Differential Galois Theory through Riemann-Hilbert Correspondence

An Elementary Introduction

GRADUATE STUDIES
IN MATHEMATICS **177**

Differential Galois Theory through Riemann-Hilbert Correspondence

An Elementary Introduction

Jacques Sauloy

American Mathematical Society
Providence, Rhode Island

The photographs that appear on the dedication page are reprinted with permission.

2010 *Mathematics Subject Classification.* Primary 12H05, 30F99, 34-01, 34A30, 34Mxx.

For additional information and updates on this book, visit
www.ams.org/bookpages/gsm-177

Library of Congress Cataloging-in-Publication Data

Names: Sauloy, Jacques.
Title: Differential Galois theory through Riemann-Hilbert correspondence : an elementary intro-
    duction / Jacques Sauloy.
Description: Providence, Rhode Island : American Mathematical Society, [2016] | Series: Gradu-
    ate studies in mathematics ; volume 177 | Includes bibliographical references and index.
Identifiers: LCCN 2016033241
Subjects: LCSH: Galois theory. | Riemann-Hilbert problems. | AMS: Field theory and polynomi-
    als – Differential and difference algebra – Differential algebra. msc | Functions of a complex
    variable – Riemann surfaces – None of the above, but in this section. msc | Ordinary differ-
    ential equations – Instructional exposition (textbooks, tutorial papers, etc.). msc | Ordinary
    differential equations – General theory – Linear equations and systems, general. msc | Ordinary
    differential equations – Differential equations in the complex domain – Differential equations
    in the complex domain. msc
Classification: LCC QA214 .S28 2016 | DDC 512/.32–dc23
    LC record available at https://lccn.loc.gov/2016033241

This book is dedicated to the Department of Mathematics of Wuda, in particular to its director Chen Hua, and to Wenyi Chen for giving me the occasion to teach this course and Jean-Pierre Ramis for giving me the ability to do so.

The math buildings at Wuda (2012) and at Toulouse University (2013)

Contents

Foreword

Nowadays differential Galois theory is a topic appearing more and more in graduate courses. There are several reasons. It mixes fundamental objects from very different areas of mathematics and uses several interesting theories. Moreover, during the last decades, mathematicians have discovered a lot of important and powerful applications of differential Galois theory. Some are quite surprising, such as some criteria for the classical problem of integrability of dynamical systems in mechanics and physics.

If differential Galois theory is a beautiful subject for a learning mathematician, it is not an easy one. There are several reference books on this topic but they are too difficult for a beginner. Jacques Sauloy's book is a wonderful success because even if it is accessible to a graduate (or a motivated undergraduate) student, it nevertheless introduces sophisticated tools in a remarkably accessible way and gives an excellent and self-contained introduction to the modern developments of the subject. Until now, I thought that such a pedagogical presentation was impossible and I envy the young beginners for having the possibility to learn the subject in such a fantastic book.

When the French mathematician Émile Picard created differential Galois theory at the end of the nineteenth century he started from some analogies between the roots of an algebraic equation and the solutions of an algebraic linear differential equation in the complex domain, which, in general, are many-valued analytic functions. Picard built on preceding works, in particular on the work of B. Riemann on the famous hypergeometric functions of Euler and Gauss, introducing the notion of monodromy representation. After Picard and his follower Vessiot, differential Galois theory was successfully algebraized by Ellis Kolchin. Today the reference books use the purely

algebraic approach of Kolchin. Jacques Sauloy returns to the origin, working only in the complex domain and making essential use of transcendental
tools as the monodromy representation and the Riemann-Hilbert correspondence. He does not at all avoid the use of algebra and of modern tools such
as categories, functors, sheaves, algebraic groups, etc., but, sewing a transcendental flesh on the algebraic bones, he throws a very interesting light
on the algebra. Moreover his point of view allows for a rather elementary
presentation of the subject with concrete approaches and a lot of interesting
examples.

The last chapter of the book (more difficult for a beginner) gives several
directions where one could extend the topics presented before and gives a
flavor of the richness of the theory.

Jean-Pierre Ramis, member of the French Academy of Sciences

Preface

In 2012, the University of Wuhan (nicknamed Wuda by its friends) asked me to give a course on differential Galois theory. Books have been published on the subject and courses have been given, all based on differential algebra and Picard-Vessiot theory and always with a very algebraic flavor (or sometimes even computer-algebra oriented). I did not feel competent to offer anything new in this direction. However, as a student of Jean-Pierre Ramis, I have been deeply influenced by his point of view that tagging algebraic objects with transcendental information enriches our understanding and brings not only new points of view but also new solutions. This certainly was illustrated by Ramis' solution to the inverse problem in differential Galois theory! Note that although there has been an overwhelming trend to algebraization during the twentieth century, the opposite tendency never died, whether in algebraic geometry or in number theory. There still are function theoretic methods in some active areas of Langlands' program! And pedagogically, I found the functional approach a nice shortcut to Tannaka theory.

On a more personal level, I remember my first advisor, Jean Giraud, an algebraist if there ever was one (he worked on the resolution of singularities in positive characteristics), sending me the beautiful 1976 article of Griffith "Variations on a theorem of Abel", which, if my memory doesn't fail me, conveyed a similar message. I explain better in the introduction why it could be a sensible choice to organize my course around such ideas.

The important point is that Chen Hua (the director of the math department at Wuda) and Wenyi Chen (who organized my stay there) accepted my proposition, and that the course took existence not only on paper. To Chen

Hua and Wenyi Chen my greatest thanks for that, and also to the whole math department of Wuda, in particular to the students who attended the course. There is no greater thrill than to teach to Chinese students. When I came back to France, I got the opportunity to teach the same course to master students. The background of French students is somewhat different from that of Chinese students, so this required some adaptation: I insisted less on complex function theory, that these students knew well (while my very young Chinese students had not yet studied it), and I used the time thus saved by giving more abstract algebraic formalism (for which French students have a strong taste).

In Wuda, the task of leading the exercise sessions with students was separated from the course proper. Luo Zhuangchu, a colleague and friend, took it in charge as well as the organization of the final exam. Xiexie Luo!

When my 2012 course in Wuhan ended, there was a two-week conference on differential equations, one of whose organizers was Changgui Zhang, who first had introduced me to the colleagues at Wuda. As a friend, as a coauthor and for having brought me first to China, I thank Changgui very much.

That conference was attended (as a speaker) by Michael Singer, also a friend and a colleague. Michael had an opportunity to see my course and found it interesting enough to suggest publication by the AMS. He put me in touch with Ina Mette of the AMS, who also encouraged me. So thanks to them both for giving me the motivation to transform my set of notes into a book. Ina Mette, along with another AMS staff member, Marcia C. Almeida, helped me all along the production process, for which I am grateful to them both. My thanks extend to production editor Mike Saitas, who handled the final publication steps. The drawings were first done by hand by my son Louis, then redrawn to professional standards by Stephen Gossmann, one of the wizards of the AMS production team led by Mary Letourneau. Thanks to all!

Last, my dearest thanks go to Jean-Pierre Ramis who, to a great extent, forged my understanding of mathematics.

Introduction

The course will involve only *complex analytic linear differential equations*. Thus, it will not be based on the general algebraic formalism of differential algebra, but rather on complex function theory.

There are two main approaches to differential Galois theory. The first one, usually called Picard-Vessiot theory, and mainly developed by Kolchin, is in some sense a transposition of the Galois theory of algebraic equations in the form it was given by the German algebraists: to a (linear) differential equation, one attaches an extension of *differential* fields and one defines the Galois group of the equation as the group of automorphisms compatible with the differential structure. This group is automatically endowed with a structure of an algebraic group, and one must take in account that structure to get information on the differential equation. This approach has been extensively developed, it has given rise to computational tools (efficient algorithms and software) and it is well documented in a huge literature.

A more recent approach is based on so-called "tannakian duality". It is very powerful and can be extended to situations where the Picard-Vessiot approach is not easily extended (like q-difference Galois theory). There is less literature and it has a reputation of being very abstract. However, in some sense, the tannakian approach can be understood as an algebraic transposition of the Riemann-Hilbert correspondence. In this way, it is rooted in very concrete and down to-earth processes: the analytic continuation of power series solutions obtained by the Cauchy theorem and the ambiguity introduced by the multivaluedness of the solutions. This is expressed by the *monodromy group*, a precursor of the differential Galois group, and by the *monodromy representation*. The Riemann-Hilbert correspondence is the other big galoisian theory of the nineteenth century, and it is likely that Picard had it in mind when he started to create differential Galois theory. The moral of this seems to be that, as understood and/or emphasized by such masters as Galois, Riemann, Birkhoff, etc., ambiguities give rise to the action of groups and the objects subject to these ambiguities are governed and can be classified by representations of groups.

Therefore, I intend to devote the first two parts[1] of the course to the study of the monodromy theory of complex analytic linear differential equations and of the Riemann-Hilbert correspondence, which is, anyhow, a must for anyone who wants to work with complex differential equations. In the third part of the course, I introduce (almost from scratch) the basic tools required for using algebraic groups in differential Galois theory, whatever the approach (Picard-Vessiot or tannakian). Last, I shall show how to attach algebraic groups and their representations to complex analytic linear differential equations. Some algebraic and functorial formalism is explained when needed, without any attempt at a systematic presentation.

The course is centered on the local analytic setting and restricted to the case of regular singular (or fuchsian) equations. One can consider it as a first-semester course. To define a second-semester course following this one would depend even more on one's personal tastes; I indicate some possibilities in Chapter 17, which gives some hints of what lies beyond, with sufficient bibliographical references. In the appendices that follow, I give some algebraic complements to facilitate the reader's work and to avoid having him or her delve into the literature without necessity. Also, the course having been taught twice before a class has given rise to two written examinations and some oral complementary tests. These are reproduced in the last appendix.

Prerequisites. The main prerequisites are: linear algebra (mostly reduction of matrices); elementary knowledge of groups and of polynomials in many variables; elementary calculus in n variables, including topology of the real euclidean spaces $\mathbf{R}^n$. Each time a more advanced result will be needed, it will be precisely stated and explained and an easily accessible reference will be given.

---

[1]The first audience of this course consisted of young enthusiastic and gifted Chinese students who had been especially prepared, except for complex function theory. Therefore I added a "crash course" on analytic functions at the beginning, which the reader may skip, although it also serves as an introduction to the more global aspects of the theory which are seldom taught at an elementary level.

Exercises. Two kinds of exercises are presented: some are inserted in the main text and serve as an illustration (*e.g.*, examples, counterexamples, explicit calculations, etc). Some come at the end of the chapters and may be either application exercises or deepening or extensions of the main text. Some solutions or hints will be posted on `www.ams.org/bookpages/gsm-177`.

Errata. I cannot hope to have corrected all the typographical and more substantial errors that appeared in the long process of making this book. A list of errata will be maintained on the same webpage mentioned above.

Notational conventions. Notation $A := B$ means that the term A is defined by formula B. New terminology is written in *emphatic style* when first defined. Note that a definition can appear in the course of a theorem, an example, an exercise, etc.

Example 0.1. The *monodromy group* of $\mathcal{F}$ at the base point a is the image of the monodromy representation:

$$\mathrm{Mon}(\mathcal{F}, a) := \mathrm{Im}\ \rho_{\mathcal{F},a} \subset \mathrm{GL}(\mathcal{F}_a).$$

We mark the end of a proof, or its absence, by the symbol $\square$.

We use commutative diagrams. For instance, to say that the diagram

$$
\begin{array}{ccc}
\mathcal{F}(U) & \xrightarrow{\ \phi_U\ } & \mathcal{F}'(U) \\
{\scriptstyle\rho^U_V}\downarrow & & \downarrow{\scriptstyle\rho'^U_V} \\
\mathcal{F}(V) & \xrightarrow[\ \phi_V\ ]{} & \mathcal{F}'(V)
\end{array}
$$

is commutative means that $\phi_V \circ \rho^U_V = \rho'^U_V \circ \phi_U$.

Index of notation

General notation.

---

[2]Sometimes, the spectrum is seen as a plain set and, for instance, writing 0_n as the null $n \times n$ matrix, $\mathrm{Sp}\ 0_n = \{0\}$. Sometimes, it is seen as a *multiset* (elements have multiplicities) and then $\mathrm{Sp}\ 0_n = \{0, \dots, 0\}$ (counted n times).

ln, e^x .... Usual logarithm and exponential

$\mathrm{Diag}(a_1,\ldots,a_n)$ .... Diagonal matrix

$\mu_n \subset \mathbf{C}^*$ .... Group of nth roots of unity

$\dim_{\mathbf{C}}$ .... Dimension of a complex space

$\overline{D}$ .... Closure of a set

N_m .... Elementary nilpotent Jordan block

$\langle x_1,\ldots,x_r\rangle$ .... Group generated by $x_1,\ldots,x_r$

$[A,B]$ .... Commutator $AB-BA$ of matrices

Specific notation. They are more competely defined in the text.

Appearing in Chapter 1.

e^z, $\exp(z)$ .... Complex exponential function

$JF(x,y)$ .... Jacobian matrix

$I(a,\gamma)$ .... Index of a loop relative to a point

e^A, $\exp(A)$ .... Exponential of a square matrix

Appearing in Chapter 2.

$\mathbf{C}[[z]]$ .... Ring of formal power series

$v_0(f)$ .... Valuation, or order, of a power series

$\mathbf{C}((z))$ .... Field of formal Laurent series

$\mathcal{O}_0 = \mathbf{C}\{z\}$ .. Ring of convergent power series, holomorphic germs at 0

Appearing in Chapter 3.

$\mathcal{O}_a = \mathbf{C}\{z-a\}$. Ring of convergent power series in $z-a$, holomorphic germs at a

$\mathcal{O}(\Omega)$ .... Ring of holomorphic functions over Ω

B_n .... Bernoulli numbers

$v_{z_0}(f)$ .... Valuation, or order, at z_0

$\mathcal{M}(\Omega)$ .... Field of meromorphic functions over a domain Ω

$I(z_0,\gamma)$ .... Index of a loop around a point

Appearing in Chapter 4.

log .... Complex logarithm (principal determination)

M_n .... Nilpotent component of a matrix

M_s .... Semi-simple component of a matrix

M_u .... Unipotent component of an invertible matrix

Appearing in Chapter 5.

Appearing in Chapter 6.

Appearing in Chapter 7.

$\rho_{\underline{a},z_0}$, $\mathrm{Mon}(E_{\underline{a}}, z_0)$ Monodromy representation and group for the scalar equation $E_{\underline{a}}$

ρ_{A,z_0}, $\mathrm{Mon}(S_A, z_0)$ ...... Monodromy representation and group for the vectorial equation S_A

$F[A]$ .... Transform of A by the gauge matrix F

$\mathcal{X}^\lambda$, $M_{[\lambda]}$ .... Analytic continuation of a fundamental matricial solution along the loop λ and corresponding monodromy matrix

$\underset{h}{\sim}$, $\underset{m}{\sim}$ .... Holomorphic and meromorphic equivalence

Appearing in Chapter 8.

$\mathfrak{Ds}$ .... Category of differential systems

$\mathfrak{Ls}$ .... Category of local systems

$\mathfrak{Sh}$ .... Category of sheaves

$\mathrm{Ob}(\mathcal{C})$, $\mathrm{Mor}_{\mathcal{C}}(X,Y)$, Id_X . Class of objects, sets of morphisms, identity morphisms in a category

$\underline{E}$ .... Constant sheaf

$\rightsquigarrow$ .... Funny arrow reserved for functors

$\mathfrak{Rep}_{\mathbf{C}}(G)$, $\mathfrak{Rep}_{\mathbf{C}}^f(G)$ .... Category of complex representations, of finite-dimensional complex representations, of G

Appearing in Chapter 9.

D, δ .... Usual and Euler differential operators

$(\alpha)_n$ .... Pochhammer symbols

$F(\alpha, \beta, \gamma; z)$ .... Hypergeometric series

Appearing in Chapter 10.

$\mathcal{E}^{(0)}$, $\mathcal{E}_f^{(0)}$, $\mathcal{E}_{f,R}^{(0)}$, $\mathcal{E}_{f,\infty}^{(0)}$ .... Various categories of differential systems

$\mathcal{M}_0 = \mathbf{C}(\{z\})$ .... Meromorphic germs at 0

Appearing in Chapter 11.

$\mathcal{B}^\lambda$ .... Analytic continuation of a basis of solutions

$HG_{\alpha,\beta,\gamma}$, $HG'_{\alpha,\beta,\gamma}$ .... Hypergeometric equation using δ or D

Appearing in Chapter 12.

RS .... Regular singular (abbreviation used only in this chapter)

Appearing in Chapter 13.

K .... Differential field, usually $\mathcal{M}_0$

$\mathcal{A}(A, z_0) = K[\mathcal{X}]$ .... Differential, algebra generated by solutions

$\mathrm{Gal}(A, z_0)$ .... Differential Galois group computed at z_0

Part 1

A Quick Introduction to Complex Analytic Functions

This "crash course" will include almost no proofs; I'll give them only if they may serve as a training for the following parts. The reader may look for further information in [**Ahl78**, **Car63**, **Rud87**].

Chapter 1

The complex exponential function

This is a *very important* function!

1.1. The series

For any $z \in \mathbf{C}$, we define:

$$\exp(z) := \sum_{n \geq 0} \frac{1}{n!} z^n = 1 + z + \frac{z^2}{2} + \frac{z^3}{6} + \frac{z^4}{24} + \cdots .$$

On the closed disk

$$\overline{\mathrm{D}}(0, R) := \{z \in \mathbf{C} \mid |z| \leq R\},$$

one has $\left| \frac{1}{n!} z^n \right| \leq \frac{1}{n!} R^n$ and we know that the series $\sum_{n \geq 0} \frac{1}{n!} R^n$ converges for any $R > 0$. Therefore, $\exp(z)$ is a normally convergent series of continuous functions on $\overline{\mathrm{D}}(0, R)$, and $z \mapsto \exp(z)$ is *a continuous function from* $\mathbf{C}$ *to* $\mathbf{C}$.

Theorem 1.1. *For any $a, b \in \mathbf{C}$, one has $\exp(a + b) = \exp(a) \exp(b)$.*

Proof. We just show the calculation, but this should be justified by arguments from real analysis (absolute convergence implies commutative convergence):

$$\exp(a+b) = \sum_{n\geq 0} \frac{1}{n!}(a+b)^n$$

$$= \sum_{n\geq 0} \frac{1}{n!} \sum_{k+l=n} \frac{(k+l)!}{k!l!}a^k b^l$$

$$= \sum_{\substack{n\geq 0 \\ k+l=n}} \frac{1}{n!}\frac{(k+l)!}{k!l!}a^k b^l$$

$$= \sum_{k,l\geq 0} \frac{1}{k!l!}a^k b^l$$

$$= \exp(a)\,\exp(b).$$

$\square$

We now give a list of basic, easily proved properties. First, the effect of complex conjugation:

$$\forall z \in \mathbf{C}\ ,\ \overline{\exp(z)} = \exp(\overline{z}).$$

Since obviously $\exp(0) = 1$, one obtains from the previous theorem:

$$\forall z \in \mathbf{C}\ ,\ \exp(z) \in \mathbf{C}^* \ \text{and}\ \exp(-z) = \frac{1}{\exp(z)}.$$

Also, $z \in \mathbf{R} \Rightarrow \exp(z) \in \mathbf{R}^*$ and then, writing $\exp(z) = \big(\exp(z/2)\big)^2$, one sees that $\exp(z) \in \mathbf{R}_+^*$.

Last, if $z \in i\mathbf{R}$ (pure imaginary), then $\overline{z} = -z$, so putting $w := \exp(z)$, one has $\overline{w} = w^{-1}$ so that $|w| = 1$. In other words, $\exp$ sends $i\mathbf{R}$ to the unit circle $\mathbf{U} := \{z \in \mathbf{C} \mid |z| = 1\}$.

Summarizing, if $x := \Re(z)$ and $y := \Im(z)$, then $\exp$ sends z to $\exp(z) = \exp(x)\exp(iy)$, where $\exp(x) \in \mathbf{R}_+^*$ and $\exp(iy) \in \mathbf{U}$.

1.2. The function $\exp$ is C-derivable

Lemma 1.2. *If* $|z| \leq R$, *then* $|\exp(z) - 1 - z| \leq \dfrac{e^R}{2}|z|^2$.

Proof. $|\exp(z) - 1 - z| = \dfrac{z^2}{2}\left(1 + \dfrac{z}{3} + \dfrac{z^2}{12} + \cdots\right)$ and $\left|1 + \dfrac{z}{3} + \dfrac{z^2}{12} + \cdots\right|$

$\le 1 + \dfrac{R}{3} + \dfrac{R^2}{12} + \cdots \le e^R.$ $\qquad\square$

Theorem 1.3. *For any fixed $z_0 \in \mathbf{C}$:*

$$\lim_{h \to 0} \frac{\exp(z_0 + h) - \exp(z_0)}{h} = \exp(z_0).$$

Proof. $\dfrac{\exp(z_0 + h) - \exp(z_0)}{h} = \exp(z_0)\,\dfrac{\exp(h) - 1}{h}$ and, after the lemma,

$\dfrac{\exp(h) - 1}{h} \to 1$ when $h \to 0$. $\qquad\square$

Therefore, exp is derivable with respect to the complex variable: we say that it is **C**-derivable (we shall change terminology later) and that its **C**-derivative is itself, which we write

$$\frac{d\exp(z)}{dz} = \exp(z) \text{ or } \exp' = \exp.$$

Corollary 1.4. *On $\mathbf{R}$, exp restricts to the usual real exponential function; that is, for $x \in \mathbf{R}$, $\exp(x) = e^x$.*

Proof. The restricted function $\exp : \mathbf{R} \to \mathbf{R}$ sends 0 to 1 and it is its own derivative, so it is the usual real exponential function. $\qquad\square$

For this reason, for now on, we shall put $e^z := \exp(z)$ when z is an arbitrary complex number.

Corollary 1.5. *For $y \in \mathbf{R}$, one has $\exp(\mathrm{i}y) = \cos(y) + \mathrm{i}\sin(y)$.*

Proof. Put $f(y) := \exp(\mathrm{i}y)$ and $g(y) := \cos(y) + \mathrm{i}\sin(y)$. These functions satisfy $f(0) = g(0) = 1$ and $f' = \mathrm{i}f$, $g' = \mathrm{i}g$. Therefore the function $h := f/g$, which is well defined from $\mathbf{R}$ to $\mathbf{C}$, satisfies $h(0) = 1$ and $h' = 0$, so that it is constant equal to 1. $\qquad\square$

Note that this implies the famous formula of Euler $e^{\mathrm{i}\pi} = -1$.

Corollary 1.6. *For $x, y \in \mathbf{R}$, one has $e^{x+\mathrm{i}y} = e^x(\cos y + \mathrm{i}\sin y)$.*

Corollary 1.7. *The exponential map $\exp : \mathbf{C} \to \mathbf{C}^*$ is surjective.*

Proof. Any $w \in \mathbf{C}^*$ can be written $w = r(\cos\theta + \mathrm{i}\sin\theta)$, $r > 0$ and $\theta \in \mathbf{R}$, so $w = \exp(\ln(r) + \mathrm{i}\theta)$. $\qquad\square$

The reader can find a beautiful proof which does not require any previous knowledge of trigonometric functions in the preliminary chapter of [**Rud87**].

The exponential viewed as a map $\mathbf{R}^2 \to \mathbf{R}^2$. It will be useful to consider functions $f : \mathbf{C} \to \mathbf{C}$ as functions $\mathbf{R}^2 \to \mathbf{R}^2$, under the usual identification of $\mathbf{C}$ with $\mathbf{R}^2$: $x + iy \leftrightarrow (x, y)$. In this way, f is described by

$$(x, y) \mapsto F(x, y) := \big(A(x, y), B(x, y)\big), \text{ where } \begin{cases} A(x, y) := \Re\big(f(x + iy)\big), \\ B(x, y) := \Im\big(f(x + iy)\big). \end{cases}$$

In the case where f is the exponential function exp, we easily compute:

$$\begin{cases} A(x, y) = e^x \cos(y), \\ B(x, y) = e^x \sin(y), \end{cases} \implies F(x, y) = \big(e^x \cos(y), e^x \sin(y)\big).$$

We are going to compare the differential of the map F with the $\mathbf{C}$-derivative of the exponential map. We use the terminology of differential calculus [**Car97, Spi65**]. On the one hand, the differential $dF(x, y)$ is the linear map defined by the relation

$$F(x + u, y + v) = F(x, y) + dF(x, y)(u, v) + o(u, v),$$

where $o(u, v)$ is small compared to the norm of (u, v) when $(u, v) \to (0, 0)$. Actually, $dF(x, y)$ can be expressed using partial derivatives:

$$dF(x, y)(u, v) = \Big(\frac{\partial A(x, y)}{\partial x} u + \frac{\partial A(x, y)}{\partial y} v, \frac{\partial B(x, y)}{\partial x} u + \frac{\partial B(x, y)}{\partial y} v\Big).$$

Therefore, it is described by the Jacobian matrix:

$$JF(x, y) = \begin{pmatrix} \dfrac{\partial A(x, y)}{\partial x} & \dfrac{\partial A(x, y)}{\partial y} \\[2mm] \dfrac{\partial B(x, y)}{\partial x} & \dfrac{\partial B(x, y)}{\partial y} \end{pmatrix}.$$

On the side of the complex function $f := \exp$, putting $z := x + iy$ and $h := u + iv$, we write:

$$f(z+h) = f(z)+hf'(z)+o(h), \text{ that is, } \exp(z+h) = \exp(z)+h\exp(z)+o(h).$$

Here, the linear part is $f'(z)h = \exp(z)h$, so we draw the conclusion that (under our correspondence of $\mathbf{C}$ with $\mathbf{R}^2$):

$$hf'(z) \longleftrightarrow dF(x, y)(u, v),$$

that is, comparing real and imaginary parts:

$$\begin{cases} \dfrac{\partial A(x, y)}{\partial x} u + \dfrac{\partial A(x, y)}{\partial y} v = \Re(f'(z))u - \Im(f'(z))v, \\[3mm] \dfrac{\partial B(x, y)}{\partial x} u + \dfrac{\partial B(x, y)}{\partial y} v = \Im(f'(z))u + \Re(f'(z))v. \end{cases}$$

Since this must be true for all u, v, we conclude that:

$$JF(x,y) = \begin{pmatrix} \dfrac{\partial A(x,y)}{\partial x} & \dfrac{\partial A(x,y)}{\partial y} \\ \dfrac{\partial B(x,y)}{\partial x} & \dfrac{\partial B(x,y)}{\partial y} \end{pmatrix} = \begin{pmatrix} \Re(f'(z)) & -\Im(f'(z)) \\ \Im(f'(z)) & \Re(f'(z)) \end{pmatrix}.$$

As a consequence, the Jacobian determinant $\det JF(x, y)$ is equal to $|f'(z)|^2$ and thus vanishes if, and only if, $f'(z) = 0$; in the case of the exponential function, it vanishes nowhere.

Exercise 1.8. Verify these formulas when $A(x, y) = e^x \cos(y)$, $B(x, y) = e^x \sin(y)$ and $f'(z) = \exp(x + iy)$.

1.3. The exponential function as a covering map

From equation $e^{x+iy} = e^x (\cos y + i \sin y)$, one sees that $e^z = 1 \Leftrightarrow z \in 2i\pi \mathbf{Z}$, i.e., $\exists k \in \mathbf{Z} : z = 2i\pi k$. It follows that $e^{z_1} = e^{z_2} \Leftrightarrow e^{z_2 - z_1} = 1 \Leftrightarrow z_2 - z_1 \in 2i\pi \mathbf{Z}$, i.e., $\exists k \in \mathbf{Z} : z_2 = z_1 + 2i\pi k$. We shall write this relation: $z_2 \equiv z_1 \pmod{2i\pi \mathbf{Z}}$ or for short $z_2 \equiv z_1 \pmod{2i\pi}$.

Theorem 1.9. *The map* $\exp : \mathbf{C} \to \mathbf{C}^*$ *is a covering map. That is, for any* $w \in \mathbf{C}^*$, *there is a neighborhood* $V \subset \mathbf{C}^*$ *of* w *such that* $\exp^{-1}(V) = \bigsqcup U_k$ *(disjoint union), where each* $U_k \subset \mathbf{C}$ *is an open set and* $\exp : U_k \to V$ *is a homeomorphism (a bicontinuous bijection).*

Proof. Choose a particular $z_0 \in \mathbf{C}$ such that $\exp(z_0) = w$. Choose an open neighborhood U_0 of z_0 such that, for any $z', z'' \in U_0$, one has $|\Im z'' - \Im z'| < 2\pi$. Then $\exp$ bijectively maps U_0 to $V := \exp(U_0)$. Moreover, one has $\exp^{-1}(V) = \bigsqcup U_k$, where k runs in $\mathbf{Z}$ and the $U_k = U_0 + 2i\pi k$ are pairwise disjoint open sets. It remains to show that V is an open set. The most generalizable way is to use the local inversion theorem [**Car97**, **Spi65**], since the Jacobian determinant vanishes nowhere. Another way is to choose an open set as in exercise 2 to this chapter. $\square$

The fact that $\exp$ is a covering map is a very important topological property and it has many consequences.

Corollary 1.10 (Path lifting property). *Let* $a < b$ *in* $\mathbf{R}$ *and let* $\gamma : [a, b] \to \mathbf{C}^*$ *be a continuous path with origin* $\gamma(a) = w_0 \in \mathbf{C}^*$. *Let* $z_0 \in \mathbf{C}$ *be such that* $\exp(z_0) = w_0$. *Then there exists a unique lifting, i.e., a continuous path* $\overline{\gamma} : [a, b] \to \mathbf{C}^*$ *such that* $\forall t \in [a, b]$, $\exp \overline{\gamma}(t) = \gamma(t)$ *and subject to the initial condition* $\overline{\gamma}(a) = z_0$.

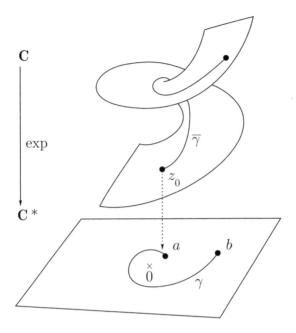

Figure 1.1. Path lifting property

Corollary 1.11 (Index of a loop with respect to a point). *Let $\gamma : [a,b] \to \mathbf{C}^*$ be a continuous loop, that is, $\gamma(a) = \gamma(b) = w_0 \in \mathbf{C}^*$. Then, for any lifting $\overline{\gamma}$ of γ, one has $\overline{\gamma}(b) - \overline{\gamma}(a) = 2i\pi n$ for some $n \in \mathbf{Z}$. The number n is the same for all the liftings and it depends only on the loop γ: it is the* index of γ around 0, *written $I(0, \gamma)$.*

Actually, another property of covering maps (the "homotopy lifting property"; see for instance [**Ful95**, Proposition 11.8]) allows one to conclude that $I(0, \gamma)$ does not change if γ is continuously deformed within $\mathbf{C}^*$: it only depends on the "homotopy class" of γ (see the last exercise at the end of this chapter).

Example 1.12. If $\gamma(t) = e^{nit}$ on $[0, 2\pi]$, then all liftings of γ have the form $\overline{\gamma}(t) = nit + 2i\pi k$ for some $k \in \mathbf{Z}$ and one finds $I(0, \gamma) = n$.

1.4. The exponential of a matrix

For a complex vector $X = \begin{pmatrix} x_1 \\ \vdots \\ x_n \end{pmatrix} \in \mathbf{C}^n$, we define $\|X\|_\infty := \max_{1 \le i \le n} (|x_i|)$.

Then, for a complex square matrix $A = (a_{i,j})_{1 \le i,j \le n} \in \mathrm{Mat}_n(\mathbf{C})$, define the subordinate norm:[1]

$$\|A\|_{\infty} := \sup_{\substack{X \in \mathbf{C}^n \\ X \neq 0}} \frac{\|AX\|_{\infty}}{\|X\|_{\infty}} = \max_{1 \leq i \leq n} \sum_{j=1}^{n} |a_{i,j}|.$$

Then, for the identity matrix, $\|I_n\|_{\infty} = 1$; and, for a product, $\|AB\|_{\infty} \leq \|A\|_{\infty} \|B\|_{\infty}$. It follows easily that $\left\| \frac{1}{k!} A^k \right\|_{\infty} \leq \frac{1}{k!} \|A\|_{\infty}^k$ for all $k \in \mathbf{N}$, so that the series $\sum_{k \geq 0} \frac{1}{k!} A^k$ converges absolutely for any $A \in \mathrm{Mat}_n(\mathbf{C})$. It actually converges normally on all compacts and therefore defines a continuous map $\exp : \mathrm{Mat}_n(\mathbf{C}) \to \mathrm{Mat}_n(\mathbf{C})$, $A \mapsto \sum_{k \geq 0} \frac{1}{k!} A^k$. We shall also write for short $e^A := \exp(A)$. In the case $n = 1$, the notation is consistent.

Examples 1.13. (i) For a diagonal matrix $A := \mathrm{Diag}(\lambda_1, \ldots, \lambda_n)$, one has $\frac{1}{k!} A^k = \mathrm{Diag}(\lambda_1^k/k!, \ldots, \lambda_n^k/k!)$, so that $\exp(A) = \mathrm{Diag}(e^{\lambda_1}, \ldots, e^{\lambda_n})$.

(ii) If A is an upper triangular matrix with diagonal $D := \mathrm{Diag}(\lambda_1, \ldots, \lambda_n)$, then $\frac{1}{k!} A^k$ is an upper triangular matrix with diagonal $\frac{1}{k!} D^k$, so that $\exp(A)$ is an upper triangular matrix with diagonal $\exp(D) = \mathrm{Diag}(e^{\lambda_1}, \ldots, e^{\lambda_n})$. Similar relations hold for lower triangular matrices.

(iii) Take $A := \begin{pmatrix} 0 & 1 \\ 1 & 0 \end{pmatrix}$. Then $A^2 = I_2$, so that $\exp(A) = aI_2 + bA = \begin{pmatrix} a & b \\ b & a \end{pmatrix}$, where $a = \sum_{k \geq 0} \frac{1}{(2k)!}$ and $b = \sum_{k \geq 0} \frac{1}{(2k+1)!}$.

Exercise 1.14. Can you recognize the values of a and b?

The same kind of calculations as for the exponential map give the rules:

$$\exp(0_n) = I_n; \quad \exp(\overline{A}) = \overline{\exp(A)},$$

and

$$AB = BA \implies \exp(A + B) = \exp(A) \exp(B) = \exp(B) \exp(A).$$

Remark 1.15. The condition $AB = BA$ is required to use the Newton binomial formula. If we take for instance $A := \begin{pmatrix} 0 & 1 \\ 0 & 0 \end{pmatrix}$ and $B := \begin{pmatrix} 0 & 0 \\ 1 & 0 \end{pmatrix}$, then $AB \neq BA$. We have $A^2 = B^2 = 0$, so that $\exp(A) = I_2 + A = \begin{pmatrix} 1 & 1 \\ 0 & 1 \end{pmatrix}$

---

[1] We shall use the same notation for a norm in the complex space V and for the corresponding subordinate norm in $\mathrm{End}(V)$ (that is, $\mathrm{Mat}_n(\mathbf{C})$ whenever $V = \mathbf{C}^n$).

and $\exp(B) = I_2 + B = \begin{pmatrix} 1 & 0 \\ 1 & 1 \end{pmatrix}$, thus $\exp(A)\exp(B) = \begin{pmatrix} 2 & 1 \\ 1 & 1 \end{pmatrix}$. On the

other hand, $A + B = \begin{pmatrix} 0 & 1 \\ 1 & 0 \end{pmatrix}$ and the previous example gave the value of

$\exp(A + B)$, which was clearly different.

It follows from the previous rules that $\exp(-A) = \big(\exp(A)\big)^{-1}$ so that exp actually sends $\mathrm{Mat}_n(\mathbf{C})$ to $\mathrm{GL}_n(\mathbf{C})$. Now there are rules more specific to matrices. For the transpose, using the fact that $^t(A^k) = (^tA)^k$, and also the continuity of $A \mapsto {}^tA$ (this is required to go to the limit in the infinite sum), we see that $\exp(^tA) = {}^t(\exp(A))$. Last, if $P \in \mathrm{GL}_n(\mathbf{C})$, from the relation $(PAP^{-1})^n = PA^nP^{-1}$ (and also from the continuity of $A \mapsto PAP^{-1}$), we deduce the very useful equality:

$$P\exp(A)P^{-1} = \exp(PAP^{-1}).$$

Now any complex matrix A is conjugate to an upper triangular matrix T having the eigenvalues of A on the diagonal; using the examples above, one concludes that if A has eigenvalues $\lambda_1, \ldots, \lambda_n$, then $\exp(A)$ has eigenvalues $e^{\lambda_1}, \ldots, e^{\lambda_n}$ with the corresponding multiplicities:[2]

$$\mathrm{Sp}(e^A) = e^{\mathrm{Sp}(A)}.$$

Note that this implies in particular the formula:

$$\det(e^A) = e^{\mathrm{Tr}A}.$$

Example 1.16. Let $A := \begin{pmatrix} 0 & -\pi \\ \pi & 0 \end{pmatrix}$. Then A is diagonalizable with spectrum $\mathrm{Sp}(A) = \{\mathrm{i}\pi, -\mathrm{i}\pi\}$. Thus, $\exp(A)$ is diagonalizable with spectrum $\{-1, -1\}$. Therefore, $\exp(A) = -I_2$.

1.5. Application to differential equations

Let $A \in \mathrm{Mat}_n(\mathbf{C})$ be fixed. Then $z \mapsto e^{zA}$ is a $\mathbf{C}$-derivable function from $\mathbf{C}$ to the complex linear space $\mathrm{Mat}_n(\mathbf{C})$; this simply means that each coefficient is a $\mathbf{C}$-derivable function from $\mathbf{C}$ to itself. Derivating our matrix-valued function coefficientwise, we find:

$$\frac{d}{dz}e^{zA} = Ae^{zA} = e^{zA}A.$$

Indeed, $\dfrac{e^{(z+h)A} - e^{zA}}{h} = e^{zA}\dfrac{e^{hA} - I_n}{h} = \dfrac{e^{hA} - I_n}{h}e^{zA}$ and $\dfrac{e^{hA} - I_n}{h} = A +$

$\dfrac{h}{2}A^2 + \cdots$.

---

[2]Most of the time, it is convenient to consider the spectrum as a *multiset*, that is, a set whose elements have multiplicities. In the same way, the roots of a polynomial make up a multiset.

Now consider the vectorial differential equation:

$$\frac{d}{dz}X(z) = AX(z),$$

where $X : \mathbf{C} \to \mathbf{C}^n$ is looked at as a $\mathbf{C}$-derivable vector-valued function, and again derivation is performed coefficientwise. We solve this by change of unknown function: $X(z) = e^{zA}Y(z)$. Then, applying Leibniz' rule for derivation, $(fg)' = f'g + fg'$ (it works the same for $\mathbf{C}$-derivation), we find:

$$X' = AX \Longrightarrow e^{zA}Y' + Ae^{zA}Y = Ae^{zA}Y \Longrightarrow e^{zA}Y' = 0 \Longrightarrow Y' = 0.$$

Therefore, $Y(z)$ is a constant function. (Again, we admit a property of $\mathbf{C}$-derivation: that $f' = 0 \Rightarrow f$ constant.) If we now fix $z_0 \in \mathbf{C}$, $X_0 \in \mathbf{C}^n$ and we address the Cauchy problem:

$$\begin{cases} \dfrac{d}{dz}X(z) = AX(z), \\ X(z_0) = X_0, \end{cases}$$

we see that the unique solution is $X(z) := e^{(z-z_0)A}X_0$.

An important theoretical consequence is the following. Call $Sol(A)$ the set of solutions of $\dfrac{d}{dz}X(z) = AX(z)$. This is obviously a complex linear space. What we proved is that the map $X \mapsto X(z_0)$ from $Sol(A)$ to $\mathbf{C}^n$, which is obviously linear, is also bijective. Therefore, it is an isomorphism of $Sol(A)$ with $\mathbf{C}^n$. This is a very particular case of the Cauchy theorem for *complex* differential equations that we shall encounter in Section 7.3.

Example 1.17. To solve the linear homogeneous second-order scalar equation (with constant coefficients) $f'' + pf' + qf = 0$ ($p, q \in \mathbf{C}$), we introduce the vector-valued function $X(z) := \begin{pmatrix} f(z) \\ f'(z) \end{pmatrix}$ and find that our scalar equation is actually equivalent to the vector equation:

$$X' = AX, \text{ where } A := \begin{pmatrix} 0 & 1 \\ -q & p \end{pmatrix}.$$

Therefore, the solution will be searched in the form $X(z) := e^{(z-z_0)A}X_0$, where z_0 may be chosen at will or else imposed by initial conditions.

Exercises

(1) For $z \in \mathbf{C}$, define

$$\cos(z) := \frac{\exp(\mathrm{i}z) + \exp(-\mathrm{i}z)}{2} \quad \text{and} \quad \sin(z) := \frac{\exp(\mathrm{i}z) - \exp(-\mathrm{i}z)}{2\mathrm{i}},$$

so that cos is an even function, sin is an odd function and $\exp(z) = \cos(z) + \mathrm{i}\sin(z)$. Translate the property of Theorem 1.1 into properties of cos and sin.

(2) Let $a, b > 0$ and $U := \{z \in \mathbf{C} \mid -a < \Re(z) < a \text{ and } -b < \Im(z) < b\}$ (thus, an open rectangle under the identification of $\mathbf{C}$ with $\mathbf{R}^2$). Assuming $b < \pi$, describe the image $V := \exp(U) \subset \mathbf{C}^*$ and define an inverse map $V \to U$.

(3) If in Corollary 1.10 one chooses another $z_0' \in \mathbf{C}$ such that $\exp(z_0') = w_0$, one gets another lifting $\overline{\gamma}' : [a, b] \to \mathbf{C}^*$ such that $\forall t \in [a, b]$, $\exp \overline{\gamma}'(t) = \gamma(t)$ and subject to the initial condition $\overline{\gamma}'(a) = z_0'$. Show that there is some *constant* $k \in \mathbf{Z}$ such that $\forall t \in [a, b]$, $\overline{\gamma}'(t) = \overline{\gamma}(t) + 2\mathrm{i}\pi k$.

(4) (i) Compute $\exp \begin{pmatrix} 0 & -b \\ b & 0 \end{pmatrix}$ in two ways: by direct calculation as in Example 1.13 (iii); by diagonalization as in Example 1.16.

   (ii) Deduce from this the value of $\exp \begin{pmatrix} a & -b \\ b & a \end{pmatrix}$.

(5) In Example 1.17 compute $e^{(z-z_0)A}$ and solve the problem with initial conditions $f(0) = a$, $f'(0) = b$. There will be a discussion according to whether $p^2 - 4q = 0$ or $\neq 0$.

(6) This exercise aims at introducing you to the concept of homotopy, which shall play a crucial role in all of the course. Ideally, you should read Chapters 11 and 12 of [**Ful95**] or content yourself with what [**Ahl78**, **Car63**, **Rud87**] say about homotopies and coverings. However, it might be even better to try to figure out by yourself the fundamental properties.

   Let Ω be a domain in $\mathbf{C}$ (although the notion is much more general). Two continuous paths $\gamma_1, \gamma_2 : [0; 1] \to \Omega$ with same origin $\gamma_1(0) = \gamma_2(0) = x$ and same extremity $\gamma_1(1) = \gamma_2(1) = y$ are said to be *homotopic* if there is a continuous map (called a *homotopy* from γ_1 to γ_2)

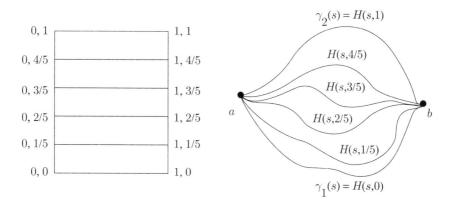

Figure 1.2. Homotopic paths

$H : [0; 1] \times [0; 1] \to \Omega$ such that:

$$\forall s \in [0; 1] , H(s, 0) = \gamma_1(s),$$
$$\forall s \in [0; 1] , H(s, 1) = \gamma_2(s),$$
$$\forall t \in [0; 1] , H(0, t) = x,$$
$$\forall t \in [0; 1] , H(1, t) = y.$$

(i) Show that this is an equivalence relation among paths from x to y.
(ii) Define the composition of a path γ from x to y with a path γ' from y to z as the path $\gamma.\gamma'$ from x to z given by the formula:

$$\gamma.\gamma'(s) = \begin{cases} \gamma(2s) \text{ if } 0 \leq s \leq 1/2, \\ \gamma'(2s - 1) \text{ if } 1/2 \leq s \leq 1. \end{cases}$$

Show that this is not associative but that (when everything is defined) $(\gamma.\gamma').\gamma''$ and $\gamma.(\gamma'.\gamma'')$ are homotopic.
(iii) More generally, show that homotopy is compatible with composition (you will have to formulate this precisely first).
(iv) Prove the homotopy lifting property.
(v) Extend all this to paths $[a; b] \to \Omega$.

Remark 1.18. We use paths with good differential properties many times, for instance such that γ' exists and is piecewise continuous. A fundamental theorem of differential topology says that when two such paths are homotopic, a homotopy H can be chosen such that all intermediate paths $\gamma_t : s \mapsto H(s, t)$ share the same good differential properties. More on this in the footnote at the bottom of page 61.

Chapter 2

Power series

2.1. Formal power series

Formal power series are actually purely algebraic objects, some kind of "polynomials of infinite degree":

$$\mathbf{C}[[z]] := \left\{ \sum_{n \geq 0} a_n z^n \mid \forall n \in \mathbf{N} , \ a_n \in \mathbf{C} \right\}.$$

This means that we do not attach (for the moment) any meaning to the "sum", do not consider z as a number and do not see a formal power series as a function. We agree to say that the formal power series $\sum_{n \geq 0} a_n z^n$ and $\sum_{n \geq 0} b_n z^n$ are equal if, and only if, they have the same coefficients.

Let $\lambda \in \mathbf{C}$ and let $\sum_{n \geq 0} a_n z^n \in \mathbf{C}[[z]]$ and $\sum_{n \geq 0} b_n z^n \in \mathbf{C}[[z]]$. Then we define the following operations:

$$\lambda . \left(\sum_{n \geq 0} a_n z^n \right) := \sum_{n \geq 0} (\lambda . a_n) z^n,$$

$$\sum_{n \geq 0} a_n z^n + \sum_{n \geq 0} b_n z^n := \sum_{n \geq 0} (a_n + b_n) z^n,$$

$$\left(\sum_{n \geq 0} a_n z^n \right) . \left(\sum_{n \geq 0} b_n z^n \right) := \sum_{n \geq 0} c_n z^n, \text{ where } \forall n \in \mathbf{N}, \ c_n := \sum_{i+j=n} a_i b_j.$$

With the first and second operations, we make $\mathbf{C}[[z]]$ into a linear space over the complex numbers. With the second and last operations, we make it into a commutative ring. Its zero element is $\sum_{n \geq 0} 0z^n$, written for short 0; its unit element is $1 + \sum_{n \geq 1} 0.z^n$, written for short 1. Altogether, we say that $\mathbf{C}[[z]]$ is a $\mathbf{C}$-algebra.

Polynomials can be considered as formal power series, with almost all their coefficients being zero. The operations are the same, so we identify $\mathbf{C}[z] \subset \mathbf{C}[[z]]$ as a sub-algebra (sub-linear space and sub-ring). Among polynomials are the constants $\mathbf{C} \subset \mathbf{C}[z] \subset \mathbf{C}[[z]]$, and so we identify $a \in \mathbf{C}$ with $a + \sum_{n \geq 1} 0.z^n \in \mathbf{C}[[z]]$.

Remark 2.1. Although it has no meaning for the moment to substitute a complex number $z_0 \in \mathbf{C}$ for the formal indeterminate z (*i.e.*, $f(z_0)$ is not defined), we will allow ourselves to write $f(0) := a_0$ when $f := \sum_{n \geq 0} a_n z^n \in \mathbf{C}[[z]]$ and call it the constant term. It has the natural properties that $(\lambda.f)(0) = \lambda.f(0)$, $(f + g)(0) = f(0) + g(0)$ and $(f.g)(0) = f(0)g(0)$.

Invertible elements. In $\mathbf{C}[z]$, only nonzero constants are invertible, but in $\mathbf{C}[[z]]$ we can perform calculations such as $\dfrac{1}{1+z} = 1 - z + z^2 - z^3 + \cdots$. Remember that we do not attach any numerical meaning to this equality, it only means that performing the product $(1+z)(1-z+z^2-z^3+\cdots)$ according to the rules above yields the result $1 + \sum_{n \geq 1} 0.z^n = 1$. More generally,

$$\left(\sum_{n \geq 0} a_n z^n\right) \cdot \left(\sum_{n \geq 0} b_n z^n\right) = 1 \iff a_0 b_0 = 1 \text{ and } \forall n \geq 1, \sum_{i+j=n} a_i b_j = 0,$$

which in turn is equivalent to

$$b_0 = 1/a_0 \text{ and } \forall n \geq 1, \ b_n = -(a_n b_0 + \cdots + a_1 b_{n-1})/a_0.$$

Therefore, $\sum_{n \geq 0} a_n z^n$ is invertible if, and only if, $a_0 \neq 0$. The coefficients of its inverse are then calculated by the above recursive formulas. The *group of units* (*invertible elements*) of the ring $\mathbf{C}[[z]]$ is:

$$\mathbf{C}[[z]]^* = \left\{\sum_{n \geq 0} a_n z^n \in \mathbf{C}[[z]] \mid a_0 \neq 0\right\} = \{f \in \mathbf{C}[[z]] \mid f(0) \neq 0\}.$$

Valuation. For $f := \sum_{n \geq 0} a_n z^n \in \mathbf{C}[[z]]$, we define $v_0(f) := \min\{n \in \mathbf{N} \mid a_n \neq 0\}$. This is called the *valuation of f* or the *order of f at 0*. By convention,[1] $v_0(0) := +\infty$. Thus, if $v_0(f) = k \in \mathbf{N}$, then $f = a_k z^k + a_{k+1} z^{k+1} + \cdots$ and $a_k \neq 0$, and therefore $f = z^k u$, where $u \in \mathbf{C}[[z]]^*$. It easily follows that $f|g$, meaning "f divides g" (*i.e.*, $\exists h \in \mathbf{C}[[z]] : g = fh$) if, and only if, $v_0(f) \leq v_0(g)$. Likewise, $f \in \mathbf{C}[[z]]^* \Leftrightarrow v_0(f) = 0$. Other useful rules are: $v_0(f+g) \geq \min(v_0(f), v_0(g))$ and $v_0(fg) = v_0(f) + v_0(g)$. An easy consequence is that $\mathbf{C}[[z]]$ is an integral ring, *i.e.*, $fg = 0 \Rightarrow f = 0$ or $g = 0$ (actually a "discrete valuation ring"; see [**Lan02**]).

Field of fractions. Let $f, g \neq 0$. If $f|g$, then $g/f \in \mathbf{C}[[z]]$. Otherwise, writing $f = z^k u$ and $g = z^l v$ with $k, l \in \mathbf{N}$ and $u, v \in \mathbf{C}[[z]]^*$ (thus $k = v_0(f)$ and $l = v_0(g)$), we have $g/f = z^{l-k}(v/u)$, where $l - k < 0$ (since f does not divide g) and $v/u \in \mathbf{C}[[z]]^*$. For example:

$$1/(z + z^2) = (1 - z + z^2 - z^3 + \cdots)/z = z^{-1} - 1 + z - z^2 + \cdots.$$

This means that quotients of formal power series are "extended" formal power series with a finite number of negative powers. We therefore define the set of *formal Laurent series*:

$$\mathbf{C}((z)) := \mathbf{C}[[z]][z^{-1}] = \mathbf{C}[[z]] + z^{-1}\mathbf{C}[z^{-1}].$$

With the same operations as in $\mathbf{C}[[z]]$, it actually becomes a field: indeed, it is the field of fractions of $\mathbf{C}[[z]]$. Beware that the elements of $\mathbf{C}((z))$ have the form $\sum_{n \gg -\infty,} a_n z^n$; the symbol $n \gg -\infty$ meaning "$n \geq n_0$ for some $n_0 \in \mathbf{Z}$".

Formal derivation. For $f := \sum_{n \geq 0} a_n z^n \in \mathbf{C}[[z]]$, we define its derivative:

$$f' := \sum_{n \geq 0} n a_n z^{n-1} = \sum_{n \geq 0} (n+1) a_{n+1} z^n.$$

We also write it df/dz. For the moment, this has no analytical meaning, it is just an algebraic operation on the coefficients. However, it satisfies the usual rules: if $\lambda \in \mathbf{C}$, then $(\lambda f)' = \lambda f'$; for any f, g, $(f+g)' = f' + g'$ and $(fg)' = fg' + f'g$ (Leibniz rule). Last, $f' = 0 \Leftrightarrow f \in \mathbf{C}$. Actually, the definition above as well as the rules can be extended without problem to formal Laurent series $f \subset \mathbf{C}((z))$, and we have two more rules: $(1/f)' = -f'/f^2$ and $(f/g)' = (f'g - fg')/g^2$. If we introduce the *logarithmic derivative* f'/f, we conclude that $(fg)'/(fg) = f'/f + g'/g$ and that $(f/g)'/(f/g) = f'/f - g'/g$.

---

[1]It is actually a general convention to set $\min \emptyset := +\infty$ and $\max \emptyset := -\infty$ when dealing with sets of real numbers. These seemingly strange values fit very nicely into general rules for the manipulation of min and max.

Example 2.2 (Newton binomial formula). The above rules sometimes allow one to transform an algebraic equation into a differential one, which may be easier to deal with, as we shall presently see. Let $p \in \mathbf{Z}$ and $q \in \mathbf{N}$, and assume they are coprime and $q > 1$. Therefore, $r := p/q \in \mathbf{Q} \setminus \mathbf{Z}$ is a rational in reduced form and not an integer. We are going to define $f := (1+z)^r$ by requiring that $f = a_0 + a_1 z + \cdots$ has constant term $f(0) = a_0 = 1$ and that $f^q = (1+z)^p$. Now, using the fact that $h = 1$ is equivalent to $h'/h = 0$ and $h(0) = 1$, one has the logical equivalences:

$$f^q = (1+z)^p \iff f^q/(1+z)^p = 1$$
$$\iff qf'/f = p/(1+z) \text{ and } a_0^q = 1$$
$$\iff (1+z)f' = rf,$$

since we have already required that $a_0 = 1$. Thus, we have a kind of Cauchy problem: a differential equation and an initial condition. Now, $(1+z)f' = \sum_{n \geq 0}((n+1)a_{n+1} + na_n)z^n$, so, by identification of coefficients, we see that our Cauchy problem is equivalent to $a_0 = 1$ and $\forall n \geq 0$, $(n+1)a_{n+1} + na_n = ra_n$, which in turn is equivalent to:

$$a_0 = 1 \text{ and } \forall n \geq 0, (n+1)a_{n+1} + na_n = ra_n \iff$$
$$a_0 = 1 \text{ and } \forall n \geq 0, (n+1)a_{n+1} = (r-n)a_n \iff$$
$$a_0 = 1 \text{ and } \forall n \geq 0, a_{n+1} = \frac{r-n}{n+1}a_n \iff \forall n \geq 0, a_n = \binom{r}{n},$$

where we have defined the *generalized binomial coefficients*:

$$\binom{r}{0} := 1 \text{ and } \forall n \geq 1, \binom{r}{n} := \frac{r(r-1)\cdots(r-n+1)}{n!}.$$

This gives the *generalized Newton binomial formula*:

$$(1+z)^r = \sum_{n \geq 0} \binom{r}{n} z^n.$$

Note that the right-hand side makes sense for any $r \in \mathbf{C}$. After the study of the series $\log(1+z)$, we shall be able to see that $(1+z)^r = \exp(r\log(1+z))$.

Substitution (or composition). Let $f, g \in \mathbf{C}[[z]]$. If they were functions, we could define the composition $g \circ f$. We shall define a formal analogue under some conditions, but rather call it *substitution of z by f in g*. The restrictive assumption is that $k := v_0(f) \geq 1$, *i.e.*, $f(0) = 0$. On the other hand, we can authorize g to be a formal Laurent series. If we write $g = \sum_{n \geq n_0} b_n z^n$, we consider first its truncated series: $g_N := \sum_{n=n_0}^{N} b_n z^n$. These are

"Laurent polynomials", that is, polynomials with some negative exponents. There are only a finite number of terms, so that it makes sense to define

$$g_N \circ f := \sum_{n=n_0}^{N} b_n f^n.$$

Now, two successive composita $g_N \circ f$ and $g_{N-1} \circ f$ differ by the term $b_N f^N$, which has order Nk (or vanishes if $b_N = 0$). Therefore, the terms up to degree Nk are the same for all $g_{N'} \circ f$ for $N' \geq N$. In this way, we see that the coefficients "stabilize": for any fixed $n \geq 0$, there is an N_0 such that all power series $g_N \circ f$, $N \geq N_0$, have the same coefficient of index n. Therefore we can define $g \circ f$ as the limit coefficientwise of the $g_N \circ f$. The composition of formal power series satisfies the usual rules for functions. For instance, $(g \circ f)(0) = g(0)$ (this only makes sense if $g \in \mathbf{C}[[z]]$), $(g \circ f_1) \circ f_2 = g \circ (f_1 \circ f_2)$, $(g_1 + g_2) \circ f = (g_1 \circ f) + (g_2 \circ f)$, $(g_1 g_2) \circ f = (g_1 \circ f)(g_2 \circ f)$ and $(g \circ f)' = (g' \circ f)f'$.

Examples 2.3. (i) If $g = z + z^2$, writing $f = a_1 z + a_2 z^2 + \cdots$:

$$g \circ f = f + f^2$$
$$= \sum_{n \geq 1} \left(a_n + \sum_{i+j=n} a_i a_j \right) z^n \text{ and}$$
$$f \circ g = f(z + z^2)$$
$$= \sum_{n \geq 1} a_n (z + z^2)^n$$
$$= \sum_{n \geq 1} \left(\sum_{i+j=n} \binom{i}{j} a_i \right) z^n.$$

(ii) Take $g = 1 + z + z^2/2 + z^3/6 + \cdots$ (the formal series with the same coefficients as exp) and $f = z - z^2/2 + z^3/3 - z^4/4 + \cdots$ (this one corresponds in some sense to $\log(1+z)$). Then the very serious reader could calculate the first terms of $g \circ f$ and find $1 + z$ and then all terms seem to vanish. This can also be seen in the following way: clearly, $g' = g$ and $f' = 1/(1+z)$, so that putting $h := g \circ f$, one has $h' = h/(1+z)$ and $h(0) = 1$, from which one deduces easily that $h = 1 + z$.

Exercise 2.4. Write down this easy deduction.

Reciprocation. It is obvious that $g \circ z = g$ and that, when defined, $z \circ f = f$. So z is a kind of neutral element for the (noncommutative) law $\circ$. One can therefore look for an inverse: g being given, does there exist f such that $g \circ f = f \circ g = z$? To distinguish this process for plain inversion $1/f$, we shall call it *reciprocation*. For these conditions to make sense, one must require that $v_0(f), v_0(g) \geq 1$. Then one sees easily that $v_0(g \circ f) = v_0(g)v_0(f)$, so that g can be reciprocated only if $v_0(g) = 1$. This necessary condition is actually sufficient, and the solution of any of the two problems $g \circ f = z$ or $f \circ g = z$ is unique, and it is a solution of the other problem; but this is rather complicated to prove (see [**Car63**]).

Exercise 2.5. Solve the two reciprocation problems when $f = z + z^2$, i.e., look for f such that $g(z + z^2) = z$, resp. $g + g^2 = z$.

2.2. Convergent power series

Theorem 2.6. *Suppose that the series* $f := \sum_{n \geq 0} a_n z_0^n$ *converges in* $\mathbf{C}$ *for some nonzero value* $z_0 \in \mathbf{C}$. *Let* $R := |z_0|$. *Then the series* $\sum_{n \geq 0} a_n z^n$ *is normally convergent on any closed disk* $\overline{\mathrm{D}}(0, R')$ *with* $0 < R' < R$.

Proof. The $|a_n z_0^n| = |a_n| R^n$ are bounded by some $M > 0$ and, on $\overline{\mathrm{D}}(0, R')$, one has $|a_n z^n| \leq M(R'/R)^n$.   □

Corollary 2.7. *The map* $z \mapsto \sum_{n \geq 0} a_n z^n$ *defines a continuous function on the open disk* $\overset{\circ}{\mathrm{D}}(0, R)$, *with values in* $\mathbf{C}$.

Definition 2.8. The *radius of convergence* (improperly abbreviated as "r.o.c.") of the series $\sum_{n \geq 0} a_n z^n$ is defined as $\sup\{|z_0| \mid \sum_{n \geq 0} a_n z_0^n \text{ converges}\}$. If the radius of convergence of $f := \sum_{n \geq 0} a_n z^n$ is strictly positive, we call f a *power series*. If necessary, we emphasize: convergent power series. The set of power series is written $\mathbf{C}\{z\}$. It is a subset of $\mathbf{C}[[z]]$ and it contains $\mathbf{C}[z]$.

Examples 2.9. The r.o.c. of $\sum z^n$ is 1. The r.o.c. of $\sum z^n/n!$ is $+\infty$. The r.o.c. of $\sum n! z^n$ is 0.

Corollary 2.10. *Let* r *be the r.o.c. of* $\sum_{n \geq 0} a_n z^n$. *If* $|z_0| < r$, *then* $\sum_{n \geq 0} a_n z_0^n$ *is absolutely convergent. If* $|z_0| > r$, *then* $\sum_{n \geq 0} a_n z_0^n$ *diverges.*

The open disk $\overset{\circ}{\mathrm{D}}(0, r)$ is called the *disk of convergence*. Its boundary, the circle $\partial \mathrm{D}(0, r)$, is called the *circle of indeterminacy*.

Examples 2.11. Let $k \in \mathbf{Z}$ (actually, what follows works for $k \in \mathbf{R}$). The series $\sum_{n \geq 1} z^n/n^k$ converges absolutely for $|z| < 1$ and it diverges for $|z| > 1$, so its r.o.c. is 1.

For $k \leq 0$, it converges at no point of the circle of indeterminacy.

For $k > 1$, it converges at all points of the circle of indeterminacy.

For $0 < k \leq 1$, it diverges at $z = 1$. But it converges at all the other points of the circle of indeterminacy.

Exercise 2.12. Formulate and prove the corresponding properties for $\sum_{n \geq 1} z^n / n^s$, where $s \in \mathbf{R}$.

Rules for computing the radius of convergence. The first rule is the easiest to use; it is a direct consequence of d'Alembert criterion for series. If $a_n \neq 0$ for n big enough and if $\lim_{n \to +\infty} \left| \dfrac{a_{n+1}}{a_n} \right| = l$, then the r.o.c. is $1/l$.

The second rule is due to Hadamard. It is more complicated, but more general: it always applies if the first rule applies, but it also applies in other cases. In its simplest form, it says that if $\lim_{n \to +\infty} \sqrt[n]{|a_n|} = l$, then the r.o.c. is $1/l$. (There is a more complete form using $\overline{\lim}$ but we shall not need it.)

What happens near the circle of indeterminacy. The power series $f := \sum_{n \geq 0} a_n z^n$ with r.o.c. r defines a continuous function on $\overset{\circ}{\mathrm{D}}(0, r)$. We shall admit:

Theorem 2.13 (Abel's radial theorem). *Suppose that f converges at some $z_0 \in \partial \mathrm{D}(0, r)$. Then:*

$$f(z_0) = \lim_{\substack{t \to 1 \\ t < 1}} f(tz_0).$$

$\square$

Example 2.14. Take $f(z) := \sum_{n \geq 1} \dfrac{(-1)^{n-1}}{n} z^n$. Its r.o.c. is 1. By the standard criterion for alternating series, it converges at $z = 1$. For $0 < t < 1$, f can be derived termwise to give $\dfrac{1}{1+t}$, so that $f(t) = \ln(1+t)$. Therefore:

$$\sum_{n \geq 1} \frac{(-1)^{n-1}}{n} = \lim_{\substack{t \to 1 \\ t < 1}} \sum_{n \geq 1} \frac{(-1)^{n-1}}{n} t^n = \lim_{\substack{t \to 1 \\ t < 1}} \ln(1+t) = \ln 2.$$

Remark 2.15. The converse of Abel's theorem is not generally true. For instance, if we take $f(z) := \sum_{n \geq 0} (-1)^n z^n$, we see that for $0 < t < 1$, one has $f(t) = \dfrac{1}{1+t}$ which tends to $1/2$ as $t \to 1$, $t < 1$. But of course $f(1)$ does not converge. (When a converse is proved to be true under additional assumption, it is called a "Tauberian theorem".)

2.3. The ring of power series

To each power series f with strictly positive r.o.c., we associate a continuous function on some open neighborhood of 0 (actually a disk), which we also write f. The neighborhood is not the same for all power series and all the associated functions.

Lemma 2.16. *If two power series define the same function in some neighborhood of 0, then they are equal, i.e., they have the same coefficients.*

Proof. Suppose $\sum_{n\geq 0} a_n z^n = \sum_{n\geq 0} b_n z^n$ for all z such that $|z| < r$, for some $r > 0$. Then putting $z = 0$ we have $a_0 = b_0$; then dividing by z, we see that $\sum_{n\geq 0} a_{n+1} z^n = \sum_{n\geq 0} b_{n+1} z^n$ for all z such that $0 < |z| < r$, hence also for $z = 0$ by continuity. Therefore, we can iterate the process. $\square$

Therefore, in order to determine the power series f, it is enough to know the function f in some undetermined neighborhood of 0. We shall say that two functions defined in some neighborhoods of 0 (maybe not the same neighborhood) define the same *germ* at 0 if they are equal in some neighborhood of 0 (maybe strictly smaller than the intersection of the neighborhoods we began with). A more formal definition is given in exercise 1 to this chapter. So, to each power series is associated a germ at 0 and the process is injective. The set of germs obtained in this way (that is, coming from convergent power series) will be written $\mathcal{O}_0$. Therefore, we can identify the power series f with the associated germ and the set $\mathbf{C}\{z\}$ with the set $\mathcal{O}_0$.

Let us temporarily write $f_1 \sim f_2$ if f_1, f_2 are functions in some neighborhoods of 0 and if they define the same germ in $\mathcal{O}_0$. Then the following rules are easily established (with obvious notation): $f_1 \sim f_2 \Rightarrow \lambda f_1 \sim \lambda f_2$; $f_1 \sim f_2$ and $g_1 \sim g_2 \Rightarrow f_1 + g_1 \sim f_2 + g_2$; $f_1 \sim f_2$ and $g_1 \sim g_2 \Rightarrow f_1 g_1 \sim f_2 g_2$. We deduce from these rules that germs can be multiplied by scalars, and added and multiplied among themselves. Clearly, they form a $\mathbf{C}$-algebra.

On the other hand, it is not difficult to see that for power series, if f has r.o.c. r and defines the germ ϕ, then λf has r.o.c. r (or maybe $+\infty$ if $\lambda = 0$) and it defines the germ $\lambda \phi$. Likewise, if f and g respectively have r.o.c. r and s and define the germs ϕ and γ, then $f + g$ has r.o.c. $\geq r$ and defines the germ $\phi + \gamma$; and fg has r.o.c. $\geq r$ and defines the germ $\phi\gamma$. If f has r.o.c. r and $f(0) \neq 0$, then its inverse series in $\mathbf{C}[[z]]$ is a convergent power series. It will follow from the next chapter on analytic functions that the r.o.c. of $1/f$ is the smallest $|z_0|$ for $f(z_0) = 0$; or is at least that of f if f has no zero in its disk of convergence. An easy example is that of $f(z) = 1 - z$, $z_0 = 1$: the inverse $1 + z + z^2 + \cdots$ has r.o.c. 1.

We conclude that $\mathbf{C}\{z\}$ is a subalgebra of $\mathbf{C}[[z]]$ and that it is isomorphic to $\mathcal{O}_0$. Its invertible elements are those such that $f(0) \neq 0$.

Now let $f, g \in \mathbf{C}\{z\}$. If $f(0) = 0$, then one can compose the series. One can prove that $g \circ f$ is a convergent power series (see the book of Cartan [**Car63**] for details) and that the associated germ is the composition of the germs associated to f and g. In the same way, the reciprocation processes for power series and for germs of functions correspond to each other.

2.4. **C**-derivability of power series

First, when h is small, $(z + h)^n = z^n + nz^{n-1}h + O(h^2)$, so that

$$\lim_{h \to 0} \frac{(z+h)^n - z^n}{h} = nz^{n-1}.$$

For a power series $f := \sum_{n \geq 0} a_n z^n$, we can therefore formally calculate the **C**-derivative:

$$\begin{aligned}
\lim_{h \to 0} \frac{f(z+h) - f(z)}{h} &= \lim_{h \to 0} \sum_{n \geq 0} a_n \frac{(z+h)^n - z^n}{h} \\
&= \sum_{n \geq 0} \lim_{h \to 0} a_n \frac{(z+h)^n - z^n}{h} \\
&= \sum_{n \geq 0} a_n (nz^{n-1}) \\
&= \sum_{n \geq 0} (n+1) a_{n+1} z^n.
\end{aligned}$$

The interchange $\lim_{h \to 0} \sum_{n \geq 0} = \sum_{n \geq 0} \lim_{h \to 0}$ can be justified on the open disk of convergence by the fact that the result converges normally in every strictly smaller closed disk. We conclude that convergent power series are **C**-derivable and that the **C**-derivation is computed in the same way as the formal derivation. Note that one cannot conclude on the circle of indeterminacy, as shown by the example of the series $\sum_{n \geq 1} z^n/n^2$.

Theorem 2.17. *A power series of r.o.c. r defines on $\overset{\circ}{\mathrm{D}}(0, r)$ an indefinitely* **C***-derivable function which is equal to its Taylor expansion at 0.*

Proof. By iterating the argument above, one finds that the kth derivative is $f^{(k)}(z) = \sum_{n \geq 0} \frac{(n+k)!}{n!} a_{n+k} z^n$, whence $a_k = \frac{f^{(k)}(0)}{k!}$. $\square$

By exactly the same computation as in the case of the exponential, we obtain:

Corollary 2.18. *The associated function $F(x, y) = (A(x, y), B(x, y))$ from $\overset{\circ}{D}(0, r)$ (viewed as an open disk in $\mathbf{R}^2$) to $\mathbf{R}^2$ is indefinitely differentiable.[2] Its Jacobian matrix is given by the formula:*

$$JF(x, y) = \begin{pmatrix} \dfrac{\partial A(x,y)}{\partial x} & \dfrac{\partial A(x,y)}{\partial y} \\ \dfrac{\partial B(x,y)}{\partial x} & \dfrac{\partial B(x,y)}{\partial y} \end{pmatrix} = \begin{pmatrix} \Re(f'(z)) & -\Im(f'(z)) \\ \Im(f'(z)) & \Re(f'(z)) \end{pmatrix}.$$

In particular, we have the Cauchy-Riemann formulas:

$$\frac{\partial A(x,y)}{\partial x} = \frac{\partial B(x,y)}{\partial y},$$
$$\frac{\partial A(x,y)}{\partial y} = -\frac{\partial B(x,y)}{\partial x},$$

which are often summarized as:

$$\frac{\partial f(z)}{\partial y} = i\frac{\partial f(z)}{\partial x}.$$

Since the Jacobian determinant is $|f'(z)|^2$, the local inversion theorem allows us to deduce:

Corollary 2.19. *At all points of the disk of convergence such that $f'(z) \neq 0$, the map f is "locally invertible" and its inverse (or reciprocal[3]) is holomorphic.*

Later, we shall prove that, if f is not constant, the zeros of f are isolated. With the corollary above, this implies:

Corollary 2.20. *If f is not constant, it defines an open map on $\overset{\circ}{D}(0, r)$. (This means that it transforms open subsets of $\overset{\cap}{D}(0, r)$ into open sets.)*

Exercise 2.21. Let $k \geq 2$. Prove using the above corollaries that there exist an open disk U and a power series f defined on U such that $f^k = 1 + z$. Deduce from this that, for $g \in \mathbf{C}\{z\}$ to be the kth power of a power series, it is necessary and sufficient that $v_0(g)$ is a multiple of k.

---

[2]In the next chapter (see Section 3.1), we shall delve again into that differential calculus point of view on complex functions.

[3]We are speaking here of the existence of a function g such that $f(g(z)) = z$ and $g(f(z)) = z$, so the right wording is reciprocal, reciprocable. But this kind of result is traditionally called "local inversion"; one should keep in mind that $1/f$ is not relevant here.

2.5. Expansion of a power series at a point $\neq 0$

Let $f := \sum\limits_{n \geq 0} a_n z^n$ with r.o.c. $r > 0$ and let $z_0 \in \overset{\circ}{\mathrm{D}}(0, r)$. We formally compute the expansion of f near z_0 as follows:

$$f(z_0 + z) = \sum_{n \geq 0} a_n (z_0 + z)^n$$

$$= \sum_{n \geq 0} a_n \sum_{l+m=n} \frac{(l+m)!}{l!m!} z_0^l z^m$$

$$= \sum_{m \geq 0} \left(\sum_{l \geq 0} \frac{(l+m)!}{l!m!} a_{l+m} z_0^l \right) z^m$$

$$= \sum_{m \geq 0} \frac{f^{(m)}(z_0)}{m!} z^m,$$

since we already know that $f^{(m)}(z_0) = \sum\limits_{l \geq 0} \frac{(l+m)!}{l!} a_{l+m} z_0^l$. This calculation can be rigorously justified, and one can prove (see for example the book of Cartan [**Car63**]):

Theorem 2.22. *The Taylor series* $\sum\limits_{m \geq 0} \frac{f^{(m)}(z_0)}{m!} z^m$ *of the function f at z_0 is convergent. Its r.o.c. is at least equal to $r - |z_0|$. The function $g(z)$ it defines is equal to $f(z_0 + z)$ on $\overset{\circ}{\mathrm{D}}(0, r - |z_0|)$.* $\qquad\square$

Example 2.23. Let $f(z) = 1 + z + z^2 + \cdots = \dfrac{1}{1-z}$ (for $|z| < 1$) and let $|z_0| < 1$. Then:

$$f^{(n)}(z) = \sum_{k \geq 0} \frac{(n+k)!}{k!} z^k = \frac{n!}{(1-z)^{n+1}}.$$

Therefore:

$$\sum_{n \geq 0} \frac{f^{(n)}(z_0)}{n!} z^n = \sum_{n \geq 0} \frac{z^n}{(1-z_0)^{n+1}}.$$

This has r.o.c. $|1 - z_0|$. For $|z| < |1 - z_0|$, one has:

$$\sum_{n \geq 0} \frac{z^n}{(1-z_0)^{n+1}} = \frac{1}{1-z_0} \frac{1}{1 - \frac{z}{1-z_0}} = \frac{1}{1 - z_0 - z} = f(z_0 + z).$$

The very last equality makes sense because $|z| < |1 - z_0| \Rightarrow |z_0 + z| < 1$, so that $z_0 + z$ is indeed in the disk of convergence of f.

Exercise 2.24. In the example above, for what values of z_0 is the r.o.c. of the new power series bigger than $|1 - z_0|$? Draw the corresponding disk to see how the domain of f has been extended.

Generally speaking, calling r' the new r.o.c., either $\overset{\circ}{D}(z_0, r')$ goes beyond the boundary of $\overset{\circ}{D}(0, r)$, or not. The points of the circle of indeterminacy which cannot be crossed out in this way are "boundary points". It can be proved that there are always boundary points on the circle of indeterminacy. (This is a consequence of "Cauchy theory".) For some special power series, like $\sum_{n \geq 0} z^{2^n}$, "Hadamard's theorem on lacunary series" (see [**Rud87**, Theorem 16.6]) implies that all the points of the circle of indeterminacy are boundary points.

2.6. Power series with values in a linear space

Let $V := \mathbf{C}^d$ and let $X(z) := \begin{pmatrix} f_1(z) \\ \vdots \\ f_d(z) \end{pmatrix}$, where the $f_i \in \mathbf{C}\{z\}$ have r.o.c. r_i.

The vector-valued function X is defined and continuous on $\overset{\circ}{D}(0, r)$, where $r := \min(r_1, \ldots, r_d) > 0$.

Defining the **C**-derivative of $X(z)$ as $X'(z) := \lim_{h \to 0} \dfrac{X(z + h) - X(z)}{h}$

(also written $dX(z)/dz$), we see that it is indeed **C**-derivable on $\overset{\circ}{D}(0, r)$ and that $X'(z) = \begin{pmatrix} f_1'(z) \\ \vdots \\ f_d'(z) \end{pmatrix}$.

We can also group the power series expansions $f_i(z) = \sum a_{i,n} z^n$ in the form $X = \sum X_n z^n$, where $X_n := \begin{pmatrix} a_{1,n} \\ \vdots \\ a_{d,n} \end{pmatrix} \in V$. One can prove that if, for an arbitrary norm on V, one has $\lim_{n \to +\infty} \sqrt[n]{\|X_n\|} = l$, then the r.o.c. of $X(z)$ is $1/l$.

The derivation of vector-valued functions is **C**-linear: $(\lambda.X)' = \lambda.X'$ and $(X + Y)' = X' + Y'$. The Leibniz rule takes the form: $(f.X)' = f'.X + f.X'$.

Exercises

(1) *Formal definition of germs.* We state this in the context of continuous functions of a complex variable; adaptation to other contexts is easy and left to the reader. Let U be an open subset of $\mathbf{C}$ and $a \in U$. We consider pairs (f, V), where $V \subset U$ is an open neighborhood of a and f is continuous on V. We say that $(f_1, V_1) \sim (f_2, V_2)$ if there is an open neighborhood W of a such that $W \subset V_1 \cap V_2$ and that f_1, f_2 have the same restriction to W. Check that this is an equivalence relation. The class of a pair (f, V) is called the germ of f at a, and f is called a representative of that germ. Then prove that germs at a of continuous functions can be added and multiplied and that they form a $\mathbf{C}$-algebra.

(2) Prove the existence of a reciprocal (Corollary 2.19) by solving a fixed point problem. For instance, if $f = az + \phi$, $a \neq 0$ and $v_0(\phi) \geq 2$, you want to solve $f(g(z)) = z$ in the form:

$$z = ag(z) + \phi(g(z)) \iff g = \Phi(g), \text{ where } \Phi(g) := \frac{1}{a}(z - \phi \circ g).$$

(3) Prove the convergence properties of $\sum_{n \geq 1} z^n / n^k$ on its circle of indeterminacy when $0 < k \leq 1$. (This uses Abel's transformation: putting $S_n := 1 + z + \cdots + z^n$, one has

$$\sum_{n=1}^{N} z^n / n^k = \sum_{n=1}^{N} (S_n - S_{n-1}) / n^k = \sum_{n=1}^{N} S_n (1/n^k - 1/(n+1)^k) + R_N,$$

where $R_N \to 0$ and, if $|z| = 1$, $z \neq 1$, the S_n are bounded.)

(4) Taking $g = 1 + z + z^2/2 + z^3/6 + \cdots$ and $f(z) = z - z^2/2 + z^3/3 - z^4/4 + \cdots$, use the relation $g \circ f = 1 + z$ (proved in the section on formal power series) to compute $\sum_{n \geq 1} i^n / n$. Then deduce the formulas:

$1 - 1/3 + 1/5 - 1/7 + \cdots = \pi/4$ and $1/2 - 1/4 + 1/6 - 1/8 + \cdots = (\ln 2)/2$.

(5) Define matrix-valued functions $A(z) := (a_{i,j}(z))_{1 \leq i,j \leq d}$ taking values in $\mathrm{Mat}_d(\mathbf{C})$, such that all $a_{i,j} \in \mathbf{C}\{z\}$. Write their $\mathbf{C}$-derivatives, the associated rules, the power series expansions. With A and X as described, what can be said of AX?

Chapter 3

Analytic functions

3.1. Analytic and holomorphic functions

Definition 3.1. (i) Let f be a function or a germ (remember that germs were defined in Section 2.3 and exercise 1 above). We say that f *admits a power series expansion at* $z_0 \in \mathbf{C}$ if there is a (convergent) power series $\sum_{n \geq 0} a_n z^n \in \mathbf{C}\{z\}$ such that, for z in some neighborhood of 0, one has: $f(z_0 + z) = \sum_{n \geq 0} a_n z^n$. We shall then rather write that, for z in some neighborhood of z_0, one has: $f(z) = \sum_{n \geq 0} a_n (z - z_0)^n \in \mathcal{O}_{z_0} = \mathbf{C}\{z - z_0\}$. For conciseness, we shall say that f is *power series expandable* at z_0.

(ii) Let f be a function on an open set $\Omega \subset \mathbf{C}$. We say that f *is analytic on* Ω if f admits a power series expansion at all points $z_0 \in \Omega$. An *analytic germ* is the germ of an analytic function.

(iii) A function analytic on the whole of $\mathbf{C}$ is said to be *entire*.

Examples 3.2. (i) The function $e^z = \sum_{n \geq 0} \frac{e^{z_0}}{n!}(z - z_0)^n$ is power series expandable at any $z_0 \in \mathbf{C}$, so it is entire.

(ii) The function $\frac{1}{z} = \frac{1}{z_0(1 + \frac{z - z_0}{z_0})} = \sum_{n \geq 0} \frac{(-1)^n}{z_0^{n+1}}(z - z_0)^n$ is power series expandable at any $z_0 \neq 0$, therefore it is analytic on $\mathbf{C}^*$. However, no power series describes it on the whole of $\mathbf{C}^*$.

(iii) If f is power series expandable at z_0 with an r.o.c. r, then it is analytic on $\overset{\circ}{\mathrm{D}}(z_0, r)$ (Theorem 2.22); in particular, if $r = +\infty$, then f is an entire function.

29

Definition 3.3. (i) We say that the function or germ f is **C**-*derivable at* z_0 if the limit $\lim\limits_{h\to 0} \dfrac{f(z_0 + h) - f(z_0)}{h}$ exists. This limit is called *the* **C**-*derivative of* f *at* z_0 and written $f'(z_0)$ or $\dfrac{df}{dz}(z_0)$. From now on, we shall simply say "derivable, derivative" instead of "**C**-derivable, **C**-derivative".

(ii) A function f defined on an open set $\Omega \subset \mathbf{C}$ is said to be *holomorphic on* Ω if it is **C**-derivable at every point of Ω. We then write f' or df/dz for the function $z_0 \mapsto f'(z_0)$.

If we identify f with a function $F(x, y) = (A(x, y), B(x, y))$ (with real variables and with values in $\mathbf{R}^2$), then a necessary and sufficient condition for f to be holomorphic is that F be differentiable and that it satisfies the Cauchy-Riemann conditions:

$$\frac{\partial A(x, y)}{\partial x} = \frac{\partial B(x, y)}{\partial y} \text{ and } \frac{\partial A(x, y)}{\partial y} = -\frac{\partial B(x, y)}{\partial x},$$

or, in a more compact form:

$$\frac{\partial f(z)}{\partial y} = i\frac{\partial f(z)}{\partial x}.$$

Then the Jacobian matrix is that of a direct similitude:[1]

$$JF(x, y) = \begin{pmatrix} \dfrac{\partial A(x, y)}{\partial x} & \dfrac{\partial A(x, y)}{\partial y} \\ \dfrac{\partial B(x, y)}{\partial x} & \dfrac{\partial B(x, y)}{\partial y} \end{pmatrix} = \begin{pmatrix} \Re(f'(z)) & -\Im(f'(z)) \\ \Im(f'(z)) & \Re(f'(z)) \end{pmatrix}.$$

Remark 3.4. The geometric consequence is that a nonconstant holomorphic function preserves the angles between tangent vectors of curves, and also the orientation: it is *conformal*.

Theorem 3.5 (FUNDAMENTAL!). *Analyticity and holomorphicity are equivalent properties.*

Proof. The fact that an analytic function is holomorphic was proved in the previous chapter. The converse implication will be admitted; see [**Ahl78**, **Car63**, **Rud87**]. □

---

[1] Remember that a direct similitude in the real plane $\mathbf{R}^2$ has a matrix of the form $\begin{pmatrix} u & -v \\ v & u \end{pmatrix}$. It corresponds in the complex plane $\mathbf{C}$ to the map $z \mapsto wz$, where $w := u + iv$. These are the only linear maps that preserve angles and orientation.

Some basic properties.

(1) Analytic functions on an open set $\Omega \subset \mathbf{C}$ form a $\mathbf{C}$-algebra, which we write $\mathcal{O}(\Omega)$.

(2) If $f \in \mathcal{O}(\Omega)$, then $1/f \in \mathcal{O}(\Omega')$, where $\Omega' := \Omega \setminus f^{-1}(0)$. In particular, the elements of $\mathcal{O}(\Omega)^*$ are the functions $f \in \mathcal{O}(\Omega)$ which vanish nowhere.

(3) If $f \in \mathcal{O}(\Omega)$, $g \in \mathcal{O}(\Omega')$ and $f(\Omega) \subset \Omega'$, then $g \circ f \in \mathcal{O}(\Omega)$.

(4) Let $z_0 \in \Omega$ and let δ denote the distance of z_0 to the exterior of Ω (or to its boundary; it is the same): $\delta := d(z_0, \mathbf{C} \setminus \Omega) = d(z_0, \partial\Omega) > 0$. Then f is indefinitely derivable at z_0, and equal to its Taylor series expansion $\sum_{m \geq 0} \dfrac{f^{(m)}(z_0)}{m!}(z - z_0)^m$ on $\overset{\circ}{\mathrm{D}}(z_0, \delta)$ (which means implicitly that this series has r.o.c. $\geq \delta$).

The following theorem will play a central role in our course. We admit it (see [**Ahl78**, **Car63**, **Rud87**]).

Theorem 3.6 (Principle of analytic continuation). *Suppose that Ω is a domain (a connected open set). If $f \in \mathcal{O}(\Omega)$ vanishes on a nonempty open set, then $f = 0$ (the zero function on Ω).* $\square$

As a consequence, if f is not the zero function on Ω, at every $z_0 \in \Omega$ it has a nontrivial power series expansion: $f(z) = \sum_{n \geq k} a_n(z - z_0)^n$, with $a_k \neq 0$. Then $f = (z - z_0)^k g$, where g is power series expandable at z_0 and $g(z_0) \neq 0$. We shall then write $v_{z_0}(f) = k$.

This implies in particular that, in some neighborhood of z_0, g does not vanish, thus is invertible (*i.e.*, $1/g$ exists).

Corollary 3.7. *The zeros of a nontrivial analytic function on a domain are isolated.*

Remember from the previous chapter that, if f is analytic on a domain Ω, then, at all points $z_0 \in \Omega$ such that $f'(z_0) \neq 0$, the map f is locally invertible. With the corollary above, this implies:

Corollary 3.8. *If f is not constant, it defines an open map on $\overset{\circ}{\mathrm{D}}(0, r)$. (This means that it transforms open subsets of $\overset{\circ}{\mathrm{D}}(0, r)$ into open subsets of $\mathbf{C}$.)*

Corollary 3.9. *Let f be a nontrivial analytic function on a domain Ω. Then, on every compact subset of Ω, f has finitely many zeros. Altogether, the set $f^{-1}(0) \subset \Omega$ of its zeros is at most denumerable.*

3.2. Singularities

Theorem 3.10 (Riemann's theorem of removable singularities). *Let $\Omega \subset \mathbf{C}$ be an open set and let $z_0 \in \Omega$. Assume that $f \in \mathcal{O}(\Omega \setminus \{z_0\})$ is bounded on some punctured neighborhood of z_0, that is, on some $U \setminus \{z_0\}$, where U is a neighborhood of z_0. Then f admits a continuation at z_0 which makes it an analytic function on the whole of Ω.*

Proof. See [**Ahl78, Car63, Rud87**].   □

Obviously, the said continuation is unique and we shall identify it with f (and write it f).

Example 3.11. The function $f(z) := \dfrac{z}{e^z - 1}$ is analytic on $\Omega := \mathbf{C} \setminus 2i\pi\mathbf{Z}$. But since $\lim\limits_{z \to 0} f(z) = \dfrac{1}{\exp'(0)} = 1$, the function f is bounded near 0 and can be continued there by putting $f(0) := 1$.

Exercise 3.12. Show that

$$\frac{z}{e^z - 1} = 1 - z/2 - \sum_{n \geq 1} (-1)^n B_n z^{2n} / (2n)!,$$

where the *Bernoulli numbers* $B_n \in \mathbf{Q}$, and compute B_1, B_2, B_3. What is the radius of convergence of this expansion? (The Bernoulli numbers are related to the values of Riemann's zeta function at the even integers; see [**Ser78**].)

Corollary 3.13. *Let $\Omega \subset \mathbf{C}$ be an open set and let $z_0 \in \Omega$ and $f \in \mathcal{O}(\Omega \setminus \{z_0\})$. Three cases are possible:*

(1) *If f is bounded on some punctured neighborhood of z_0, we consider it as analytic on Ω.*

(2) *Otherwise, if there exists $N \geq 1$ such that $f(z) = O\left(|z - z_0|^{-N}\right)$, then there exists a unique $k \geq 1$ (actually, the smallest of all possible such N) such that $g := (z - z_0)^k f$ is analytic and $g(z_0) \neq 0$. In this case, f is said to have a pole of order k at z_0. We put $v_{z_0}(f) := -k$.*

(3) *Otherwise, we say that f has an essential singularity at z_0.*

Example 3.14. The function $\dfrac{e^{1/z}}{e^z - 1}$ has simple poles (*i.e.*, of order 1) at all points of $2i\pi\mathbf{Z}$ except 0 and an essential singularity at 0 .

If f has a pole of order k at z_0, it admits a (convergent) Laurent series expansion $f(z) = \sum\limits_{n \geq -k} a_n(z - z_0)^n$, with $a_{-k} \neq 0$. We write $\mathbf{C}(\{z - z_0\})$ as the $\mathbf{C}$-algebra of such series. It is the field of fractions of $\mathbf{C}\{z - z_0\}$ and it is actually equal to $\mathbf{C}\{z - z_0\}[1/(z - z_0)]$. Clearly, poles are isolated.

On the other hand, in the case of an essential singularity, f has a "generalized Laurent series expansion", with infinitely many negative powers; for instance, $e^{1/z} = \sum\limits_{n \geq 0} z^{-n}/n!$. Essential singularities need not be isolated, as shown in the following example:

Exercise 3.15. What are the singularities of $1/\sin(1/z)$? Which ones are poles, resp. essential singularities?

Definition 3.16. The function f is said to be *meromorphic* on the open set Ω if there is a discrete subset $X \subset \Omega$ such that f is analytic on $\Omega \setminus X$ and has poles on X.

The following is easy to prove:

Theorem 3.17. *Meromorphic functions on a domain Ω form a field $\mathcal{M}(\Omega)$. In particular, if $f, g \in \mathcal{O}(\Omega)$ and $g \neq 0$, then $f/g \in \mathcal{M}(\Omega)$.*

Much more difficult is the theorem that all meromorphic functions are quotients of holomorphic functions (see [**Rud87**, Theorem 15.12]).

Examples 3.18. (i) Rational functions $f := P/Q \in \mathbf{C}(z)$ are meromorphic on $\mathbf{C}$. If $P, Q \in \mathbf{C}[z]$ are coprime polynomials and if z_0 is a root of order k of P, then $v_{z_0}(f) = k$. If z_0 is a root of order k of Q, then it is a pole of order k of f and $v_{z_0}(f) = -k$. If z_0 is a root of neither P nor Q, then $v_{z_0}(f) = 0$.

(ii) If Ω is a domain and $f \in \mathcal{M}(\Omega)$, $f \neq 0$, then $f'/f \in \mathcal{M}(\Omega)$. The poles of f'/f are the zeros and the poles of f. They are all simple. All this comes from the fact that if $f := (z - z_0)^k g$ with g analytic at z_0 and $g(z_0) \neq 0$, then $f'/f = k/(z - z_0) + g'/g$, and g'/g is analytic at z_0.

3.3. Cauchy theory

Let Ω be a domain, let $f \in \mathcal{O}(\Omega)$ and let $\gamma : [a, b] \to \Omega$ be a continuous path ($a, b \in \mathbf{R}$ and $a < b$). We shall define:

$$\int_\gamma f(z)\, dz := \int_a^b f(\gamma(t))\gamma'(t)\, dt.$$

For this definition to make sense, we shall require the path γ to be of class $\mathcal{C}^1$, that is, continuously differentiable. Note however that the weaker assumption piecewise continuously differentiable (and continuous) would be sufficient. One can easily check that, in the above formula, reparameterizing the path (that is, using $\gamma(\phi(s))$, where $\phi : [a', b'] \to [a, b]$ is a change of parameters) does not change the integral.

Note that, if f has a primitive F on Ω (that is, $F \in \mathcal{O}(\Omega)$ and $F' = f$), then:

$$\int_\gamma f(z)\, dz = F(\gamma(b)) - F(\gamma(a)).$$

In particular, if γ is a loop, then (still assuming the existence of a primitive) $\int_\gamma f(z)\, dz = 0$.

Examples 3.19. (i) Let $k \in \mathbf{Z}$ and let $\gamma_1 : [0, 2\pi] \to \mathbf{C}^*$, $t \mapsto e^{ikt}$ and $\gamma_2 : [0, 1] \to \mathbf{C}^*$, $t \mapsto e^{2i\pi kt}$. Let $f(z) := z^n$, $n \in \mathbf{Z}$. Then:

$$\int_{\gamma_1} f(z)\, dz = \int_{\gamma_2} f(z)\, dz = \begin{cases} 2i\pi k & \text{if } n = -1, \\ 0 & \text{otherwise.} \end{cases}$$

Indeed:

$$\int_{\gamma_1} z^n\, dz = \int_0^{2\pi} ik e^{ik(n+1)t}\, dt \text{ and } \int_{\gamma_2} z^n\, dz = \int_0^1 2i\pi k e^{2i\pi k(n+1)t}\, dt.$$

(ii) From this, by elementary computation, one finds that if f has a generalized Laurent series expansion (be it holomorphic, meromorphic or essentially singular) $\sum\limits_{n \in \mathbf{Z}} a_n z^n$ at 0 and if its domain of convergence contains the unit circle, then:

$$\int_{\gamma_1} f(z)\, dz = \int_{\gamma_2} f(z)\, dz = 2i\pi k a_{-1}.$$

The following important theorems due to Cauchy are proved in [**Ahl78**, **Car63**, **Rud87**].

Theorem 3.20 (Cauchy). *If $f \in \mathcal{O}(\Omega)$ and $\gamma_1, \gamma_2 : [a, b] \to \Omega$ are two homotopic paths of class $\mathcal{C}^1$ (that is, they can be continuously deformed into each other within Ω; see exercise 6 in Chapter 1), then:*

$$\int_{\gamma_1} f(z)\, dz = \int_{\gamma_2} f(z)\, dz.$$

$\square$

Using the calculations in the examples, one deduces:

Corollary 3.21. *If $f(z) = \sum\limits_{n \in \mathbf{Z}} a_n(z - z_0)^n$ and if the loop γ has its image in the domain of convergence of f, then:*

$$\int_\gamma f(z)\, dz = 2i\pi\, I(z_0, \gamma)\, a_{-1}.$$

Remember from Corollary 1.11 that the index $I(z_0, \gamma)$ is the number of times that the loop γ turns around z_0 in the positive sense. Actually, the corollary above yields a possible alternative definition of the index.

Definition 3.22. If $f(z) = \sum\limits_{n \in \mathbf{Z}} a_n(z - z_0)^n$, the complex number a_{-1} is called the *residue of f at z_0* and written $\mathrm{Res}_{z_0}(f)$.

Cauchy formulas in simply-connected domains. From now on, in this section (Cauchy theory), we shall assume[2] the domain Ω to be *simply connected*, that is, all loops can be continuously shrunk to a point. The precise definition rests on homotopy theory (see exercise 5 at the end of the chapter), but the reader can think intuitively that there is no "hole" in Ω; in particular, convex domains are simply connected.

Theorem 3.23 (Cauchy residue formula). *Let f have a finite number of singularities in Ω and let γ be a loop in Ω avoiding all these singularities. Then:*

$$\int_\gamma f(z)\,dz = 2\mathrm{i}\pi \sum_{z_0} I(z_0, \gamma)\mathrm{Res}_{z_0}(f),$$

the sum being taken for all singularities z_0. □

Corollary 3.24. *Suppose $f \in \mathcal{M}(\Omega)$ (a simply-connected domain), $f \neq 0$, and γ is a loop in Ω avoiding all zeros and poles of f. Then:*

$$\int_\gamma (f'/f)(z)\,dz = 2\mathrm{i}\pi \sum_{z_0} I(z_0, \gamma)v_{z_0}(f),$$

the sum being taken for all zeros and poles.

Corollary 3.25 (Cauchy formula). *Suppose $f \in \mathcal{O}(\Omega)$ and γ is a loop in Ω avoiding $z_0 \in \Omega$. Then:*

$$\int_\gamma \frac{f(z)}{(z - z_0)^{k+1}}\,dz = 2\mathrm{i}\pi\, I(z_0, \gamma)\, \frac{f^{(k)}(z_0)}{k!}.$$

Remark 3.26. The standard application of the Cauchy residue formula and of its corollaries is when the integration contour loops exactly once around the set of all singularities (it is homotopic to a circle surrounding them and followed in the positive sense): then all indexes $I(z_0, \gamma)$ equal 1 and we get equalities such as $\int_\gamma f(z)\,dz = 2\mathrm{i}\pi \sum_{z_0} \mathrm{Res}_{z_0}(f)$, etc.

---

[2]Rudin in [**Rud87**] requires a weaker condition, but such generality is of no use for our purposes.

Primitives. Curvilinear integrals (*i.e.*, integrals on paths) can serve to compute primitives (or prove they do not exist: see the remark below and the next chapter for this). Let Ω be a simply-connected domain. Then by the Cauchy theorem on homotopy invariance (Theorem 3.20), if $f \in \mathcal{O}(\Omega)$ and $\gamma_1, \gamma_2 : [a, b] \to \Omega$ are any two paths of class $\mathcal{C}^1$ with $\gamma_1(a) = \gamma_2(a)$ and $\gamma_1(b) = \gamma_2(b)$, we have:

$$\int_{\gamma_1} f(z)\, dz = \int_{\gamma_2} f(z)\, dz.$$

Now fix $a = z_0$ and consider $b = z$ as a variable. Then the above integrals define a same function $F(z)$ on Ω. We would then write:

$$F(z) = \int_{z_0}^{z} f(t)\, dt.$$

Theorem 3.27. *This function is the unique primitive of f (that is, $F' = f$) such that $F(z_0) = 0$.* $\qquad\qquad\qquad\qquad\qquad\qquad\qquad\qquad\qquad\square$

Remark 3.28. If Ω is not simply connected, some functions may have no primitives on Ω. For instance, if $a \notin \Omega$ and there is a loop γ in Ω such that $I(a, \gamma) \neq 0$, then the function $f(z) := \dfrac{1}{z - a}$ is holomorphic on Ω but has no primitive; indeed, $\int_{\gamma} f(z)\, dz = I(a, \gamma)$.

If Ω is not assumed to be simply connected, values of integrals of f on loops are called "periods" of f, and $\int_{z_0}^{z} f(t)\, dt$ is defined up to a period.

3.4. Our first differential algebras

Let Ω a nonempty domain in $\mathbf{C}$. The map $D : f \mapsto f'$ maps the $\mathbf{C}$-algebra $\mathcal{O}(\Omega)$ into itself, it is $\mathbf{C}$-linear, and it satisfies the *Leibniz rule*:

$$\forall f, g \in \mathcal{O}(\Omega)\,,\ D(fg) = f D(g) + D(f) g.$$

We say that D is a ($\mathbf{C}$-linear) *derivation* and that $\mathcal{O}(\Omega)$, along with that derivation, is a *differential algebra* over $\mathbf{C}$.

If $a \in \Omega$, the natural map $\mathcal{O}(\Omega) \to \mathcal{O}_a$ sending a function to its germ at a is injective as a consequence of the principle of analytic continuation. Identifying $\mathcal{O}(\Omega)$ with a sub-$\mathbf{C}$-algebra of $\mathcal{O}_a$, we see that the derivation D on $\mathcal{O}(\Omega)$ can be extended to a derivation on $\mathcal{O}_a$, making it a differential algebra. Conversely, we can see $\mathcal{O}(\Omega)$ as a sub-algebra of the differential algebra $\mathcal{O}_a$ stable under the derivation of $\mathcal{O}_a$, *i.e.*, a *sub-differential algebra* of the differential algebra $\mathcal{O}_a$.

Exercises

(1) Find the zeros of $\sin(\pi/z)$. Do they accumulate? Does this contradict the above results on isolated zeros?

(2) Show that Bernoulli numbers are related to the volumes of spheres in $\mathbf{R}^n$.

(3) Let $\gamma(t) := Re^{it}$ on $[0, 2\pi]$, where R is "big" and $R \notin 2i\pi\mathbf{N}$. For $k \in \mathbf{N}$, compute $\displaystyle\int_\gamma z^{-k} \frac{z}{e^z - 1}\, dz$.

(4) Check that the function $1/z(1-z)$, which is holomorphic on the domain $\Omega \setminus \{0, 1\}$, has no primitive and compute the resulting periods.

(5) Say that a domain Ω is *simply connected* if every loop in Ω is homotopic (exercise 6 in Chapter 1) to a constant loop (*i.e.*, one whose image is a point). Prove that every convex domain is simply connected. More generally, any star-shaped domain (*i.e.*, one such that there exists a point that can be linked to any point of the domain along a straight line) is simply connected. Draw other examples. Prove that an annulus is not simply connected. Prove that, if Ω is an arbitrary domain and $a \in \Omega$, then $\Omega \setminus \{a\}$ is not simply connected.

(6) Prove that if Ω is simply connected and $a \notin \Omega$, then no loop in Ω winds around a. (This last statement is the condition used in [**Rud87**] as a foundation for Cauchy theory.)

(7) In the notation of Section 3.4, prove that $\mathcal{O}(\Omega)$ can be identified with a sub-algebra of O_a and that the derivation on $\mathcal{O}(\Omega)$ can be naturally extended to $\mathcal{O}_a$.

Chapter 4

The complex logarithm

4.1. Can one invert the complex exponential function?

We know that the real exponential function $x \mapsto e^x$ from $\mathbf{R}$ to $\mathbf{R}_+^*$ can be inverted by $\ln : \mathbf{R}_+^* \to \mathbf{R}$ in the sense that $\forall x \in \mathbf{R}$, $\ln(e^x) = x$ and $\forall y \in \mathbf{R}_+^*$, $e^{\ln(y)} = y$. Moreover, $\ln$ is a rather "good" function: it is continuous, derivable, etc. We are going to *try* to extend this process to $\mathbf{C}$, that is, to invert the complex exponential function $\exp : \mathbf{C} \to \mathbf{C}^*$. (However, we shall use the notation $\ln$ only for the real logarithm mentioned above.) It is impossible to have a function $L : \mathbf{C} \to \mathbf{C}^*$ such that $\forall z \in \mathbf{C}$, $L(e^z) = z$. Indeed, since $e^{z+2i\pi} = e^z$, this would imply $z + 2i\pi = z$. Clearly, the impossibility stems from the fact that $\exp$ is not injective. However, we know that $\exp$ is surjective, so that for each $z \in \mathbf{C}^*$ there exists a (nonunique) complex number, say $L(z)$, such that $e^{L(z)} = z$. In this way, we can build a function $L : \mathbf{C}^* \to \mathbf{C}$ such that $\forall z \in \mathbf{C}^*$, $e^{L(z)} = z$. Now, the values $L(z)$ having been chosen at random (each time among infinitely many choices), it is not clear that one can get in this way a "good" function. Indeed one cannot:

Lemma 4.1. *There is no continuous function* $L : \mathbf{C}^* \to \mathbf{C}$ *such that* $\forall z \in \mathbf{C}^*$, $e^{L(z)} = z$.

Proof. Actually, it is not even possible to find such a continuous function on the unit circle $\mathbf{U}$. Assume indeed by contradiction that there was such a function $L : \mathbf{U} \to \mathbf{C}$ and put, for all $t \in \mathbf{R}$, $f(t) := L(e^{it}) - it$. Then f is a continuous function from $\mathbf{R}$ to $\mathbf{C}$. Since $e^{L(e^{it})} = e^{it}$, we see that $e^{f(t)} = 1$ for all t. Therefore the continuous function f sends the connected set $\mathbf{R}$ to the discrete set $2i\pi\mathbf{Z}$; this is only possible if it is constant. Thus, there exists a fixed $k \in \mathbf{Z}$ such that: $\forall t \in \mathbf{R}$, $L(e^{it}) = it + 2i\pi k$. Now, writing this for $t := 0$ and $t := 2\pi$ yields the desired contradiction. $\qquad\square$

Therefore, we are going to look for *local determinations of the logarithm*: this means a continuous function $L : \Omega \to \mathbf{C}$, where Ω is some open subset of $\mathbf{C}^*$, such that $\forall z \in \Omega$, $e^{L(z)} = z$. This will not be possible for arbitrary Ω.

Lemma 4.2. *Let $\Omega \subset \mathbf{C}^*$ be a domain (a connected open set). Then any two determinations of the logarithm on Ω differ by a constant. (Of course, there may exist no such determination at all!)*

Proof. If L_1 and L_2 are two determinations of the logarithm on Ω, then $\forall z \in \Omega$, $e^{L_2(z) - L_1(z)} = 1$, so that the continuous function $L_2 - L_1$ sends the connected set Ω to the discrete set $2i\pi\mathbf{Z}$, so it is constant. $\qquad\square$

Therefore, if there is at least one determination of the logarithm on Ω, there is a denumerable family of them differing by constants $2i\pi k$, $k \in \mathbf{Z}$. If one wants to specify one of them, one uses an *initial condition*: for some $z_0 \in \Omega$, one chooses a particular $w_0 \in \mathbf{C}$ such that $e^{w_0} = z_0$, and one knows that there is a unique determination such that $L(z_0) = w_0$.

4.2. The complex logarithm via trigonometry

We fix $\theta_0 \in \mathbf{R}$ arbitrary, indicating a direction in the plane, that is, a half-line $\mathbf{R}_+ e^{i\theta_0}$. We define the "cut plane" $\Omega := \mathbf{C} \setminus \mathbf{R}_- e^{i\theta_0}$ (that is, we think that we have "cut off" the "prohibited half-line" $\mathbf{R}_- e^{i\theta_0}$); it is an open subset of $\mathbf{C}^*$. Then, for all $z \in \Omega$, there is a unique pair $(r, \theta) \in \mathbf{R}_+^* \times]\theta_0 - \pi, \theta_0 + \pi[$ such that $z = re^{i\theta}$. Moreover, r and θ are continuous functions of z. Therefore, putting $L_{\theta_0}(z) := \ln(r) + i\theta$, we get a continuous function $L_{\theta_0} : \Omega \to \mathbf{C}$. This is clearly a determination of the logarithm on Ω, characterized by the initial condition $L_{\theta_0}(e^{i\theta_0}) = i\theta_0$.

If we take $\theta_0 := 0$, we get the *principal determination of the logarithm*, which we write log. It is defined on the cut plane $\mathbf{C} \setminus \mathbf{R}_- \subset \mathbf{C}^*$ and characterized by the initial condition $\log(1) = 0$. Its restriction to $\mathbf{R}_+^*$ is ln.

Remark 4.3. This nice function cannot be continuously extended to $\mathbf{R}_-$. Indeed, if for instance $z \in \mathbf{C} \setminus \mathbf{R}_-$ tends to -1, then it can be written $z = re^{i\theta}$, where $r > 0$ and $-\pi < \theta < \pi$. One has $r \to 1$ and $\theta \to \pm\pi$: if z approaches -1 by above, then $\theta \to +\pi$; if z approaches -1 by below, then $\theta \to -\pi$. In the first case, $\log(z) \to i\pi$; in the second case, $\log(z) \to -i\pi$. In full generality, z could alternate above and below and then $\log(z)$ would tend to nothing.

Remark 4.4. If we change the argument θ_0 by $\theta_0 + 2\pi$, the open set Ω does not change, but the determination of the logarithm does. We know that L_{θ_0} and $L_{\theta_0+2\pi}$ differ by a constant, so we just have to test them on the initial conditions. Since $L_{\theta_0}(e^{i\theta_0}) = i\theta_0$ and $L_{\theta_0+2\pi}(e^{i(\theta_0+2\pi)}) = i(\theta_0 + 2\pi)$, and since $e^{i(\theta_0+2\pi)} = e^{i\theta_0}$, we conclude that $L_{\theta_0+2\pi} = L_{\theta_0} + 2i\pi$.

Exercise 4.5. Under what condition on θ_0 does the open set Ω contain the positive real half-line $\mathbf{R}_+^*$? Assuming this, under which supplementary condition does one have $L_{\theta_0}(1) = 0$? Again assuming this, show that the restriction of L_{θ_0} to $\mathbf{R}_+^*$ is ln.

4.3. The complex logarithm as an analytic function

Let $L(z) := \sum_{n \geq 1} \dfrac{(-1)^{n-1}}{n}(z-1)^n$. Then L is analytic on $\overset{\circ}{\mathrm{D}}(1,1)$ and satisfies the initial condition $L(1) = 0$. Moreover, $L'(z) = \sum_{n \geq 0}(-1)^n(z-1)^n = \dfrac{1}{z}$ on $\overset{\circ}{\mathrm{D}}(1,1)$, from which one obtains $(e^L)' = e^L/z$; then $(e^L/z)' = 0$ and one concludes that $e^{L(z)} = z$. Therefore, L is a determination of the logarithm on $\overset{\circ}{\mathrm{D}}(1,1)$. Since $\overset{\circ}{\mathrm{D}}(1,1) \subset \mathbf{C} \setminus \mathbf{R}_-$, we deduce from the initial condition at 1 that L is the restriction of $\log$ (the principal determination) to $\overset{\circ}{\mathrm{D}}(1,1)$:

$$\forall z \in \overset{\circ}{\mathrm{D}}(1,1), \; \log(z) = \sum_{n \geq 1}\frac{(-1)^{n-1}}{n}(z-1)^n.$$

Proposition 4.6. *Let $z_0 := re^{i\theta_0}$ and $w_0 := \ln(r) + i\theta_0$. Then:*

$$\forall z \in \overset{\circ}{\mathrm{D}}(z_0, |z_0|), \; L_{\theta_0}(z) = w_0 + \sum_{n \geq 1}\frac{(-1)^{n-1}}{nz_0^n}(z-z_0)^n.$$

Proof. The function $M(z) := w_0 + L(z/z_0)$ is analytic on $\overset{\circ}{\mathrm{D}}(z_0, |z_0|)$ and satisfies the relations $M(z_0) = w_0$ and, since w_0 is a logarithm[1] of z_0, $e^{M(z)} = z$. Therefore it is the (unique) determination of the logarithm on $\overset{\circ}{\mathrm{D}}(z_0, |z_0|)$ satisfying the same initial condition as L_{θ_0}. $\qquad\square$

Corollary 4.7. *All determinations of the logarithm are analytic functions.*

Proof. Indeed, at any point, there is one determination on some disk which is an analytic function (the one in the proposition), and they all differ by constants. $\qquad\square$

---

[1] We say *a* logarithm of $z \in \mathbf{C}^*$ for a complex number w such that $e^w = z$; there are infinitely many logarithms of z. We shall try not to confuse them with the "determinations of *the* logarithm", which are functions.

Primitives. We have already used the following argument: $L' = 1/z \implies (e^L)' = e^L/z \implies (e^L/z)' = 0$. Therefore, if $L' = 1/z$ on a domain $\Omega \subset \mathbf{C}^*$ and, for some $z_0 \in \Omega$, $L(z_0)$ is a logarithm of z_0, then L is a determination of the logarithm on Ω. Therefore, we can define L in the following way.

Proposition 4.8. *Let Ω be a domain in $\mathbf{C}^*$ such that for any loop γ in Ω one has $I(0, \gamma) = 0$. (This is true for instance if Ω is simply connected.) Fix z_0 in Ω and fix w_0 a logarithm of z_0. Then the function*

$$L(z) := w_0 + \int_\gamma \frac{dz}{z},$$

where γ is any path from z_0 to z, is well defined and it is the unique determination of the logarithm on Ω satisfying the initial condition $L(z_0) = w_0$.
□

Exercise 4.9. Prove that, conversely, if there is a determination of the logarithm on Ω, then for any loop γ in Ω one has $I(0, \gamma) = 0$.

4.4. The logarithm of an invertible matrix

We know that if $A \in \mathrm{Mat}_n(\mathbf{C})$, then $\exp(A) \in \mathrm{GL}_n(\mathbf{C})$. We shall now prove that the exponential map $\exp : \mathrm{Mat}_n(\mathbf{C}) \to \mathrm{GL}_n(\mathbf{C})$ is surjective. This fact will be needed in the sequel, but not its proof; the reader may skip it. But the proof is not easily found in books, so it is given here. A rather different proof is proposed in Appendix A and yet another construction of the logarithm of a matrix in Appendix B. However, we consider the path followed in the present section as the most relevant to our purposes.

Semi-simple matrices. On $\mathbf{C}$, semi-simple matrices are the same as diagonalizable matrices; but the notion is more general, and this terminology is more similar to that of algebraic groups, which we shall tackle later. So we consider a matrix $S := P\mathrm{Diag}(\mu_1, \ldots, \mu_n)P^{-1} \in \mathrm{GL}_n(\mathbf{C})$. Since S is invertible, its eigenvalues are nonzero: $\mu_i \in \mathbf{C}^*$, and we can choose logarithms $\lambda_i \in \mathbf{C}$ such that $e^{\lambda_i} = \mu_i$ for $i = 1, \ldots, n$. Then $A := P\mathrm{Diag}(\lambda_1, \ldots, \lambda_n)P^{-1}$ satisfies $\exp(A) = S$.

However, for technical reasons that will appear in the proof of the theorem below, we want more. So we make a refined choice of the logarithms λ_i. Precisely, we choose them so that whenever $\mu_i = \mu_j$, then $\lambda_i = \lambda_j$. Using polynomial interpolation (for instance, Lagrange interpolation polynomials) we see that there exists a polynomial $F \in \mathbf{C}[z]$ such that $F(\mu_i) = \lambda_i$ for $i = 1, \ldots, n$. Then:

$$F(S) = F\left(\mathrm{PDiag}(\mu_1, \ldots, \mu_n)P^{-1}\right) = PF\left(\mathrm{Diag}(\mu_1, \ldots, \mu_n)\right)P^{-1}$$

$$= P\mathrm{Diag}(F(\mu_1), \ldots, F(\mu_n))P^{-1} = P\mathrm{Diag}(\lambda_1, \ldots, \lambda_n)P^{-1} = A.$$

Unipotent matrices. Let $U \in \mathrm{GL}_n(\mathbf{C})$ be such that $U - I_n$ is nilpotent; then we know that $(U - I_n)^n = 0_n$. Define:

$$N := \sum_{1 \leq k < n} \frac{(-1)^{k-1}}{k}(U - I_n)^k.$$

Then $\exp(N) = U$. Indeed, this follows from the composition of the formal series for exp and log, except that here we truncated the latter, taking off the vanishing terms. Algebraically, the argument takes the following form. Consider the formal power series $E := \sum_{k \geq 0} z^k/k!$ and $L := \sum_{k \geq 1} (-1)^{k-1} z^k/k$. They are respectively the power series expansions of $\exp(z)$ and $\log(1 + z)$, so $E \circ L$ is the power series expansion of $\exp(\log(1 + z))$, so $E \circ L = 1 + z$. Defining the truncated series $L_n := \sum_{1 \leq k < n} (-1)^{k-1} \frac{z^k}{k}$, we see that (obviously) $L \equiv L_n \pmod{z^n}$ (meaning as is usual in algebra that $L - L_n$ is a multiple of z^n) and then, by computation (and less obviously), that $E \circ L \equiv 1 + z \pmod{z^n}$. Since $(U - I_n)^n = 0_n$, the equality $\exp(N) = U$ follows. Note that again the "logarithm" N of U is obtained in the form $G(U)$ for some polynomial $G \in \mathbf{C}[z]$.

Jordan decomposition. It is a classical fact in linear algebra that every matrix $M \in \mathrm{Mat}_n(\mathbf{C})$ admits a unique decomposition in the form $M = M_s + M_n$, where M_s is semi-simple, M_n is nilpotent and they commute: $M_s M_n = M_n M_s$. This is sometimes called the (additive) Dunford decomposition. Moreover, M_s has the same spectrum as M. Therefore, if we take $M \in \mathrm{GL}_n(\mathbf{C})$, then $M_s \in \mathrm{GL}_n(\mathbf{C})$ and we can write $M = M_s M_u = M_u M_s$, where $M_u := I_n + M_s^{-1}M_n = I_n + M_n M_s^{-1}$ is unipotent. This is the *Jordan decomposition of M*; it is of course also unique. (In France this is sometimes called "multiplicative Dunford decomposition".) A complete proof of all these facts is given in Appendix C.

Theorem 4.10. *Let $B \in \mathrm{GL}_n(\mathbf{C})$. Then there exists $A \in \mathrm{Mat}_n(\mathbf{C})$ such that $\exp(A) = B$.*

Proof. Write $B = B_s B_u$ as the Jordan decomposition. Find a semi-simple matrix $A_s = F(B_s)$ such that $\exp(A_s) = B_s$ and a nilpotent matrix $A_n = G(B_u)$ such that $\exp(A_n) = B_n$, where $F, G \in \mathbf{C}[z]$ are polynomials. Last, define $A := A_s + A_n$. Then $B_s B_u = B_u B_s \Rightarrow F(B_s)G(B_u) = G(B_u)F(B_s)$, i.e., $A_s A_n = A_n A_s$, so that $\exp(A) = \exp(A_s + A_n) = \exp(A_s)\exp(A_n) = B_s B_u = B$. □

Note that $A := A_s + A_n$ is the Dunford decomposition of A.

Exercises

(1) (i) When does one have $\log(e^z) = z$?
     (ii) Compare $\log(ab)$ with $\log(a) + \log(b)$.

(2) (i) Suppose $\theta \in \,]\theta_0 - \pi, \theta_0 + \pi[$ and also $\theta \in \,]\theta_1 - \pi, \theta_1 + \pi[$. Then compare $L_{\theta_0}(e^{i\theta})$ with $L_{\theta_1}(e^{i\theta})$.
     (ii) Let $\Omega_0 := \mathbf{C} \setminus \mathbf{R}_- e^{i\theta_0}$ and $\Omega_1 := \mathbf{C} \setminus \mathbf{R}_- e^{i\theta_1}$. Describe the intersection $\Omega_0 \cap \Omega_1$ and compare L_{θ_0} with L_{θ_1} on this set.

(3) Prove that $\exp(A) = I_n$ is equivalent to: A is semi-simple and all its eigenvalues are in $2i\pi\mathbf{Z}$. Can such an A be upper triangular?

Chapter 5

From the local to the global

5.1. Analytic continuation

We saw that there are various incarnations of the logarithm in various regions of the plane. This is a very general (and fundamental) phenomenon regarding analytic functions. We shall formalize it as a *process of analytic continuation along a path*. (You may have a look at Figure 6.3 on page 73.)

The data.

(1) Let $a \in \mathbf{C}$ and let f be analytic in some neighborhood of a. The neighborhood does not matter, so we consider f as a *germ* (remember this notion from Section 2.3 and from exercise 1 in Chapter 2) at a and write $f \in \mathcal{O}_a = \mathbf{C}\{z - a\}$.

(2) Let $b \in \mathbf{C}$ and let $\gamma : [0, 1] \to \mathbf{C}$ be a path from $\gamma(0) = a$ to $\gamma(1) = b$. We require that γ be continuous, nothing more. Of course, we could take another interval as a source for γ.

(3) Let $0 = t_0 < t_1 < \cdots < t_n = 1$ be a *subdivision* of $[0, 1]$.

(4) We cover the image curve $\gamma([0, 1]) \subset \mathbf{C}$ by open disks $D_i := \overset{\circ}{\mathrm{D}}(z_i, r_i)$, where, for $i = 0, \ldots, n$, we have $z_i = \gamma(t_i)$ (thus a point on the curve) and $r_i > 0$. Note that the first and last disk are respectively centered at $z_0 = \gamma(0) = a$ and at $z_n = \gamma(1) = b$. We assume that, for $i = 1, \ldots, n$ the disks D_i and D_{i-1} have a nonempty intersection: $D_i \cap D_{i-1} \neq \emptyset$.

Definition 5.1. Suppose that there are functions $f_i \in \mathcal{O}(D_i)$ for $i = 0, \ldots, n$ such that f is the germ of f_0 and that, for $i = 1, \ldots, n$, the functions f_i and f_{i-1} have the same restriction on $D_i \cap D_{i-1}$. Call $g \in \mathcal{O}_b$ the germ of f_n. Then g is called *the result of the analytic continuation of f along γ.*

Note that, the data above being fixed, this result is necessarily unique. Indeed, from the principle of analytic continuation (Theorem 3.6) and since the D_i are domains, f_0 is uniquely determined by its germ f; and, for $i = 1, \ldots, n$, each f_i is uniquely determined by its restriction to $D_i \cap D_{i-1}$, thus by f_{i-1}. Moreover, with some combinatorial and geometrical reasoning, one can see that the choice of the t_i, the z_i and the r_i does not change the result: that is why it is sound, in the definition, to speak of γ alone and not of the other data. Actually, the process is even much more invariant as shown in the following essential result (the proof of which can be found in [**Ahl78**]).

Theorem 5.2 (Principle of monodromy). *Let Ω be a domain, $a, b \in \Omega$ and $f \in \mathcal{O}_a$ an analytic germ at a. Assume that, for all paths from a to b in Ω, analytic continuation of f along γ is possible. Then, if γ_1 and γ_2 are two paths from a to b which are homotopic in Ω (that is, they can be continuously deformed into each other within Ω; see exercise 6 in Chapter 1), then the result of the analytic continuation of f along γ_1 or γ_2 is the same.* $\square$

Example 5.3. Let $f_0(z) := \sum_{n \geq 0} \binom{1/2}{n}(z-1)^n = (1 + (z-1))^{1/2}$, which, for obvious reasons, we write $\sqrt{z}$: it is an analytic function on $\overset{\circ}{D}(1,1)$. It can be defined trigonometrically by the formula $f_0(re^{i\theta}) = \sqrt{r}e^{i\theta/2}$ for $-\pi < \theta < \pi$. (Argument: both functions are continuous on a domain, with same square and same initial value at 1.) Put $\gamma(t) = e^{it}$ on the segment $[0, 2\pi]$, thus a loop: the most interesting case! Take $n = 4$ and the subdivision of the $t_k = k\pi/2$ for $k = 0, \ldots, 4$, so that $z_k = i^k$. Take all radii $r_k := 1$. We have $z_0 = z_4 = 1$, the base point of the loop; and the circle is covered by four disks, because the first and last disks are equal: $D_0 = D_4$.

Now we define functions similar to f_0 in the following way: the function f_k will be defined on D_k: $f_k(re^{i\theta}) = \sqrt{r}e^{i\theta/2}$ for $\theta \in \,]k\pi/2 - \pi, k\pi/2 + \pi[$. Thus, each f_k is a continuous determination of the square root on D_k. Thus, on each $D_k \cap D_{k-1}$ (which are nonempty domains) the functions f_k and f_{k-1} are equal or opposite (because the quotient of the two functions is continuous with values in $\{+1, -1\}$). To check that they are equal, one initial condition is enough. It can be found each time by using the point $i^{k-1}e^{i\pi/4} \in D_k \cap D_{k-1}$.

The conclusion is that the germ g of f_4 at 1 is the analytic continuation of the germ f of f_0 at 1 along γ. But $f_4 = -f_0$: the square root $\sqrt{z}$ has been transformed into its opposite.

Exercise 5.4. If a domain $\Omega \subset \mathbf{C}^*$ contains the image of the above loop, show that there is no continuous function f on Ω such that $f(z)^2 = z$. (Consider the function $f(e^{it})e^{-it/2}$.)

Example 5.5. We use the same loop, subdivision and disks than in the previous example and look for the analytic continuation of the germ at 1 of the function $f_0(z) := \sum_{n \geq 1} \dfrac{(-1)^{n-1}}{n}(z-1)^n$, the principal determination of the logarithm. We know that $f_0(re^{i\theta}) = \ln(r) + i\theta$ for $-\pi < \theta < \pi$. We define f_k on D_k by $f_k(re^{i\theta}) = \ln(r) + i\theta$ for $\theta \in]k\pi/2 - \pi, k\pi/2 + \pi[$. These are also determinations of the logarithms, so in their common domains they differ by constants in $2i\pi\mathbf{Z}$. Using the same points as before as initial conditions, we find that f_k and f_{k-1} are equal on $D_k \cap D_{k-1}$ and therefore that the germ g of f_4 at 1 is the analytic continuation of the germ f of f_0 at 1 along γ. But $f_4 = f_0 + 2i\pi$: the logarithm $\log(z)$ has been transformed into $\log(z) + 2i\pi$.

Exercise 5.6. Deduce from this a new proof that there is no determination of the logarithm on the whole of $\mathbf{C}^*$. (If there was one, it would be equal to f_k on D_k.)

Remark 5.7. Analytic continuation is not always possible: for instance, the germ at 0 defined by the lacunary series $\sum z^{2^n}$ admits no analytic continuation out of $\overset{\circ}{D}(0,1)$; see [**Rud87**].

5.2. Monodromy

The principle of "monodromy", after the greek "mono" for unique and "dromos" for path, means that the result of analytic continuation does not depend on the path — except if something prevents deformation. We shall see that, for differential equations, what prevents deformations is usually the presence of singularities, and we shall consider monodromy as the effect of these singularities on the changes of values by analytic continuations. In some sense, we shall rather try to understand the multiplicity than the unicity!

Now, with some algebraic formalism, we shall give power to the principle of monodromy. It will be useful to have a notation[1] for the result of analytic continuation along a path. So if $f \in \mathcal{O}_a$ and if the path γ goes from a to b, then we write f^γ as the result of the analytic continuation of f along γ if it

---

[1]The power of algebra often rests on using good notation!

exists: thus the notation is only partially defined, it may represent nothing in some cases. Here are the successive steps of the algebraic formalization. We suppose that a domain Ω has been fixed and everything (points, paths, homotopies, arguments of functions, etc.) lives there. For $a \in \Omega$, we shall write $\tilde{\mathcal{O}}_a$ as the subset of $\mathcal{O}_a$ made of germs which admit analytic continuation along any path in Ω starting from a. Beware that the notation is not really local at a, since it refers implicitly to the domain Ω.

(1) Suppose that $f, g \in \mathcal{O}_a$ admit an analytical continuation along the path γ from a to b. Then adding them, multiplying them and derivating them yield the same relations and operations between intermediate functions, and therefore between the results:

$$(\lambda f + \mu g)^\gamma = \lambda f^\gamma + \mu g^\gamma,$$
$$(fg)^\gamma = f^\gamma g^\gamma,$$
$$(f')^\gamma = (f^\gamma)'.$$

Therefore, the subset of $\mathcal{O}_a$ formed by germs which admit an analytic continuation along γ is a sub-**C**-algebra of $\mathcal{O}_a$ and it is moreover stable under derivation: it is a sub-differential algebra of the differential algebra $\mathcal{O}_a$ (this notion was defined in Section 3.4).

Of course, $\tilde{\mathcal{O}}_a$ is itself a sub-differential algebra of this differential algebra. Moreover, $f \mapsto f^\gamma$ is a morphism of differential algebras (it is linear, a morphism of rings, and it commutes with derivation). Altogether, these facts are called "principle of conservation of algebraic and differential relations". They are a natural property of monodromy, true for transcendental reasons (based on analysis) but their algebraization is the basis of differential Galois theory; see Chapter 13.

(2) From now on, we are going to use the terminology of paths and homotopy that was introduced in exercise 6 at the end of Chapter 1, so you should refer to it in case of doubt. If γ_1 goes from a to b and γ_2 goes from b to c, then we write $\gamma_1.\gamma_2$ as the composite path from a to c. Then, if $f \in \mathcal{O}_a$ is continued to $g \in \mathcal{O}_b$ along γ_1 and g is continued to $h \in \mathcal{O}_c$ along γ_2, it is clear that f is continued to h along $\gamma_1.\gamma_2$:

$$f^{\gamma_1.\gamma_2} = (f^{\gamma_1})^{\gamma_2},$$

meaning that, if one side of the equality is meaningful, so is the other and then they are equal.

(3) Let γ_1 and γ_2 be two paths from a to b and suppose that they are homotopic (in Ω by convention). We write $\gamma_1 \sim \gamma_2$ to express this relation. Then we know that, for functions satisfying the

assumptions of the principle of monodromy (Theorem 5.2), one has $f^{\gamma_1} = f^{\gamma_2}$:

$$\gamma_1 \sim \gamma_2 \Longrightarrow f^{\gamma_1} = f^{\gamma_2}.$$

As a consequence, f^γ only depends on the *homotopy class* $[\gamma] \in \Pi_1(\Omega; a, b)$ of γ; of course, $\Pi_1(\Omega; a, b)$ denotes the set of all those homotopy classes. (Remember this notion was intuitively introduced just after Corollary 1.11 and then formally defined in exercise 6 at the end of Chapter 1.) Therefore, we can define:

$$f^{[\gamma]} := f^\gamma.$$

Then the principle of conservation of algebraic and differential relations says that we have a map:

$$\Pi_1(\Omega; a, b) \to \mathrm{Iso}_{\mathbf{C}-algdiff}(\tilde{\mathcal{O}}_a, \tilde{\mathcal{O}}_b).$$

The target of this map is the set of all isomorphisms of differential **C**-algebras $\phi : \tilde{\mathcal{O}}_a \to \tilde{\mathcal{O}}_b$, meaning that ϕ is a **C**-algebra isomorphism such that $\phi(f') = (\phi(f))'$. We have added something to the principle of conservation here. First, that analytic continuation sends $\tilde{\mathcal{O}}_a$ to $\tilde{\mathcal{O}}_b$; second, that it is bijective. Both statements come from the fact that analytic continuation can be reversed by going along the inverse path.

(4) Remember from the course of topology that homotopy is compatible with the composition of paths, so that for paths γ_1 from a to b and γ_2 from b to c, and for their homotopy classes $[\gamma_1] \in \Pi_1(\Omega; a, b)$ and $[\gamma_2] \in \Pi_1(\Omega; b, c)$, one can define the product $[\gamma_1].[\gamma_2] \in \Pi_1(\Omega; a, c)$ in such a way that:

$$[\gamma_1].[\gamma_2] = [\gamma_1.\gamma_2].$$

Then the previous relation on the effect of composition of paths becomes:

$$f^{[\gamma_1.\gamma_2]} = (f^{[\gamma_1]})^{[\gamma_2]}.$$

(5) Suppose now that $a = b$: the loop case is the most interesting of all. Then $\Pi_1(\Omega; a, b)$ is $\pi_1(\Omega; a)$, the *fundamental group of Ω with base point a*, and we have a map:

$$\phi : \pi_1(\Omega; a) \to \mathrm{Aut}_{\mathbf{C}-algdiff}(\tilde{\mathcal{O}}_a),$$

the group of automorphisms of the differential algebra $\tilde{\mathcal{O}}_a$. Moreover, the equality $f^{[\gamma_1.\gamma_2]} = (f^{[\gamma_1]})^{[\gamma_2]}$ can be translated as: $\phi(xy) = \phi(y)\phi(x)$ (taking $x := [\gamma_1]$ and $y := [\gamma_2]$). Therefore, ϕ is an antimorphism of groups.

The result can be summarized as follows:

Theorem 5.8. *The group $\pi_1(\Omega; a)$ operates on the right on the differential algebra $\tilde{\mathcal{O}}_a$.* □

Remark 5.9. The fact that ϕ is an anti-morphism instead of a morphism is unavoidable if one wants to keep intuitive notation. Some books write $\gamma_2.\gamma_1$ where we have written $\gamma_1.\gamma_2$ and then they have a morphism. With their convention, the result of analytic continuation of f along γ is written $\gamma.f$ and one has: $[\gamma_2.\gamma_1].f = [\gamma_2]([\gamma_1].f)$, that is, the fundamental group operates on the left, which is more usual. But this notation is awkward for the composition of paths.

Corollary 5.10. *The fixed set of the operation of $\pi_1(\Omega; a)$ on $\tilde{\mathcal{O}}_a$ is $\mathcal{O}(\Omega)$.*

Proof. Indeed, if a germ can be continued everywhere without ambiguity, it defines a *global* analytic function on Ω. □

Note how this result looks like a theorem from Galois theory! We have a small domain $\mathcal{O}(\Omega)$ (analogous to the base field of an extension in algebra), a big domain $\tilde{\mathcal{O}}_a$ (analogous to the extension field) and a "group of ambiguity" $\pi_1(\Omega; a)$ (analogous to the group of automorphisms of the bigger field relative to the smaller field; the set of fixed points of the bigger field under the action of this group being exactly the smaller field). This is quite the framework of the Galois theory of fields. This analogy will be pursued in Remark 7.37.

5.3. A first look at differential equations with a singularity

We have already used, in Section 1.5, the exponential of a matrix to solve the differential equation with constant coefficients $X' = AX$, where $A \in \mathrm{Mat}_n(\mathbf{C})$. We shall presently use the logarithm function to solve the (very simple) *singular* equation $zX' = AX$. It is said to be singular because $X' = z^{-1}AX$ and $z^{-1}A$ is not defined at 0. We shall solve it on $\mathbf{C}^*$, that is, we shall look for an analytic solution $X : \mathbf{C}^* \to \mathbf{C}^n$.

Lemma 5.11. *Let $z_0 \in \mathbf{C}^*$ and let L be a determination of the logarithm in a domain Ω containing z_0. Then the matrix-valued function $\mathcal{X}(z) := e^{(L(z)-L(z_0))A}$ defined in Ω and with values in $\mathrm{Mat}_n(\mathbf{C})$ (actually in $\mathrm{GL}_n(\mathbf{C})$) satisfies the equation:*

$$\mathcal{X}'(z) = (z^{-1}A)\mathcal{X}(z) = \mathcal{X}(z)(z^{-1}A) \text{ with initial condition } \mathcal{X}(z_0) = I_n.$$

Proof. This is an immediate consequence of the following general fact. □

Lemma 5.12. *Let $M(z)$ be a matrix-valued analytic function on Ω such that, for all $z \in \Omega$, $M(z)M'(z) = M'(z)M(z)$ (for instance, $M(z) = f(z)A$, where $A \in \mathrm{GL}_n(\mathbf{C})$ is fixed and where $f(z)$ is a scalar analytic function). Then e^M is analytic on Ω and $(e^M)' = e^M M' = M'e^M$.*

Proof. Since $MM' = M'M$, the Leibniz formula applied to M^k gives $(M^k)' = kM^{k-1}M' = M'(kM^{k-1})$. The rest of the proof is standard. $\square$

Returning to our differential equation, we conclude again by a particular case of the Cauchy theorem for complex analytic differential equations (which we shall meet in Section 7.3):

Theorem 5.13. *Let $Sol(z^{-1}A, \Omega) \subset \mathcal{O}(\Omega)^n$ be the set of solutions of our differential equation in Ω. Then the map $X \mapsto X(z_0)$ from $Sol(z^{-1}A, \Omega)$ to $\mathbf{C}^n$ is an isomorphism of linear spaces.*

Proof. It is clear that $Sol(z^{-1}A, \Omega)$ is a linear subspace of $\mathcal{O}(\Omega)^n$ and that the map $X \mapsto X(z_0)$ is $\mathbf{C}$-linear. We are going to prove that its inverse is the map $X_0 \mapsto \mathcal{X}X_0$ from $\mathbf{C}^n$ to $\mathcal{O}(\Omega)^n$.

Indeed, an immediate calculation shows that $\mathcal{X}X_0$ is a solution of the Cauchy problem $\begin{cases} X' = (z^{-1}A)X, \\ X(z_0) = X_0, \end{cases}$ and we have to see that it is the only one. But any solution X can be written $X = \mathcal{X}Y$ (since $\mathcal{X}(z)$ is invertible for every z) and the Leibniz rule, along with Lemma 5.11 and the equality $\mathcal{X}(z_0) = I_n$, then gives $Y' = 0$ and $Y(z_0) = X_0$. $\square$

Note that this applies to all domains $\Omega \subset \mathbf{C}^*$ around z_0 on which there is a determination of the logarithm, in particular, to all simply-connected domains. Therefore, all the corresponding spaces of solutions $Sol(z^{-1}A, \Omega)$ are isomorphic to each other.

Corollary 5.14. *The space $Sol(z^{-1}A)_{z_0} \subset \mathcal{O}_{z_0}^n$ of germs at z_0 of solutions of the differential equation is isomorphic to the space $\mathbf{C}^n$ of initial conditions, through the map $X \mapsto X(z_0)$.*

Example 5.15. To solve $\begin{cases} f' = 1/z \\ f(1) = 0 \end{cases}$, we solve the equivalent Cauchy problem $\begin{cases} zf'' + f' = 0 \\ f(1) = 0 \\ f'(1) = 1 \end{cases}$ (because $zf'' + f' = (zf')'$). We make it into a vectorial differential equation of order 1 by putting $X := \begin{pmatrix} f \\ zf' \end{pmatrix}$. Our problem then boils down to $\begin{cases} X' = (z^{-1}A)X \\ X(1) = X_0 \end{cases}$, where $A := \begin{pmatrix} 0 & 1 \\ 0 & 0 \end{pmatrix}$ and $X_0 := \begin{pmatrix} 0 \\ 1 \end{pmatrix}$. Here

$A^2 = 0_2$, so that $e^{A \log z} = I_2 + A \log z = \begin{pmatrix} 1 & \log z \\ 0 & 1 \end{pmatrix}$, which gives in the end

$f = \log z$ and $z f' = 1$: this is correct and consistent.

Monodromy of the solutions. With the same notation as above, let γ be a loop in $\mathbf{C}^*$ based at z_0. Then the result of the analytic continuation of $L(z)$ along γ is $L(z) + 2i\pi k$, where $k := I(0, \gamma)$; the result of the analytic continuation of $\mathcal{X}(z)$ along γ is therefore $\mathcal{X}(z)e^{2i\pi kA} = e^{2i\pi kA}\mathcal{X}(z)$; and the result of the analytic continuation of a solution $X(z) = \mathcal{X}(z)X_0$ along γ is $\mathcal{X}(z)e^{2i\pi kA}X_0$, that is, the solution with initial condition $e^{2i\pi kA}X_0$. We express this fact by a *commutative diagram*:

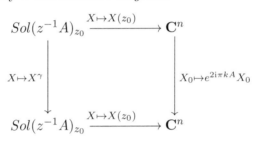

Example 5.16. In the case of the last example, we get $e^{2i\pi kA} = \begin{pmatrix} 1 & 2i\pi k \\ 0 & 1 \end{pmatrix}$. Therefore, the solution $f = a + b \log z$ with initial conditions $f(1) = a$ and $(zf')(1) = b$ is transformed into the solution with initial conditions $\begin{pmatrix} 1 & 2i\pi k \\ 0 & 1 \end{pmatrix}\begin{pmatrix} a \\ b \end{pmatrix} = \begin{pmatrix} a + 2i\pi kb \\ b \end{pmatrix}$, that is, the function $f^\gamma = (a + 2i\pi kb) + b \log z$: this is consistent.

Exercise 5.17. Solve in the same way $z^2 f'' + z f' + f = 0$ and find its monodromy.

Exercises

(1) Take $\Omega := \mathbf{C}^*$ and $a := 1$, so that $\pi_1(\Omega; a)$ is isomorphic to $\mathbf{Z}$. Show that the linear space generated by 1 and $\log$ is stable under the operation of $\pi_1(\Omega; a)$ and describe the induced action.

(2) It is customary to write z^A as the value of $\mathcal{X}(z)$, the fundamental matricial solution of $\mathcal{X}'(z) = (z^{-1}A)\mathcal{X}(z)$ on $\mathbf{C} \setminus \mathbf{R}_-$ with initial condition $\mathcal{X}(1) = I_n$. Check that $z^A = \exp(A \log z)$ and give a way to compute it explicitly. Which kind of functions appear as coefficients?

(3) Although there is generally no simple expression for the derivative of the matrix function $e^{M(z)}$, one can explicitly describe the differential of the map $M \mapsto e^M$ from the normed space $\mathrm{Mat}_n(\mathbf{C})$ to itself. We have to write:

$$e^{M+H} = e^M + \Lambda_M(H) + o(\|H\|),$$

where $\Lambda_M : \mathrm{Mat}_n(\mathbf{C}) \to \mathrm{Mat}_n(\mathbf{C})$ is a linear map, actually the differential $d\exp(M)$ of $\exp : \mathrm{Mat}_n(\mathbf{C}) \to \mathrm{Mat}_n(\mathbf{C})$ at M.

Clearly $\Lambda_M(H)$ is the part of $\sum_{k \geq 0} (M+H)^k/k!$ that involves H exactly at the first power. It can be explicitly computed with the help of the operators $L_M : H \mapsto MH$ and $R_M : H \mapsto HM$. For instance, the part of $(M+H)^3$ that involves H exactly at the first power is $M^2H + MHM + HM^2 = (L_M^2 + L_M R_M + R_M^2)(H)$. Do the complete computation and deduce from general differential calculus a (not very simple) expression for the derivative of $e^{M(z)}$.

Part 2

Complex Linear Differential Equations and their Monodromy

Chapter 6

Two basic equations
and their monodromy

The first section of this chapter was written to be the introduction to the
whole course, so the reader who is already conversant with complex analytic
functions can begin here; on the other hand, if you have carefully read Part
1, you will find a lot of redundancy in it, so maybe you should browse it
rather quickly. However, the following section (Section 6.2) is *not* redundant
and should be studied carefully.

6.1. The "characters" z^α

Let $\alpha \in \mathbf{Q}$ (the rational numbers) and assume that α is not a rational integer: $\alpha \notin \mathbf{Z}$. How can we define z^α for complex values $z \in \mathbf{C}$? If z is
a strictly positive real, $z \in \mathbf{R}_+^*$, then one can use the real logarithm and
complex exponential functions and put $z^\alpha := \exp(\alpha \ln z)$. In the complex
domain, however, the logarithm function is not globally defined (as explained
in Chapter 4), so we expect to obtain a "multivalued" function, that is, local
determinations and monodromy. In order to introduce the present chapter
on complex differential equations, and also in order to define a family of
useful basic functions for the sequel, we shall tackle the problem differently.

If $\alpha = p/q$ with $p \in \mathbf{Z}$ and $q \in \mathbf{N}^*$ (a nonzero natural integer), we can
assume that p and q are relatively prime; and, since $\alpha \notin \mathbf{Z}$, we know that
$q \geq 2$. Then z^α must be a complex number $w \in \mathbf{C}$ such that $w^q = z^p$.
If $z \neq 0$, writing $z = re^{i\theta}$ with $r > 0$ and $\theta \in \mathbf{R}$, one finds that there
are q such q^{th} roots of z^p; these are the complex numbers $r^\alpha e^{i\alpha\theta}j$, where

$r^\alpha := \exp(\alpha \ln r)$ (since $r \in \mathbf{R}_+^*$, this makes sense!) and where j is a qth root of unity:

$$j \in \mu_q := \{e^{2k\mathrm{i}\pi/q} \mid 0 \le k \le q-1\}.$$

Note that r^α can also be defined in a more elementary way as $\sqrt[q]{r^p}$.

Thus, for each particular nonzero complex number z, the fractional power z^α must be chosen among q possible values. In various senses, it is generally not possible to make such choices for all $z \in \mathbf{C}^*$ in a consistent way.

Exercise 6.1. (i) Show that putting $z^\alpha := \exp(\alpha \ln z)$ for $z > 0$, one has the rule: $\forall z_1, z_2 \in \mathbf{R}_+^*,\ (z_1 z_2)^\alpha = (z^\alpha)(z_2^\alpha)$.

(ii) Taking for example $\alpha := 1/2$, show that there is no way to define $\sqrt{z} = z^{1/2}$ for all $z \in \mathbf{C}^*$ in such a way that $(\sqrt{z})^2 = z$ and that moreover $\forall z_1, z_2 \in \mathbf{R}_+^*,\ \sqrt{z_1 z_2} = \sqrt{z_1}\sqrt{z_2}$. (Hint: try to find a square root for -1.)

There is no global definition of z^α on $\mathbf{C}^*$. In this course we shall be more concerned with analytical aspects (although we intend to approach them through algebra) and we will rather insist on the impossibility to define z^α in such a way that it depends continuously on z:

Lemma 6.2. *We keep the same notation and assumptions about α, p, q. Let V be a punctured neighborhood of 0 in $\mathbf{C}$. Then there is no continuous function $f : V \to \mathbf{C}$ such that $\forall z \in V,\ (f(z))^q = z^p$.*

Proof. By definition, V is a subset of $\mathbf{C}^*$ which contains some nontrivial punctured disk centered at 0:

$$V \supset \overset{\circ}{\mathrm{D}}(0,t) \setminus \{0\}, \text{ where } t > 0.$$

Then, for some fixed $0 < s < t$ and for any $\theta \in \mathbf{R}$, one can define $g(\theta) := \dfrac{f(se^{\mathrm{i}\theta})}{s^\alpha e^{\mathrm{i}\alpha\theta}}$. This is a continuous function of θ and it satisfies:

$$\forall \theta \in \mathbf{R},\ \big(g(\theta)\big)^q = \frac{s^p e^{\mathrm{i}p\theta}}{s^p e^{\mathrm{i}p\theta}} = 1.$$

Therefore, the continuous function g maps the connected set $\mathbf{R}$ to the discrete finite set μ_q, therefore it must be constant: there exists a fixed $j \in \mu_q$ such that:

$$\forall \theta \in \mathbf{R},\ f(se^{\mathrm{i}\theta}) = js^\alpha e^{\mathrm{i}\alpha\theta}.$$

Now, replacing θ by $\theta + 2\pi$, the left-hand side does not change, while the right-hand side (which is $\ne 0$) is multiplied by $e^{2\mathrm{i}\pi\alpha}$. This implies that $e^{2\mathrm{i}\pi\alpha} = 1$, which contradicts the assumption that $\alpha \notin \mathbf{Z}$. $\qquad\square$

The lemma obviously implies that there is no way to define z^α as a continuous map on $\mathbf{C}^*$. What we are going to do is to look for local definitions, that is, continuous functions on sufficiently small neighborhoods of all nonzero complex numbers. Then we shall consider the possibility of patching together these local objects.

Transforming an algebraic equation into a differential one. We fix $z_0 \in \mathbf{C}^*$ and choose a particular $w_0 \in \mathbf{C}^*$ such that $w_0^q = z_0^p$. This w_0 will serve as a kind of "initial condition" to define the function $z \mapsto z^\alpha$ in the neighborhood of z_0.

Proposition 6.3. (i) *There exists a unique power series*

$$f(z) = \sum a_n (z - z_0)^n$$

such that $f(z_0) = w_0$ and $\big(f(z)\big)^q = z^p$. Its radius of convergence is $|z_0|$.

(ii) *Any function g defined and continuous in a connected neighborhood $V \subset \overset{\circ}{\mathrm{D}}(z_0, |z_0|)$ of z_0 and such that $\big(g(z)\big)^q = z^p$ in V is a constant multiple of f.*

Proof. (i) We first note that, for any power series f defined in a connected neighborhood U of z_0 in $\mathbf{C}^*$, one has the equivalence:

$$\begin{cases} f(z_0) = w_0, \\ \forall z \in U, \ \big(f(z)\big)^q = z^p \end{cases} \iff \begin{cases} f(z_0) = w_0, \\ \forall z \in U, \ z f'(z) = \alpha f(z). \end{cases}$$

Indeed, if the algebraic equation $\big(f(z)\big)^q = z^p$ is true, then f does not vanish anywhere on U and one can take the logarithmic derivatives on both sides, which does yield the differential equation $z f'(z) = \alpha f(z)$. Conversely, the differential equation implies that the function $\big(f(z)\big)^q / z^p$ has a trivial derivative, hence (U being connected) it is constant; the initial condition then implies that it is equal to 1, ending the proof of the logical equivalence. Now the first assertion of the theorem follows from the similar one for differential equations, which will be proved below (Theorem 6.7).

(ii) If g is such a solution, then $(g/f)^q = z^p/z^p = 1$ on V. The continuous map g/f sends the connected set V to the discrete set μ_q; therefore it is constant. $\qquad\square$

From now on, we shall therefore study the differential equation $z f' = \alpha f$, where $\alpha \in \mathbf{C}$ is an arbitrary complex number. Indeed there is no reason to restrict to rational α. We first prove again the impossibility of a global solution except in trivial cases.

Lemma 6.4. *If $\alpha \notin \mathbf{Z}$, the differential equation $z f' = \alpha f$ has no nontrivial solution in any punctured neighborhood of 0.*

Proof. Of course, if $\alpha \in \mathbf{Z}$, the solution z^α is well defined in $\mathbf{C}$ or $\mathbf{C}^*$ according to the sign of α; and for arbitrary α, there is always the trivial solution $f = 0$.

Suppose that f is a nontrivial solution in some punctured disk $\overset{\circ}{\mathrm{D}}(0, t) \setminus \{0\}$, where $t > 0$. Then, for some fixed $0 < s < t$, and for any $\theta \in \mathbf{R}$, one can define $g(\theta) := \dfrac{f(se^{i\theta})}{e^{i\alpha\theta}}$. This is a differentiable function of θ and a simple computation shows that it satisfies:

$$\forall \theta \in \mathbf{R}, \ g'(\theta) = 0.$$

Therefore, g is constant and there exists c such that:

$$\forall \theta \in \mathbf{R}, \ f(se^{i\theta}) = ce^{i\alpha\theta}.$$

Now, replacing θ by $\theta + 2\pi$, the left-hand side does not change, while the right-hand side (which is $\neq 0$) is multiplied by $e^{2i\pi\alpha}$. This implies that $e^{2i\pi\alpha} = 1$, which contradicts the assumption that $\alpha \notin \mathbf{Z}$. □

Exercise 6.5. Give another proof that $\alpha \in \mathbf{Z}$ by integrating $f'/f = \alpha/z$ on a small circle centered at 0 and by using the Cauchy residue formula (Theorem 3.23).

Local solutions of the differential equation. Before stating the theorem, let us make some preliminary remarks about the solutions of the differential equation $zf' = \alpha f$.

(1) On $\mathbf{R}_+^* =]0, +\infty[$, there is the obvious solution $z^\alpha := \exp(\alpha \ln z)$. Moreover, any solution defined on a connected open set of $\mathbf{R}_+^*$ (*i.e.*, on an interval) has to be a constant multiple of this one, because the differential equation implies that f/z^α has zero derivative.

(2) If V is any open subset of $\mathbf{C}^*$, we shall write $\mathcal{F}(V)$ as the set of solutions of the differential equation $zf' = \alpha f$. Then $\mathcal{F}(V)$ is a linear space over the field $\mathbf{C}$ of complex numbers. Indeed, if f_1, f_2 are any solutions and if λ_1, λ_2 are any complex coefficients, it is obvious that $\lambda_1 f_1 + \lambda_2 f_2$ is a solution.

(3) Suppose moreover that the open set $V \subset \mathbf{C}^*$ is connected and suppose that f_0 is a nontrivial solution on V. Then, for any solution f, the meromorphic function $g := f/f_0$ has a trivial derivative (its logarithmic derivative is $g'/g = f'/f - f_0'/f_0 = \alpha/z - \alpha/z = 0$); therefore g is constant on V. This means that all solutions f are constant multiples of f_0.

(4) As a corollary, if V is connected, there is a dichotomy: either
  - $\mathcal{F}(V) = \{0\}$, so there is no nontrivial solution on V. This happens for instance if $V = \mathbf{C}^*$, more generally if V is a punctured

neighborood of 0 (under the assumption that $\alpha \notin \mathbf{Z}$) and even more generally if it contains the image of a loop γ such that $I(0, \gamma) \neq 0$ (see Exercise 6.5[1]); or
- $\mathcal{F}(V)$ is generated by any of its nonzero elements, that is, it has dimension 1.

Remark 6.6. When V is empty, we shall take the convention that $\mathcal{F}(V) = \{0\}$, the trivial linear space. The reader can check that all our general assertions shall remain true in that degenerate case.

Theorem 6.7. *For any $z_0 \in \mathbf{C}^*$ and $w_0 \in \mathbf{C}$, the differential equation (with initial condition)*

(6.1)
$$\begin{cases} f(z_0) = w_0, \\ zf' = \alpha f \end{cases}$$

has a unique power series solution $f = \sum_{n \geq 0} a_n (z - z_0)^n$. If $w_0 \neq 0$ and $\alpha \notin \mathbf{N}$, the radius of convergence of f is exactly $|z_0|$.

Proof. The initial condition $f(z_0) = w_0$ translates to $a_0 = w_0$, which determines the first coefficient. Then, from the calculation

$$zf' = (z - z_0)f' + z_0 f' = \sum_{n \geq 0} n a_n (z - z_0)^n + z_0 \sum_{n \geq 0} (n+1) a_{n+1} (z - z_0)^n$$
$$= \sum_{n \geq 0} (n a_n + (n+1) a_{n+1} z_0)(z - z_0)^n,$$

one gets the recursive relation:

$$\forall n \in \mathbf{N}, \ n a_n + (n+1) a_{n+1} z_0 = \alpha a_n$$
$$\implies \forall n \in \mathbf{N}, \ (n+1) a_{n+1} z_0 = (\alpha - n) a_n$$
$$\implies \forall n \in \mathbf{N}, \ a_{n+1} = z_0^{-1} \frac{\alpha - n}{n+1} a_n.$$

Solving this, we get:

$$\forall n \in \mathbf{N}, \ a_n = w_0 \binom{\alpha}{n} z_0^{-n},$$

---

[1] The exercise requires a path of class $\mathcal{C}^1$, but one can prove that any continuous path with values in an open set of $\mathbf{C}^*$ is homotopic to such a path. See for instance the article by R. Vyborny, *On the use of a differentiable homotopy in the proof of the Cauchy theorem*, in the American Mathematical Monthly, 1979; or the book by Madsen and Tornehave, "From Calculus to Cohomology".

where the generalized binomial coefficients $\binom{\alpha}{n}$ have been defined in Chapter 2. Thus:

$$(6.2) \qquad f(z) = w_0 \sum_{n \geq 0} \binom{\alpha}{n} \left(\frac{z - z_0}{z_0} \right)^n.$$

One also recognizes the generalized Newton binomial formula, already met as a formal power series:

$$(6.3) \qquad (1 + u)^\alpha := \sum_{n \geq 0} \binom{\alpha}{n} u^n.$$

(As we are going to see, this is only well defined for $|u| < 1$.) Then our solution can be conveniently expressed as:

$$f(z) = w_0 \left(1 + \frac{z - z_0}{z_0} \right)^\alpha.$$

Of course, if $w_0 = 0$ this is trivial, so assume that $w_0 \neq 0$. Then, as we saw, if $\alpha \in \mathbf{N}$, this is a polynomial, so that the radius of convergence is infinite; so assume that $\alpha \notin \mathbf{N}$. Then the coefficients a_n are all nonzero, and

$$\left| \frac{\binom{\alpha}{n+1}}{\binom{\alpha}{n}} \right| = \left| \frac{\alpha - n}{n + 1} \right| \xrightarrow[n \to +\infty]{} 1,$$

so that the radius of convergence of the power series (6.3) is 1 and the radius of convergence of the power series (6.2) is $|z_0|$. $\qquad \square$

The principal determination. Now consider all the open disks $\overset{\circ}{\mathrm{D}}(z_0, |z_0|)$ with $z_0 > 0$. This is an increasing family of open subsets of the complex plane $\mathbf{C}$, with union the right ("eastern") half-plane:

$$\bigcup_{z_0 > 0} \overset{\circ}{\mathrm{D}}(z_0, |z_0|) = H_0 := \{z \in \mathbf{C} \mid \Re(z) > 0\}.$$

The elements of H_0 are the complex numbers which can be written $z = r e^{i\theta}$ with $r > 0$ and $-\pi/2 < \theta < \pi/2$.

For any disk $\overset{\circ}{\mathrm{D}}(z_0, |z_0|)$ with $z_0 > 0$, there is a unique solution of the differential equation (6.1) with initial condition $f(z_0) = z_0^\alpha := e^{\alpha \ln z_0}$; we temporarily write $f_{z_0} \in \mathcal{F}\big(\overset{\circ}{\mathrm{D}}(z_0, |z_0|)\big)$ for this solution.

Lemma 6.8. *If $0 < z_0 < z_1$, then the restriction of f_{z_1} to $\overset{\circ}{\mathrm{D}}(z_0, |z_0|) \subset$ $\overset{\circ}{\mathrm{D}}(z_1, |z_1|)$ is equal to f_{z_0}. Temporarily write f as the unique function on H_0 which, for all $z_0 > 0$, restricts to f_{z_0} on $\overset{\circ}{\mathrm{D}}(z_0, |z_0|)$ (this exists by the previous assertion). Then the restriction of f to $\mathbf{R}_+^*$ is the function $z \mapsto z^\alpha := e^{\alpha \ln z}$.*

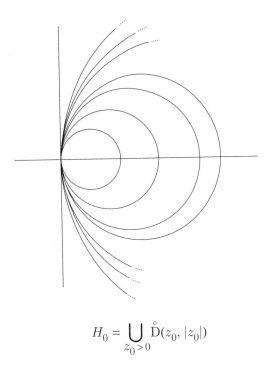

$$H_0 = \bigcup_{z_0 > 0} \overset{\circ}{\mathrm{D}}(z_0, |z_0|)$$

Figure 6.1. Covering of right half-plane by increasing disks

Proof. The restriction of f_{z_0} to $\overset{\circ}{\mathrm{D}}(z_0, |z_0|) \cap \mathbf{R}_+^* = \,]0, 2z_0[$ is equal to the function $z \mapsto z^\alpha$: this is because they satisfy the same differential equation with the same initial condition on that open interval. Therefore, if $0 < z_0 < z_1$, the functions f_{z_1} and f_{z_0} have the same value at z_0, hence they are equal on $\overset{\circ}{\mathrm{D}}(z_0, |z_0|)$ by the unicity property in Theorem 6.7. $\qquad\square$

From now on, we shall write $z^\alpha := f(z)$. This function is the unique $f \in \mathcal{F}(H_0)$ such that $f(1) = 1$.

Proposition 6.9. *Let $z \in H_0$ be written $z = re^{i\theta}$ with $r > 0$ and $-\pi/2 < \theta < \pi/2$. Then $z^\alpha = r^\alpha e^{i\alpha\theta}$.*

Proof. Let $g(re^{i\theta}) := r^\alpha e^{i\alpha\theta} = e^{\alpha(\ln r + i\theta)}$. This defines a function from H_0 to $\mathbf{C}^*$ which coincides with z^α on $\mathbf{R}_+^*$. Differentiating the relations $r^2 = x^2 + y^2$ and $\tan\theta = y/x$, one first obtains:

$$r\, dr = x\, dx + y\, dy \text{ and } (1 + \tan^2\theta)\, d\theta = \frac{x\, dy - y\, dx}{x^2},$$

whence:

$$\frac{\partial r}{\partial x} = \frac{x}{r}, \quad \frac{\partial r}{\partial y} = \frac{y}{r},$$

$$\frac{\partial \theta}{\partial x} = \frac{-y}{r^2}, \quad \frac{\partial \theta}{\partial y} = \frac{x}{r^2},$$

from which one computes:

$$\frac{\partial g}{\partial x} = \alpha \frac{x - \mathrm{i}y}{r^2} g,$$

$$\frac{\partial g}{\partial y} = \alpha \frac{y + \mathrm{i}x}{r^2} g.$$

Thus, $\dfrac{\partial g}{\partial y} = \mathrm{i}\dfrac{\partial g}{\partial x}$, which proves that g is holomorphic on $\mathbf{C}^*$. Moreover, $zg' = \alpha g$, which shows that $g \in \mathcal{F}(H_0)$. Since $g(1) = 1$, the function g is equal to the function $z \mapsto z^\alpha$. $\qquad\square$

Exercise 6.10. From the values of $\dfrac{\partial g}{\partial x}$ and $\dfrac{\partial g}{\partial y}$ given above deduce that indeed $zg' = \alpha g$.

Remark 6.11. If $\alpha \in \mathbf{Z}$, the above formula is trivial. However, if $\alpha \notin \mathbf{Z}$ the formula can only be applied if $-\pi/2 < \theta < \pi/2$. For instance, trying to compute 1^α using the representation $1 = 1.e^{2\mathrm{i}\pi}$ would yield the incorrect result $e^{2\mathrm{i}\pi\alpha} \neq 1$.

Corollary 6.12. *The unique function* $f \in \mathcal{F}\big(\overset{\circ}{\mathrm{D}}(z_0, |z_0|)\big)$ *such that* $f(z_0) = w_0$ *is defined by* $z \mapsto w_0(z/z_0)^\alpha = w_0(r/r_0)^\alpha e^{\mathrm{i}\alpha(\theta - \theta_0)}$, *where* $z = re^{\mathrm{i}\theta}$ *and* $z_0 = r_0 e^{\mathrm{i}\theta_0}$, *with* $r, r_0 > 0$ *and* $-\pi/2 < \theta - \theta_0 < \pi/2$.

Corollary 6.13. *Let* $H_{\theta_0} := e^{\mathrm{i}\theta_0} H_0 = \{z \in \mathbf{C} \mid \Re(z/e^{\mathrm{i}\theta_0}) > 0\}$. *Then the function defined by*

$$re^{\mathrm{i}\theta} \mapsto r^\alpha e^{\mathrm{i}\alpha(\theta - \theta_0)}, \text{ where } r > 0 \text{ and } -\pi/2 < \theta - \theta_0 < \pi/2,$$

generates $\mathcal{F}(H_{\theta_0})$.

How can one patch the local solutions. We shall now try to "patch together" these "local" solutions. We start with a simple example.

Example 6.14. We consider the upper half-plane (also called *Poincaré half-plane*) $H_{\pi/2} = \mathrm{i}H_0 = \{z \in \mathbf{C} \mid \Im(z) > 0\}$. Then $H_0 \cap H_{\pi/2}$ is the upper right quadrant. The restriction of $z^\alpha \in \mathcal{F}(H_0)$ to $H_0 \cap H_{\pi/2}$ has a unique extension to $H_{\pi/2}$. This extension can be described as follows: one chooses an arbitrary $z_0 \in H_0 \cap H_{\pi/2}$; then this extension is the unique $f \in \mathcal{F}(H_{\pi/2})$ satisfying the initial condition $f(z_0) = z_0^\alpha$. Note that this makes sense as an initial condition for an element of $\mathcal{F}(H_{\pi/2})$, since $z_0 \in H_{\pi/2}$; and the right-hand side z_0^α makes sense since $z_0 \in H_0$.

Also note that the particular choice of z_0 does not matter. Indeed, in any case, the unique $f \in \mathcal{F}(H_{\pi/2})$ such that $f(z_0) = z_0^\alpha$ will have to coincide with z^α on $H_0 \cap H_{\pi/2}$. (The reader should check this statement!)

To compute the function f explicitly, we first choose a nice particular value of z_0. We shall take $z_0 := \dfrac{1 + \mathrm{i}}{\sqrt{2}} = e^{\mathrm{i}\pi/4}$. Since $-\pi/2 < \pi/4 < \pi/2$, using Proposition 6.9, we see that $z_0^\alpha = e^{\mathrm{i}\alpha\pi/4}$.

Now, by Corollary 6.13, the function f on $H_{\pi/2}$ can be computed as $f(z) = w_0(z/\mathrm{i})^\alpha$, which makes sense because $z \in H_{\pi/2} \Rightarrow z/\mathrm{i} \in H_0$. We determine w_0 using the initial condition at z_0. Since $z_0/\mathrm{i} = e^{\mathrm{i}\pi/4 - \mathrm{i}\pi/2} = e^{-\mathrm{i}\pi/4}$ and since $-\pi/2 < -\pi/4 < \pi/2$, using Proposition 6.9 we see that $(z_0/\mathrm{i})^\alpha = e^{-\mathrm{i}\alpha\pi/4}$. Now, from the initial condition $f(z_0) = z_0^\alpha = e^{\mathrm{i}\alpha\pi/4} = w_0 e^{-\mathrm{i}\alpha\pi/4}$, we conclude that $w_0 = e^{\mathrm{i}\alpha\pi/2}$ and we obtain the following formula for f:

$$\forall z \in H_{\pi/2}, \ f(z) = e^{\mathrm{i}\alpha\pi/2}(z/\mathrm{i})^\alpha.$$

Exercise 6.15. Check that the unique $f \in \mathcal{F}(H_{\pi/2})$ such that $f(z_0) = z_0^\alpha$ will have to coincide with z^α on $H_0 \cap H_{\pi/2}$.

In some sense, we have tried to continue the function z^α along a path that starts at $z_0 := 1$ and that turns around the origin 0 in the positive direction (up, then left). We can go further by considering any half-plane H_θ which meets H_0, that is, any half-plane except $-H_0 = H_\pi = H_{-\pi}$. So we take any θ_0 such that $-\pi < \theta_0 < \pi$. Then $H_0 \cap H_{\theta_0}$ is connected, it is actually a sector; precisely, if $\theta_0 > 0$:

$$H_0 \cap H_{\theta_0} = \{re^{\mathrm{i}\theta} \mid r > 0 \text{ and } \theta_0 - \pi/2 < \theta < \pi/2\},$$

and if $\theta_0 < 0$:

$$H_0 \cap H_{\theta_0} = \{re^{\mathrm{i}\theta} \mid r > 0 \text{ and } -\pi/2 < \theta < \theta_0 + \pi/2\}.$$

Lemma 6.16. *Under these conditions, there is a unique function $f \in \mathcal{F}(H_{\theta_0})$ which coincides with z^α on their common domain of definition. This function is given by the formula:*

$$\forall \theta \in H_{\theta_0}, \ f(z) = e^{\mathrm{i}\alpha\theta_0}(z/e^{\mathrm{i}\theta_0})^\alpha.$$

Proof. Putting $f(z) = C(z/e^{\mathrm{i}\theta_0})^\alpha$ with $C \in \mathbf{C}^*$, we just have to replace z by a particular point of $H_0 \cap H_{\theta_0}$ to be able to determine the unknown factor C. We choose $z := e^{\mathrm{i}\theta_0/2}$ and compute:

$$z^\alpha = e^{\mathrm{i}\alpha\theta_0/2} \text{ since } -\pi/2 < \theta_0/2 < \pi/2,$$

$$(z/e^{\mathrm{i}\theta_0})^\alpha = e^{-\mathrm{i}\alpha\theta_0/2} \text{ since } -\pi/2 < -\theta_0/2 < \pi/2,$$

and we conclude that $C = e^{\mathrm{i}\alpha\theta_0}$. $\qquad\square$

So suppose we now want to compute $(-1)^\alpha$. To begin with, this is not really defined since $-1 \notin H_0$. So a natural way to proceed is to choose some θ_0 as before, to take the unique continuation of z^α into a function $f \in H_{\theta_0}$ and to evaluate $f(-1)$.

Exercise 6.17. For which values of θ_0 is this guaranteed to work?

We shall try two distinct possibilities for θ_0.

Example 6.18. Take $\theta_0 \in \,]\pi/2, \pi[$. Then $-1 = e^{i\pi}$ with $\theta_0 - \pi/2 < \pi < \theta_0 + \pi/2$, so that, from the lemma: $f(-1) = e^{\alpha i\pi}$.

Example 6.19. Take $\theta_0 \in \,]-\pi, -\pi/2[$. Then $-1 = e^{-i\pi}$ with $\theta_0 - \pi/2 < -\pi < \theta_0 + \pi/2$, so that, from the lemma: $f(-1) = e^{-\alpha i\pi}$.

Therefore, we have two candidate values for $(-1)^\alpha$, that is, $e^{\alpha i\pi}$ and $e^{-\alpha i\pi}$. If $\alpha \notin \mathbf{Z}$, these values are distinct. Different continuations of z^α from a neighborhood of 1 to a neighborhood of -1 have given different results. This can be seen as an "explanation" of the impossibility of globally defining z^α on $\mathbf{C}^*$.

Analytic continuation and differential equations. Given that we are not in general able to continue z^α to a solution of $zf' = \alpha f$ in the whole of $\mathbf{C}^*$, we might decide to relax our condition and look for functions satisfying weaker conditions. For instance, it is not too difficult to prove that the function z^α on H_0 can be extended (in many ways) to a function which is indefinitely differentiable on $\mathbf{C}^*$ in its variables x, y. (Because of the wild behavior of z^α near 0, we cannot hope for an extension to the whole of $\mathbf{C}$.)

The problem as we study it appeared in the nineteenth century, when mathematicians like Gauss, Cauchy, Riemann, etc. discovered the marvelous properties of *analytic* functions of complex variables. Now, such functions satisfy very strong "rigidity" properties. For instance, if a function f defined on a connected open set $U \subset \mathbf{C}$ satisfies an algebraic equation (like $f^q = z^p$) or a differential equation (like $zf' = \alpha f$), then all its analytic continuations satisfy the same equation. We shall now explain this sentence and prove it in a simple case; the general case for linear differential equations[2] will be tackled in the following chapters.

So assume either that $f^q = z^p$ or that $zf' = \alpha f$. Now consider an open connected set V such that $U \cap V \neq \emptyset$. Suppose there is an analytic function g on V such that f and g coincide on $U \cap V$. This implies that the function

---

[2]The case of algebraic equations is studied in all books about "algebraic functions", or in the corresponding chapter of many books on complex functions, like [**Ahl78**].

$h := g^q - z^p$ (in the first case) or $h := zg' - \alpha g$ (in the second case) has a trivial restriction to $U \cap V$. Since h is analytic by the *principle of analytic continuation*, this implies that h is trivial on V, thus that g is also a solution of the algebraic or the differential equation. The function g is called a *direct analytic continuation* of f. If we consider a direct analytic continuation of g, etc., we obtain various analytic continuations of f. All these continuations are solutions of the same algebraic or differential equation as f.

Riemann proved in two celebrated works[3] that algebraic functions and solutions of differential equations could be better understood through the properties of their analytic continuations (and also through the study of their singularities). Galois saw that the ambiguities caused by the multiplicity of analytic continuations could be used to build a "Galois theory of transcendental functions" as he had done for algebraic equations satisfied by numbers; see [**And12**, **Ram93**] for a discussion of this story.

A formal look at what has been done. We now try to understand the process of analytic continuation of the function z^α but more globally, for all functions $f \in \mathcal{F}(H_0)$ simultaneously. We recall that $\mathcal{F}(H_0)$ is a complex linear space and that it is generated by any of its nontrivial elements, for instance by z^α; so the dimension of this complex space is 1.

If we choose any $\theta_0 \neq 0$ such that $-\pi < \theta_0 < \pi$, then $H_0 \cap H_{\theta_0}$ is nonempty and connected. Restricting $f \in \mathcal{F}(H_0)$ to $H_0 \cap H_{\theta_0}$ gives a function g on $H_0 \cap H_{\theta_0}$ such that $zg' = \alpha g$, that is, an element $g \in \mathcal{F}(H_0 \cap H_{\theta_0})$. In this way, we obtain a map:

$$\begin{cases} \mathcal{F}(H_0) \to \mathcal{F}(H_0 \cap H_{\theta_0}), \\ f \mapsto f_{|H_0 \cap H_{\theta_0}}. \end{cases}$$

This map is obviously linear, and it follows from our previous arguments that it is bijective: it is an isomorphism.

In the same way, one defines an isomorphism:

$$\begin{cases} \mathcal{F}(H_{\theta_0}) \to \mathcal{F}(H_0 \cap H_{\theta_0}), \\ f \mapsto f_{|H_0 \cap H_{\theta_0}}. \end{cases}$$

---

[3] "Grundlagen für eine allgemeine Theorie der Functionen einer veränderlichen complexen Grösse" and "Beiträge sur Theorie der durch die Gauss'sche Reihe $F(\alpha, \beta, \gamma, x)$ darstellbaren Functionen", in his "Gesammelte Mathematische Werke"; the second one is the most relevant here.

Then by composition of the first isomorphism with the inverse of the second isomorphism:

$$\begin{cases} \mathcal{F}(H_0) \to \mathcal{F}(H_{\theta_0}), \\ f \mapsto \text{ the unique extension to } H_{\theta_0} \text{ of } f_{|H_0 \cap H_{\theta_0}}. \end{cases}$$

For instance, we proved that the image of z^α under this isomorphism is the function $e^{i\alpha\theta_0}(z/e^{i\theta_0})^\alpha$.

Now we can prove in the same way that there are isomorphisms:

$$\begin{cases} \mathcal{F}(H_{\theta_0}) \to \mathcal{F}(H_\pi \cap H_{\theta_0}), \\ f \mapsto f_{|H_\pi \cap H_{\theta_0}}. \end{cases}$$

$$\begin{cases} \mathcal{F}(H_\pi) \to \mathcal{F}(H_\pi \cap H_{\theta_0}), \\ f \mapsto f_{|H_\pi \cap H_{\theta_0}}. \end{cases}$$

Then by composition of the first isomorphism with the inverse of the second isomorphism:

$$\begin{cases} \mathcal{F}(H_{\theta_0}) \to \mathcal{F}(H_\pi), \\ f \mapsto \text{ the unique extension to } H_\pi \text{ of } f_{|H_\pi \cap H_{\theta_0}}. \end{cases}$$

To summarize, we obtain an isomorphism $\mathcal{F}(H_0) \to \mathcal{F}(H_\pi)$ by following a complicated path:

$$\mathcal{F}(H_0) \longrightarrow \mathcal{F}(H_0 \cap H_{\theta_0}) \longleftarrow \mathcal{F}(H_{\theta_0}) \longrightarrow \mathcal{F}(H_\pi \cap H_{\theta_0}) \longleftarrow \mathcal{F}(H_\pi).$$

All the arrows are restriction maps. The path always goes from left to right, so we follow some arrows backwards! (Of course, this is possible because they are isomorphisms.)

Now, we shall try to characterize the isomorphism $\mathcal{F}(H_0) \to \mathcal{F}(H_\pi)$ thus defined. It is sufficient to compute the image f of the generator z^α of $\mathcal{F}(H_0)$. As an element of $\mathcal{F}(H_\pi)$, the function f must have the form $f(z) = C(z/e^{i\pi})^\alpha = C(-z)^\alpha$ (remember that here we have $z \in H_\pi$, so that $-z \in H_0$). The constant C can be determined by taking $z = -1$: we have $C = f(-1)$. This value has been computed in Examples 6.18 and 6.19. There, we found that C depends on the choice of the intermediate plane:

$$\forall z \in H_\pi,\ f(z) = C(-z)^\alpha, \text{ where } C = \begin{cases} C_{up} := e^{\alpha i \pi} \text{ if } \theta_0 \in\ ]\pi/2, \pi[, \\ C_{down} := e^{-\alpha i \pi} \text{ if } \theta_0 \in\ ]-\pi, -\pi/2[. \end{cases}$$

Now we do some very simple linear algebra. We call u the generator of $\mathcal{F}(H_0)$ that we wrote until now as z^α. We call v the generator of $\mathcal{F}(H_\pi)$ that we wrote until now as $(-z)^\alpha$. We have defined two isomorphisms ϕ_{up} and ϕ_{down} from $\mathcal{F}(H_0)$ to $\mathcal{F}(H_\pi)$: one is such that $\phi_{up}(u) = C_{up}v$, the other is

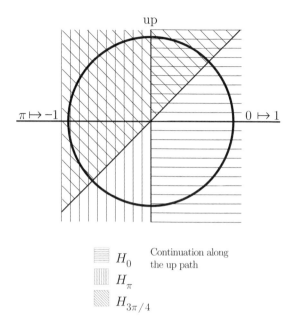

H_0 Continuation along
the up path

H_π

$H_{3\pi/4}$

Figure 6.2. Upper path through an intermediate half-plane

such that $\phi_{down}(u) = C_{down}v$. We can compose one of these isomorphisms with the inverse of the other and obtain an automorphism ψ of $\mathcal{F}(H_0)$, defined by:

$$\psi := \phi_{down}^{-1} \circ \phi_{up} : H_0 \to H_0.$$

This is characterized by the fact that $\psi(u) = \dfrac{C_{up}}{C_{down}}u = e^{2\alpha i\pi}u$. Since u is a generator of $\mathcal{F}(H_0)$, this implies:

$$\forall f \in \mathcal{F}(H_0), \ \psi(f) = e^{2\alpha i\pi}f.$$

How can we understand the automorphism ψ of $\mathcal{F}(H_0)$? It has been obtained by following an even longer complicated path:

$$\mathcal{F}(H_0) \longrightarrow \mathcal{F}(H_0 \cap H_{\theta_0'}) \longleftarrow \mathcal{F}(H_{\theta_0'}) \longrightarrow \mathcal{F}(H_\pi \cap H_{\theta_0'}) \longleftarrow \mathcal{F}(H_\pi)$$
$$\longrightarrow \cdots \longrightarrow \mathcal{F}(H_\pi \cap H_{\theta_0''}) \longleftarrow \mathcal{F}(H_{\theta_0''}) \longrightarrow \mathcal{F}(H_0 \cap H_{\theta_0''}) \longleftarrow \mathcal{F}(H_0).$$

(As before, all the arrows are restriction maps and we follow some arrows backwards.) Here we have taken two distinct intermediate half-planes, one "up", characterized by $\theta_0' \in \,]\pi/2, \pi[$, the other "down", characterized by $\theta_0'' \in \,]-\pi, -\pi/2[$. In some sense, we have turned around 0 in the positive sense (counterclockwise) and $f \in \mathcal{F}(H_0)$ has been changed in the process.

6.2. A new look at the complex logarithm

We shall now do for the logarithm what we just did for the function z^α: study the ambiguity of its definition starting from the differential equation of which it is a solution. Instead of the equation $f' = 1/z$, we shall here use its consequence:

$$(6.4) \qquad\qquad zf'' + f' = 0,$$

because it is a linear homogeneous scalar differential equation, and so its solutions form a linear space. Of course, we have greatly increased the set of solutions beyond the logarithm; formally:

$$(zf')' = zf'' + f' = 0 \Longrightarrow zf' = a \Longrightarrow f = a\log z + b,$$

and we can only fix the two constants of integration by specifying initial conditions.

We saw that putting $X := \begin{pmatrix} f \\ zf' \end{pmatrix}$, we obtain an equivalent vectorial differential equation (or system):

$$(6.5) \qquad\qquad X' = z^{-1}AX, \text{ where } A := \begin{pmatrix} 0 & 1 \\ 0 & 0 \end{pmatrix}.$$

We shall now formalize more precisely in what sense they are equivalent. For any open set $U \subset \mathbf{C}$, let us write:

$$\mathcal{F}_1(U) := \{f \in \mathcal{O}(U) \mid f \text{ is a solution of } (6.4)\}$$

and

$$\mathcal{F}_2(U) := \{X \in \mathcal{O}(U)^2 \mid X \text{ is a solution of } (6.5)\}.$$

These are two linear spaces, and the map $\phi : f \mapsto \begin{pmatrix} f \\ zf' \end{pmatrix}$ is an isomorphism from the first to the second. If we want to take into account initial conditions, we introduce the map $IC_2 : X \mapsto X(z_0)$ from $\mathcal{F}_2(U)$ to $\mathbf{C}^2$ and, for compatibility, a map IC_1 from $\mathcal{F}_1(U)$ to $\mathbf{C}^2$ defined as $f \mapsto (f(z_0), z_0 f'(z_0))$. Then we have a commutative diagram in which the horizontal arrow ϕ is an isomorphism:

$$\begin{array}{ccc}
\mathcal{F}_1(U) & \xrightarrow{\phi} & \mathcal{F}_2(U) \\
& & \\
\searrow{\scriptstyle IC_1} & & \swarrow{\scriptstyle IC_2} \\
& \mathbf{C}^2 &
\end{array}$$

In the course, we shall generally study the vectorial form (after having shown that it is equivalent to the scalar form), but in this particular case we shall rather use the scalar form (6.4). So, we simply write $\mathcal{F}(U)$ for $\mathcal{F}_1(U)$ (and IC for the initial map IC_1). We shall first study some particular cases, depending on the open set U. Then we shall see how all these spaces $\mathcal{F}(U)$ are globally related.

The singularity at 0. This is considered as a singularity because, if we write the equation in the form $f'' + p(z)f' + q(z)f$, the function $p = 1/z$ is singular at 0 (one has $q = 0$) and the Cauchy theorem on differential equations (which we shall meet later; see Section 7.3) cannot be applied. Specifically, if we try to solve (6.4) with a power series $f := \sum a_n z^n$, we see that it is equivalent to:

$$zf'' + f' = \sum (n+1)^2 a_{n+1} z^n = 0 \iff \forall n, \ (n+1)^2 a_{n+1} = 0$$

$$\iff \forall n \neq 0, \ a_n = 0.$$

Therefore, the only solutions are constants, even on very small neighborhoods of 0. Note that we said nothing about the values of n: the argument applies as well to Laurent series and even generalized Laurent series, that is, solutions which are analytic in a punctured neighborhood of 0. This is clearly not satisfying, since for an equation of order 2 we hope to obtain two arbitrary constants of integration. So, from now on, we shall only consider open sets $U \subset \mathbf{C}^*$.

Small scale properties of $\mathcal{F}(U)$. As already noted, $\mathcal{F}(U)$ is a complex linear space for any open set $U \subset \mathbf{C}^*$. We shall make the convention that $\mathcal{F}(\emptyset) = \{0\}$, the trivial linear space. If U is not connected, we can write $U = U_1 \cup U_2$, where U_1, U_2 are two nonempty disjoint open subsets. It is obvious that a solution on U is determined by independent choices of a solution on U_1 and of a solution on U_2, *i.e.*, there is an isomorphism:

$$\begin{cases} \mathcal{F}(U) \to \mathcal{F}(U_1) \times \mathcal{F}(U_2), \\ f \mapsto (f_{|U_1}, f_{|U_2}). \end{cases}$$

Whatever the number of connected components of U (even infinite), they are all open and a decomposition $U = \bigsqcup U_k$ gives rise in a similar way to an isomorphism $\mathcal{F}(U) \to \prod \mathcal{F}(U_k)$.

Therefore, the case of interest is if U is a nonempty domain. In this case, we know for sure that the constants are solutions: $\mathbf{C} \subset \mathcal{F}(U)$, so that $\dim_{\mathbf{C}} \mathcal{F}(U) \geq 1$. Now, for a domain, one also has the upper bound: $\dim_{\mathbf{C}} \mathcal{F}(U) \leq 2$. This is a particular case of the "wronskian lemma", which will be proved in Section 7.2, so we prefer not to give a proof here; but the reader can try to imagine one (see exercise 6 at the end of the chapter).

So the question is: for what kind of domain does one achieve the optimal dimension 2? We just saw that punctured neighborhoods of 0 are excluded.

So we choose $z_0 \neq 0$ and try for a power series $f(z) := \sum_{n \geq 0} a_n (z - z_0)^n$. We find:

$$z f'' + f' = \sum_{n \geq 0} \left((n+1)^2 a_{n+1} + (n+1)(n+2) a_{n+2} z_0 \right) (z - z_0)^n$$

and deduce:

$$z f'' + f' = 0 \iff \forall n \in \mathbf{N}, \ (n+1)^2 a_{n+1} + (n+1)(n+2) a_{n+2} z_0 = 0.$$

The recurrence is readily solved: there is no condition on a_0, and $a_n = (a_1 z_0) \dfrac{(-1)^{n-1}}{n z_0^n}$ for $n \geq 1$. This means that $f = a_0 + (a_1 z_0) L$, where the series L has r.o.c. $|z_0|$. Actually, we recognize the series for the determination of the logarithm around z_0. It follows that, for any disk $U := \overset{\circ}{\mathrm{D}}(z_0, |z_0|)$ (see how it carefully avoids the singular point 0?), one has the optimal dimension $\dim_{\mathbf{C}} \mathcal{F}(U) = 2$. Also note that the image of $a_0 + (a_1 z_0) L$ by the initial condition map IC is $f \mapsto (a_0, a_1 z_0)$, so that it is an isomorphism in this case. (Finding the equality of dimensions would not be by itself a sufficient argument.)

Large scale properties of $\mathcal{F}(U)$. Here, "large scale" means that we consider globally the collection of all open sets $U \subset \mathbf{C}^*$ and the collection of all linear spaces $\mathcal{F}(U)$. The main relation among these is that if $V \subset U$, there is a *restriction map* $f \mapsto f_{|V}$ from $\mathcal{F}(U)$ to $\mathcal{F}(V)$; of course, it is linear. The obvious fact that, if U is covered by open subsets V_k, then f can be uniquely recovered from the family of all the $f_k := f_{|V_k} \in \mathcal{F}(V_k)$, provided these are compatible (*i.e.*, f_k and f_l have the same restriction in $\mathcal{F}(V_k \cap V_l)$) is formulated by saying that "$\mathcal{F}$ is a sheaf" (a complete definition will be given in Section 7.4).

A property more related to the analyticity of the solutions is that, if U is a domain and if $V \subset U$ is a nonempty open subset, then the restriction map $\mathcal{F}(U) \to \mathcal{F}(V)$ is injective: this is indeed a direct consequence of the principle of analytic continuation (Theorem 3.6). As a consequence of this injectivity and of the calculation of dimensions, we find that if U is a disk centered at z_0 and if V is a domain, then the restriction map $\mathcal{F}(U) \to \mathcal{F}(V)$ is bijective. Therefore, for all domains U contained in some $\overset{\circ}{\mathrm{D}}(z_0, |z_0|)$, one has $\dim_{\mathbf{C}} \mathcal{F}(U) = 2$. Another consequence is that all the spaces $\mathcal{F}_{z_0}$ of germs of solution have dimension 2; and also that the initial condition map IC gives an isomorphism from $\mathcal{F}_{z_0}$ to $\mathbf{C}^2$.

Monodromy. We already know the rules of the game. Fix $a, b \in \mathbf{C}^*$ and a path γ from a to b within $\mathbf{C}^*$. We cover the image curve of γ by

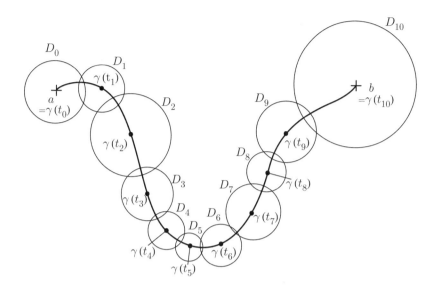

Figure 6.3. Analytic continuation along overlapping disks

disks $D_0, \ldots, D_n$ centered on the curve, the first one at a, the last one at b, and any two consecutive disks having nonempty intersection. From this, we obtain isomorphisms: first, $\mathcal{F}_a \leftarrow \mathcal{F}(D_0)$; then, for $k = 1, \ldots, n$, $\mathcal{F}(D_{k-1}) \rightarrow \mathcal{F}(D_{k-1} \cap D_k) \leftarrow \mathcal{F}(D_k)$; last, $\mathcal{F}(D_n) \rightarrow \mathcal{F}_b$. By composition, we get the isomorphism $\mathcal{F}_a \rightarrow \mathcal{F}_b$ induced by analytic continuation along the path γ. By the principle of monodromy (Theorem 5.2), the isomorphism $\mathcal{F}_a \rightarrow \mathcal{F}_b$ depends only on the homotopy class of γ in $\mathbf{C}^*$.

Taking $a = b$, this gives a map:

$$\pi_1(\mathbf{C}^*; a) \rightarrow \mathrm{GL}(\mathcal{F}_a).$$

From the algebraic rules stated in Section 5.2, this is an anti-morphism of groups. However, in this particular case, we know that $\pi_1(\mathbf{C}^*; a) \simeq \mathbf{Z}$ is commutative, so an anti-morphism is the same thing as a morphism. Actually, this morphism is totally determined from the knowledge of the image of a generator of the fundamental group, for instance the homotopy class of the positive loop $\gamma : [0, 2\pi] \rightarrow \mathbf{C}^*, t \mapsto ae^{it}$.

To see more concretely what this "monodromy representation" (this is its official name!) looks like, we shall take $a := 1$. Then a basis of $\mathcal{F}_1$ is given by (the germs of) the constant map 1 and the principal determination of the logarithm $\log$. This gives an identification of $\mathrm{GL}(\mathcal{F}_1)$ with $\mathrm{GL}_2(\mathbf{C})$ and we obtain a second description of the monodromy representation, related to the first one by a commutative diagram:

$$
\begin{array}{ccc}
\pi_1(\mathbf{C}^*; 1) & \longrightarrow & \mathrm{GL}(\mathcal{F}_1) \\
\downarrow & & \downarrow \\
\mathbf{Z} & \longrightarrow & \mathrm{GL}_2(\mathbf{C})
\end{array}
$$

The vertical maps are isomorphisms, respectively induced by the choice of a generator of $\pi_1(\mathbf{C}^*; 1)$ (the homotopy class of the loop γ) and by the choice of a basis of $\mathcal{F}_1$ (the germs of 1 and log). The horizontal maps are the monodromy representation in its abstract and in its matricial form. The lower horizontal map is characterized by the image of $1 \in \mathbf{Z}$. This corresponds to the action of the loop γ. Along this loop, the constant germ 1 is continued to itself and log to $\log +2\mathrm{i}\pi$. Therefore, our basis $(1, \log)$ is transformed to $(1, \log +2\mathrm{i}\pi)$. The matrix of this automorphism is $\begin{pmatrix} 1 & 2\mathrm{i}\pi \\ 0 & 1 \end{pmatrix}$. Therefore, we get at last the concrete description of the monodromy representation:

$$
\begin{cases}
\mathbf{Z} \to \mathrm{GL}_2(\mathbf{C}), \\
k \mapsto \begin{pmatrix} 1 & 2\mathrm{i}\pi \\ 0 & 1 \end{pmatrix}^k = \begin{pmatrix} 1 & 2\mathrm{i}\pi k \\ 0 & 1 \end{pmatrix}.
\end{cases}
$$

Note that the monodromy automorphisms are unipotent: this is characteristic of the logarithm.

Exercise 6.20. A general solution of $zf'' + f' = 0$ has the form $f = a + b\log = \mathcal{B}\begin{pmatrix} a \\ b \end{pmatrix}$ for some $a, b \in \mathbf{C}$. Thus, its analytic continuation along a loop λ is $f^\lambda = \mathcal{B}^\lambda \begin{pmatrix} a \\ b \end{pmatrix}$, where $\mathcal{B}^\lambda = \mathcal{B}\begin{pmatrix} 1 & 2\mathrm{i}\pi \\ 0 & 1 \end{pmatrix}^k$ for some $k \in \mathbf{Z}$. Deduce from this the expression of f^λ.

6.3. Back again to the first example

We shall describe again what has been found in the first example (Section 6.1) in light of the constructions of Section 6.2. Calling $\mathcal{F}_1$ the space of germs of solutions of $zf' = \alpha f$, where $\alpha \in \mathbf{C}$ is arbitrary, we saw that it is a linear space of dimension 1 and that turning around 0 once in the positive sense induced an automorphism $\psi : f \mapsto \beta f$ of $\mathcal{F}_1$, where $\beta := e^{2\alpha \mathrm{i}\pi}$. Note that an automorphism of $\mathcal{F}_1$ is always of this form, that is, we have a canonical identification $\mathrm{GL}_1(\mathcal{F}_1) \simeq \mathbf{C}^*$. Here, "canonical" means that the isomorphism does not depend on the choice of a basis of $\mathcal{F}_1$. Now if we turn k times around 0, any $f \in \mathcal{F}(H_0)$ is multiplied k times by β, that is, the corresponding automorphism is $\psi^k : f \mapsto \beta^k f$. (This also works if $k < 0$.)

In light of the study of the second example, "turning around 0" just means performing analytic continuation along the loop $\gamma : [0, 2\pi] \to \mathbf{C}^*, t \mapsto ae^{it}$ and thereby identifying $\pi_1(\mathbf{C}^*, 1)$ with $\mathbf{Z}$. Since we know that it is only the homotopy class of the loop that matters, and also composing paths, we must compose the automorphisms (using the rules of Section 5.2); we get an anti-morphism of groups from $\pi_1(\mathbf{C}^*, 1)$ to $\mathrm{GL}(\mathcal{F}_1)$. As already noted, since $\pi_1(\mathbf{C}^*, 1)$ is commutative, this is also a group morphism.

To the equation $zf' = \alpha f$, $\alpha \in \mathbf{C}$, we have therefore attached the monodromy representation:

$$\begin{cases} \mathbf{Z} \to \mathbf{C}^*, \\ k \mapsto \beta^k, \end{cases}$$

where $\beta := e^{2i\pi\alpha}$.

Exercises

(1) Use the calculation of the monodromy of $\sqrt{z}$ in Example 5.3, page 46, to give another proof that there is no global determination of $\sqrt{z}$.

(2) (i) Remember that the series (6.3) is the unique power series solution of the differential equation:

$$\begin{cases} f(0) = 1, \\ (1 + u)f' = \alpha f. \end{cases}$$

Using only this characterization, prove the following formula:

$$(1 + u)^{\alpha+\beta} = (1 + u)^\alpha (1 + u)^\beta.$$

(ii) As an application, prove the following formula:

$$\forall n \in \mathbf{N}, \quad \sum_{i+j=n} \binom{\alpha}{i}\binom{\beta}{j} = \binom{\alpha + \beta}{n}.$$

(3) Prove the following formulas: $\forall p \in \mathbf{N}^*$,

$$\frac{1}{(1 + u)^p} = \sum_{n \geq 0} (-1)^n \frac{(p + n - 1)!}{(p - 1)!n!} u^n,$$

$$\sqrt{1 + u} = 1 + \frac{1}{2}\sum_{n \geq 1} \left(\frac{-1}{4}\right)^{n-1} \frac{1}{n}\binom{2n - 2}{n - 1} u^n = 1 + \frac{u}{2} - \frac{u^2}{8} + \cdots,$$

$$\frac{1}{\sqrt{1 + u}} = \sum_{n \geq 0} \left(\frac{-1}{4}\right)^n \binom{2n}{n} u^n = 1 - \frac{u}{2} + \frac{3u^2}{8} + \cdots.$$

(4) Setting $z^\alpha := \exp(\alpha \log z)$, recover the results of Section 6.1 from those of Section 6.2.

(5) Call D_θ the open disk $\overset{\circ}{\mathrm{D}}(e^{i\theta}, 1)$. Let $\mathcal{F}$ be any of the sheaves of solutions appearing in this chapter. Define isomorphisms:

$$\mathcal{F}(D_0) \to \mathcal{F}(D_0 \cap D_{\pi/2}) \leftarrow \mathcal{F}(D_{\pi/2}) \to \mathcal{F}(D_\pi \cap D_{\pi/2}) \leftarrow \mathcal{F}(D_\pi)$$
$$\to \cdots \to \mathcal{F}(D_\pi \cap D_{3\pi/2}) \leftarrow \mathcal{F}(D_{3\pi/2}) \to \mathcal{F}(D_{2\pi} \cap D_{3\pi/2}) \leftarrow \mathcal{F}(D_{2\pi}).$$

From the fact that $D_{2\pi} = D_0$, deduce an automorphism of $\mathcal{F}(D_0)$ and explicitly describe that automorphism.

(6) Let p, q be holomorphic on the nonempty domain $\Omega \subset \mathbf{C}$ and let $f_1, f_2 \in \mathcal{O}(\Omega)$ be two solutions of the differential equation $f'' + pf' + qf = 0$. We set $w := f_1 f_2' - f_1' f_2$ (the "wronskian" of f_1 and f_2, which shall be introduced in all generality in Section 7.2).

   (i) Show that $w' = -pw$ and deduce that either w vanishes identically or it has no zeros on Ω. We now assume that $w \neq 0$.

   (ii) Let f be an arbitrary solution. Show that there exist $a_1, a_2 \in \mathcal{O}(\Omega)$ such that $f = a_1 f_1 + a_2 f_2$ and $f' = a_1 f_1' + a_2 f_2'$.

   (iii) Show that $a_1, a_2 \in \mathbf{C}$.

Chapter 7

Linear complex analytic differential equations

This chapter lays down the basic theory of linear complex analytic differential equations. (The nonlinear theory is much more difficult and we shall say nothing about it.) Since one of the ultimate goals of the whole theory is to understand the so-called "special functions", and since the study of these functions[1] shows interesting features of their asymptotic behavior at infinity, the theory is done on the "Riemann sphere" which is the complex plane augmented with a "point at infinity". Therefore, the first section here is a complement to the course on analytic functions.

7.1. The Riemann sphere

We want to study analytic functions "at infinity" as if this was a place, so that we can use geometric reasoning as we did in the complex plane $\mathbf{C}$. There are various ways[2] to do this. A most efficient way is to consider that there is a "point at infinity" that we write ∞ (without the $+$ or $-$ sign). The resulting set is the *Riemann sphere*:

$$\mathbf{S} := \mathbf{C} \cup \{\infty\}.$$

---

[1] In this course, we shall not go into any detail about special functions. The best way to get acquainted with them is by far the book [**WW27**] by Whittaker and Watson, second part.

[2] For instance, in real projective geometry, one adds a whole (real projective) line at infinity; the result is quite different from the one we shall be dealing with here.

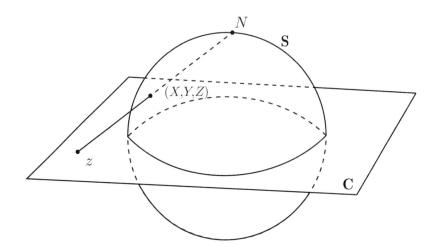

Figure 7.1. Stereographic projection from the north pole

Other names and notation for this are: the Alexandrov one-point compact-ification $\hat{\mathbf{C}}$; the complex projective line[3] $\mathbf{P}^1(\mathbf{C})$. We shall not use these names and notation.

The reason why $\mathbf{S}$ is called a sphere can be understood through a pro-cess coming from cartography, the so-called "stereographic projection" of the unit sphere $S : X^2 + Y^2 + Z^2 = 1$ in $\mathbf{R}^3$ (not to be confused with the Riemann sphere $\mathbf{S}$, although they look like twin sisters!) from its north pole $N(0, 0, 1)$ to its equatorial plane $P(Z = 0)$. To any point $A(X, Y, Z)$ in $\dot{S} := S \setminus \{N\}$, it associates the intersection $B(x, y)$ of the straight line (NA) with P.

Now we shall identify as usual B with $z := x + iy \in \mathbf{C}$, whence a map $(X, Y, Z) \mapsto z$ from $\dot{S}$ to $\mathbf{C}$ given by the equation:

$$z = \frac{X}{1 - Z} + i\frac{Y}{1 - Z}.$$

This map can easily be inverted into the map from $\mathbf{C}$ to $\dot{S}$ described by the equation:

$$z \mapsto A(z) := \left(\frac{2\Re(z)}{|z|^2 + 1}, \frac{2\Im(z)}{|z|^2 + 1}, \frac{|z|^2 - 1}{|z|^2 + 1} \right).$$

Exercise 7.1. Prove these equations.

---

[3]There are some reasons to say that adding one point to the complex plane yields a complex *line*! Indeed, although its real dimension is 2, its complex dimension is 1 and algebraic geometers view it as the simplest of all projective algebraic varieties.

So we have a homeomorphism (a bicontinuous bijection) from the plane $\mathbf{C}$ to $\dot{S}$; it is actually a diffeomorphism (differentiable with a differentiable inverse). Moreover, when z tends to infinity in the plane in the sense that $|z| \to +\infty$ (thus without any particular direction), then the corresponding point $A(z) \in \dot{S}$ tends to N. For this reason, we transport the topology and differentiable structure on $S = \dot{S} \cup \{N\}$ to $\mathbf{S} = \mathbf{C} \cup \{\infty\}$. Therefore, $\mathbf{S}$ is homeomorphic to S. In particular, *the Riemann sphere is compact, arcwise connected and simply connected.*

The open subsets of $\mathbf{S}$ for this topology are, on the one hand, the usual open subsets of $\mathbf{C}$; and, on the other hand, the open neighborhoods of ∞. These can be described as follows: they must contain a "disk centered at infinity" $\overset{\circ}{\mathrm{D}}(\infty, r) := \{z \in \mathbf{S} \mid |w| < r\}$, where $w := 1/z$ with of course the special convention that $1/\infty = 0$. There are three particular important open subsets of $\mathbf{S}$:

$$\mathbf{C}_0 := \mathbf{S} \setminus \{\infty\} = \mathbf{C}, \quad \mathbf{C}_\infty := \mathbf{S} \setminus \{0\}, \quad \mathbf{C}^* = \mathbf{S} \setminus \{0, \infty\} = \mathbf{C}_0 \cap \mathbf{C}_\infty.$$

When working with a point $z \in \mathbf{C}_0$, we may use the usual "coordinate" z; when working with a point $z \in \mathbf{C}_\infty$, we may use the new "coordinate" $w = 1/z$; and when working with a point of $z \in \mathbf{C}^*$, we may use either coordinate. This works without problem for questions of topology and of differential calculus, because, on the common domain of validity $\mathbf{C}^*$ of the two coordinates, the *changes of coordinates* $z \mapsto w = 1/z$ and $w \mapsto z = 1/w$ are homeomorphisms and even diffeomorphisms. But note that they are also both analytic, which justifies the following definition.

Definition 7.2. A function defined on a neighborhood of a point of $\mathbf{S}$ and with values in $\mathbf{C}$ is said to be *analytic*, resp. *holomorphic*, resp. *meromorphic*, if it is so when expressed as a function of whichever coordinate (z or w) is defined at this point. Functions holomorphic at all points of an open set Ω form a differential $\mathbf{C}$-algebra (see Section 3.4) written $\mathcal{O}(\Omega)$ (thus an integral domain if Ω is connected).

In the case of usual functions on $\mathbf{C}$, these definitions are obviously compatible with those given in Chapter 3. Note that here again analyticity is equivalent to holomorphy. Practically, if one wants to study $f(z)$ at infinity, one puts $g(w) := f(1/w)$ and one studies g at $w = 0$.

Example 7.3. Let $f(z) := P(z)/Q(z) \in \mathbf{C}(z)$, $P, Q \neq 0$, be a nontrivial rational function. We write $P(z) = a_0 + \cdots + a_d z^d$, with $a_d \neq 0$ (so that $\deg P = d$), and $Q(z) = b_0 + \cdots + b_e z^e$, with $b_e \neq 0$ (so that $\deg Q = e$). Moreover, we assume that P and Q have no common root. Then f is holomorphic on $\mathbf{C} \setminus Q^{-1}(0)$ and meromorphic at points of $Q^{-1}(0)$: any root of multiplicity k of Q is a pole of order k of f. To see what happens at infinity, we use $w = 1/z$ and consider:

$$g(w) := f(1/w) = \frac{a_0 + \cdots + a_d w^{-d}}{b_0 + \cdots + b_e w^{-e}} = w^{e-d} \frac{a_0 w^d + \cdots + a_d}{b_0 w^e + \cdots + b_e}.$$

Since $a_d b_e \neq 0$, we conclude that it behaves like $(a_d/b_e) w^{e-d}$ for w near 0. Otherwise said, if $e \geq d$, then f is holomorphic at infinity; if $d > e$, then ∞ is a pole of order $d - e$. In any case, f is meromorphic on the whole of $\mathbf{S}$.

For f to be holomorphic on the whole of $\mathbf{S}$, a necessary and sufficient condition is that Q has no root, that is, $e = 0$ (nonconstant complex polynomials always have roots); and that $e \geq d$, so $d = 0$. Therefore, a rational function is holomorphic on the whole of $\mathbf{S}$ if, and only if, it is a constant.

Example 7.4. To study the differential equation $zf' = \alpha f$ at infinity, we put $g(w) = f(w^{-1})$ so that $g'(w) = -f'(w^{-1})w^{-2} = -(\alpha/w^{-1})f(w^{-1})w^{-2} = -\alpha g(w)w^{-1}$, i.e., $wg' = -\alpha g$. The basic solution is, logically, $w^{-\alpha}$.

From the first example above, it follows that $\mathbf{C} \subset \mathcal{O}(\mathbf{S})$ and $\mathbf{C}(z) \subset \mathcal{M}(\mathbf{S})$. The following important theorem is proved in [**Ahl78**, **Car63**], [**Rud87**]:

Theorem 7.5. (i) *Every analytic function on* $\mathbf{S}$ *is constant:* $\mathcal{O}(\mathbf{S}) = \mathbf{C}$.

(ii) *Every meromorphic function on* $\mathbf{S}$ *is rational:* $\mathcal{M}(\mathbf{S}) = \mathbf{C}(z)$. $\square$

Description of holomorphic functions on some open sets. We now describe $\mathcal{O}(\Omega)$ for various open subsets Ω of $\mathbf{S}$.

(1) $\mathcal{O}(\mathbf{C}_0)$: these are the power series $\sum_{n \geq 0} a_n z^n$ with an infinite r.o.c., for example e^z. They are called *entire functions*. A theorem of Liouville says that for an entire function f, if $|f| = O\left(|z|^k\right)$ when $|z| \to +\infty$, then f is a polynomial. (This theorem and the second part of Theorem 7.5 are easy consequences of the first part of Theorem 7.5, which itself follows easily from the fact that every bounded entire function is constant. That last fact is less easy to prove; see [**Ahl78**, **Car63**, **Rud87**].)

(2) $\mathcal{O}(\mathbf{C}_\infty)$: these are the functions $\sum_{n \geq 0} a_n z^{-n}$, where $\sum_{n \geq 0} a_n w^n$ is entire, for example $e^{1/z}$.

(3) $\mathcal{O}(\mathbf{C}^*)$: these are "generalized Laurent series" of the form $\sum_{n \in \mathbf{Z}} a_n z^n$, where both $\sum_{n \geq 0} a_n z^n$ and $\sum_{n \geq 0} a_{-n} w^n$ are entire functions.

(4) $\mathcal{O}(\overset{\circ}{\mathrm{D}}(0,R) \setminus \{0\})$: these are generalized Laurent series of the form $\sum\limits_{n \in \mathbf{Z}} a_n z^n$, where $\sum\limits_{n \geq 0} a_{-n} w^n$ is an entire function and $\sum\limits_{n \geq 0} a_n z^n$ has r.o.c. R.

(5) $\mathcal{O}(\overset{\circ}{\mathrm{D}}(\infty,r) \setminus \{\infty\})$: it is left to the reader to describe this case from the previous one or from the case of an annulus tackled below.

The two last items are particular cases ($r = 0$ or $R = +\infty$) of the *open annulus of radii $r < R$*:

$$\mathcal{C}(r,R) := \overset{\circ}{\mathrm{D}}(0,R) \setminus \overline{\mathrm{D}}(0,r) = \{z \in \mathbf{C} \mid r < |z| < R\}.$$

Then the elements of $\mathcal{O}\left(\mathcal{C}(r,R)\right)$ are generalized Laurent series of the form $\sum\limits_{n \in \mathbf{Z}} a_n z^n$, where $\sum\limits_{n \geq 0} a_{-n} w^n$ has r.o.c. $1/r$ and $\sum\limits_{n \geq 0} a_n z^n$ has r.o.c. R.

Note that the appearance of generalized Laurent series here means that essential singularities are possible. For functions having poles at 0 or ∞, the corresponding series will be usual Laurent series. All the facts above are proved in [**Ahl78, Car63, Rud87**].

7.2. Equations of order n and systems of rank n

Let Ω be a nonempty connected subset (a domain) of $\mathbf{S}$ and let $a_1, \ldots, a_n \in \mathcal{O}(\Omega)$. We shall denote by $E_{\underline{a}}$ the following *linear homogeneous scalar differential equation of order n*:

$$f^{(n)} + a_1 f^{(n-1)} + \cdots + a_n f = 0.$$

The reader may think of the two examples $f' - (\alpha/z)f = 0$ and $f'' + (1/z)f' + 0f = 0$ of Chapter 6; there, $\Omega = \mathbf{C}^*$. For any open subset U of Ω, we write $\mathcal{F}_{\underline{a}}(U)$ as the set of solutions of $E_{\underline{a}}$ on U:

$$\mathcal{F}_{\underline{a}}(U) := \{f \in \mathcal{O}(U) \mid \forall z \in U, \ f^{(n)}(z) + a_1(z)f^{(n-1)}(z) + \cdots + a_n(z)f(z) = 0\}.$$

This is a linear space over $\mathbf{C}$. The reason to consider various open sets U is that we hope to understand better the differential equation $E_{\underline{a}}$ on the whole of Ω by studying the collection of all spaces $\mathcal{F}_{\underline{a}}(U)$. For instance, we saw examples in Chapter 6 where $\mathcal{F}_{\underline{a}}(\Omega) = \{0\}$ but nevertheless some local solutions are interesting.

By convention, we agree that $\mathcal{F}_{\underline{a}}(\emptyset) = \{0\}$, the trivial space. If U is not connected and if $U = \bigsqcup U_i$ is its decomposition in connected components, we know that the U_i are open sets and that there is an isomorphism $\mathcal{O}(U) \to \prod \mathcal{O}(U_i)$, which sends $f \in \mathcal{O}(U)$ to the family (f_i) of its restrictions $f_i := f_{|U_i}$. This clearly induces an isomorphism of $\mathcal{F}_{\underline{a}}(U)$ with $\prod \mathcal{F}_{\underline{a}}(U_i)$. For these reasons, we shall assume most of the time that U is a nonempty domain.

Let $A = (a_{i,j})_{1 \le i,j \le n} \in \mathrm{Mat}_n(\mathcal{O}(\Omega))$ be a matrix having analytic coefficients $a_{i,j} \in \mathcal{O}(\Omega)$. We shall denote by S_A the following *linear homogeneous vectorial differential equation of rank n and order* 1:

$$X' = AX.$$

The reader may think of the vectorial form of the equation of the logarithm in Chapter 6; there, $\Omega = \mathbf{C}^*$. The other standard name for a vectorial differential equation is *differential system*.

Remark 7.6. There may be difficulty when $\infty \in \Omega$. Indeed, then the coordinate $w := 1/z$ should be used. Writing $B(w) := A(1/w)$ and $Y(w) := X(1/w)$, S_A becomes $Y'(w) = -w^{-2}B(w)Y(w)$. However, it is not true that if A is analytic at ∞, then $-w^{-2}B(w)$ is analytic at $w = 0$. For instance, the differential equation $f' = f$ is not analytic at ∞, despite appearances. The totally rigorous solution to this difficulty is to write S_A in the more invariant form $dX = \Sigma X$ and to require that the "matrix of differential forms" $\Sigma := A(z)dz = B(w)dw$ be analytic on Ω. At ∞ that would translate to the correct condition that $-w^{-2}B(w)$ is analytic at $w = 0$. Except in the following exercise, we shall not tackle this problem anymore;[4] only in concrete cases when we have to.

Exercise 7.7. Prove that a differential system that is holomorphic on the whole of $\mathbf{S}$ has the form $X' = 0$.

For any open subset U of Ω, we write $\mathcal{F}_A(U)$ as the set of vectorial solutions of S_A on U:

$$\mathcal{F}_A(U) := \{X \in \mathcal{O}(U)^n \mid \forall z \in U, \ X'(z) = A(z)X(z)\}.$$

This is a linear space over $\mathbf{C}$. By convention, we agree that $\mathcal{F}_A(\emptyset) = \{0\}$. If $U = \bigsqcup U_i$ is the decomposition of U in connected components, the map which sends $X \in \mathcal{F}_A(U)$ to the family of its restrictions $X_{|U_i}$ is an isomorphism of $\mathcal{F}_A(U)$ with $\prod \mathcal{F}_A(U_i)$. We shall also sometimes consider *matricial* solutions $M \in \mathrm{Mat}_{n,p}(\mathcal{O}(U))$, that is, matrices with coefficients in $\mathcal{O}(U)$ and such that $M'(z) = A(z)M(z)$ on U; this is equivalent to saying that all the columns of M belong to $\mathcal{F}_A(U)$. For instance, if $A \in \mathrm{Mat}_n(\mathbf{C})$ (constant coefficients), then e^{zA} is a matricial solution of S_A. Note that S_A can be seen as a system of scalar differential equations, or a *differential system*; if $f_1, \ldots, f_n$ are the components of X, then:

$$S_A \iff \begin{cases} f_1' = a_{1,1}f_1 + \cdots + a_{1,n}f_n, \\ \cdots\cdots\cdots\cdots\cdots\cdots\cdots, \\ f_n' = a_{n,1}f_1 + \cdots + a_{n,n}f_n. \end{cases}$$

---

[4]Actually, this coordinate-free way to write a differential equation has the advantage of being transposable to more general Riemann surfaces than the Riemann sphere; see for instance [**Del70**] or [**Sab93**].

Proposition 7.8. *Given* $a_1, \ldots, a_n \in \mathcal{O}(\Omega)$ *and* $f \in \mathcal{O}(U)$, $U \subset \Omega$, *define:*

$$
A_{\underline{a}} := \begin{pmatrix}
0 & 1 & 0 & \cdots & 0 & 0 \\
0 & 0 & 1 & \cdots & 0 & 0 \\
\vdots & \vdots & \vdots & \ddots & \vdots & \vdots \\
0 & 0 & 0 & \cdots & 1 & 0 \\
0 & 0 & 0 & \cdots & 0 & 1 \\
-a_n & -a_{n-1} & -a_{n-2} & \cdots & -a_2 & -a_1
\end{pmatrix} \in \mathrm{Mat}_n(\mathcal{O}(\Omega))
$$

and

$$
X_f := \begin{pmatrix} f \\ f' \\ \vdots \\ f^{(n-1)} \end{pmatrix} \in \mathcal{O}(U)^n.
$$

Then the scalar differential equation $E_{\underline{a}}$ is equivalent to the vectorial differential equation S_A for $A := A_{\underline{a}}$ in the following sense: for any open subset $U \subset \Omega$, the map $f \mapsto X_f$ is an isomorphism of linear spaces from $\mathcal{F}_{\underline{a}}(U)$ to $\mathcal{F}_A(U)$.

Proof. Let $X \in \mathcal{O}(U)^n$ have components $f_1, \ldots, f_n$. Then $X' = A_{\underline{a}} X$ is equivalent to $f_1' = f_2, \ldots, f_{n-1}' = f_n$ and $f_n' + a_1 f_n + \cdots + a_n f_1 = 0$, which in turn is equivalent to $X = X_f$, where $f := f_1$ is a solution of $E_{\underline{a}}$ and the unique antecedent of X by $f \mapsto X_f$. $\qquad\square$

The wronskian.

Definition 7.9. The *wronskian matrix* of $f_1, \ldots, f_n \in \mathcal{O}(U)$ is:

$$
W_n(f_1, \ldots, f_n) := \begin{pmatrix}
f_1 & \cdots & f_j & \cdots & f_n \\
\vdots & \ddots & \vdots & \ddots & \vdots \\
f_1^{(i-1)} & \cdots & f_j^{(i-1)} & \cdots & f_n^{(i-1)} \\
\vdots & \ddots & \vdots & \ddots & \vdots \\
f_1^{(n-1)} & \cdots & f_j^{(n-1)} & \cdots & f_n^{(n-1)}
\end{pmatrix}
$$

$$
= [X_{f_1}, \ldots, X_{f_n}] \in \mathrm{Mat}_n(\mathcal{O}(U)).
$$

Beware that its (i, j)-coefficient is $f_j^{(i-1)}$. The *wronskian determinant* of $f_1, \ldots, f_n \in \mathcal{O}(U)$ is:

$$
w_n(f_1, \ldots, f_n) := \det W_n(f_1, \ldots, f_n) \in \mathcal{O}(U).
$$

It is also simply called the *wronskian* of $f_1, \ldots, f_n$.

The wronskian is an extremely useful tool in the study of differential systems. We prepare its computation with two lemmas.

Lemma 7.10. *Let* $X_1, \ldots, X_n : U \to \mathbf{C}^n$ *be vector-valued analytic functions. Then* $\det(X_1, \ldots, X_n) \in \mathcal{O}(U)$ *and*

$$(\det(X_1, \ldots, X_n))' = \det(X_1', \ldots, X_i, \ldots, X_n)$$
$$+ \cdots + \det(X_1, \ldots, X_i', \ldots, X_n)$$
$$+ \cdots + \det(X_1, \ldots, X_i, \ldots, X_n').$$

Proof. From the "small" Leibniz formula $(fg)' = f'g + fg'$, one obtains by induction the "big" Leibniz formula $(f_1 \cdots f_n)' = f_1' \cdots f_i \cdots f_n + \cdots + f_1 \cdots f_i' \cdots f_n + \cdots + f_1 \cdots f_i \cdots f_n'$. Then we apply this formula to each of the $n!$ monomials composing the determinant. $\qquad\square$

Lemma 7.11. *Let* $X_1, \ldots, X_n \in \mathbf{C}^n$ *and let* $A \in \mathrm{Mat}_n(\mathbf{C})$. *Then:*

$$\det(AX_1, \ldots, X_i, \ldots, X_n) + \cdots + \det(X_1, \ldots, AX_i, \ldots, X_n)$$
$$+ \cdots + \det(X_1, \ldots, X_i, \ldots, AX_n)$$
$$= \mathrm{Tr}(A) \det(X_1, \ldots, X_n).$$

Proof. The left-hand side is an alternated n-linear function of the X_i, so, by general multilinear algebra (see for instance [**Lan02**]), it is equal to $C \det(X_1, \ldots, X_n)$ for some constant C. Taking for $X_1, \ldots, X_n$ the canonical basis of $\mathbf{C}^n$ gives the desired result. $\qquad\square$

Proposition 7.12. *Let* $X_1, \ldots, X_n \in \mathcal{F}_A(U)$. *Write* $\mathcal{X} := [X_1, \ldots, X_n]$ *as the matrix having the* X_i *as columns. Then* $\mathcal{X} \in \mathrm{Mat}_n(\mathcal{O}(U))$, $\mathcal{X}' = A\mathcal{X}$ *and*

$$(\det \mathcal{X})' = \mathrm{Tr}(A)(\det \mathcal{X}).$$

As a consequence, if U *is a domain, either* $\det \mathcal{X}$ *vanishes nowhere on* U *or it vanishes identically.*

Proof. The fact that $\mathcal{X} \in \mathrm{Mat}_n(\mathcal{O}(U))$ is obvious; the fact that $\mathcal{X}' = A\mathcal{X}$ follows because the columns of $A\mathcal{X}$ are the $AX_i = X_i'$. For the last formula, we compute with the help of the two lemmas above:

$$(\det \mathcal{X})' = \det(X_1', \ldots, X_i, \ldots, X_n) + \cdots + \det(X_1, \ldots, X_i, \ldots, X_n')$$
$$= \det(AX_1, \ldots, X_i, \ldots, X_n) + \cdots + \det(X_1, \ldots, X_i, \ldots, AX_n)$$
$$= \mathrm{Tr}(A) \det(X_1, \ldots, X_n) = \mathrm{Tr}(A)(\det \mathcal{X}).$$

The last statement is proved as follows. Write for short $w(z) := (\det \mathcal{X})(z)$ and suppose that $w(z_0) = 0$ for some $z_0 \in U$. Let V be a nonempty simply-connected open neighborhood of z_0 in U (for instance, an open disk centered at z_0). Then the analytic function $\mathrm{Tr}(A)$ has a primitive f on V (see Section 3.3). Therefore, $w' = f'w$ on V, that is, $(e^{-f}w)' = 0$ so that $w = Ce^f$ for some constant $C \in \mathbf{C}$. Since $w(z_0) = 0$, this implies that $C = 0$ and w

vanishes on V. By the principle of analytic continuation (Theorem 3.6), U being connected, w vanishes on U. ☐

Corollary 7.13. *Let* $f_1, \ldots, f_n \in \mathcal{F}_{\underline{a}}(U)$ *and write* $w(z) := w_n(f_1, \ldots, f_n)(z)$. *Then* $w \in \mathcal{O}(U)$, $w' = -a_1 w$ *and, if* U *is a domain, either* w *vanishes nowhere on* U *or it vanishes identically.*

Proof. Indeed, $\mathrm{Tr}(A_{\underline{a}}) = -a_1$. ☐

Definition 7.14. Let U be a domain.

(i) If $X_1, \ldots, X_n \in \mathcal{F}_A(U)$ are such that $\det(X_1, \ldots, X_n)$ vanishes nowhere on U, then $\mathcal{X} := [X_1, \ldots, X_n]$ is called a *fundamental matricial solution of S_A on U.*

(ii) If $f_1, \ldots, f_n \in \mathcal{F}_{\underline{a}}(U)$ are such that their wronskian vanishes nowhere on U, then $(f_1, \ldots, f_n)$ is called a *fundamental system of solutions of $E_{\underline{a}}$ on U.*

Examples 7.15. (i) If $(f_1, \ldots, f_n)$ is a fundamental system of solutions of $E_{\underline{a}}$ on U, then $[X_{f_1}, \ldots, X_{f_n}]$ is a fundamental matricial solution of S_A on U for $A := A_{\underline{a}}$.

(ii) If $A \in \mathrm{Mat}_n(\mathbf{C})$, then e^{zA} is a fundamental matricial solution of $X' = AX$ on $\mathbf{C}$.

(iii) If $\alpha \in \mathbf{C}$, then z^α is a fundamental matricial solution (of rank 1) and also a fundamental system of solutions of $zf' = \alpha f$ on $\mathbf{C} \setminus \mathbf{R}_-$.

(iv) The pair $(1, \log)$ is a fundamental system of solutions of $zf'' + f' = 0$ on $\mathbf{C} \setminus \mathbf{R}_-$. Indeed, its wronskian matrix is $\begin{pmatrix} 1 & \log z \\ 0 & 1/z \end{pmatrix}$. Its wronskian determinant is $w(z) = 1/z$, which satisfies $w' = -a_1 w$ with $a_1 = 1/z$ here (and $a_2 = 0$).

Remark 7.16. If $\mathcal{X} \in \mathrm{Mat}_n(\mathcal{O}(U))$ is a fundamental matricial solution of S_A, then $\mathcal{X}^{-1} \in \mathrm{Mat}_n(\mathcal{O}(U))$. Indeed, the inverse of an arbitrary matrix A is computed as $\dfrac{1}{\det A}\, {}^t\mathrm{com}(A)$, where $\mathrm{com}(A)$ (the so-called "comatrix" of A) has as coefficients the minor determinants of A. In our case, $\mathrm{com}(\mathcal{X})$ and its transpose ${}^t\mathrm{com}(\mathcal{X})$ are obviously in $\mathrm{Mat}_n(\mathcal{O}(U))$; and, since $\det \mathcal{X}$ is analytic and vanishes nowhere, $\dfrac{1}{\det \mathcal{X}} \in \mathcal{O}(U)$. We shall therefore write $\mathcal{X} \in \mathrm{GL}_n(\mathcal{O}(U))$. More generally, for any commutative ring R, the matrices $A \in \mathrm{Mat}_n(R)$ which have an inverse $A^{-1} \in \mathrm{Mat}_n(R)$ are those such that $\det A$ is invertible in R. The group of such matrices is written $\mathrm{GL}_n(R)$.

Theorem 7.17. *Let* U *be a nonempty domain. Let* $X_1, \ldots, X_n \in \mathcal{F}_A(U)$ *and let* $\mathcal{X} := [X_1, \ldots, X_n] \in \mathrm{Mat}_n(\mathcal{O}(U))$. *Write* $w(z) := (\det \mathcal{X})(z)$. *If there exists* $z_0 \in U$ *such that* $w(z_0) \neq 0$, *then* w *vanishes nowhere (and* *therefore $\mathcal{X}$ is a fundamental matricial solution of S_A on U). In this case, $(X_1, \ldots, X_n)$ is a basis of $\mathcal{F}_A(U)$.*

Proof. Suppose there exists $z_0 \in U$ such that $w(z_0) \neq 0$. The fact that w vanishes nowhere is contained in the previous proposition. Then $\mathcal{X}$ is invertible. Let $X \in \mathcal{F}_A(U)$ be an arbitrary solution. Taking $Y := \mathcal{X}^{-1}X$, we can write $X = \mathcal{X}Y$ with $Y \in \mathcal{O}(U)^n$. From $X' = AX$, i.e., $(\mathcal{X}Y)' = A(\mathcal{X}Y)$, we obtain

$$\mathcal{X}Y' + \mathcal{X}'Y = A\mathcal{X}Y \implies \mathcal{X}Y' + A\mathcal{X}Y = A\mathcal{X}Y \implies \mathcal{X}Y' = 0 \implies Y' = 0,$$

so that $Y \in \mathbf{C}^n$ since U is connected (otherwise, having a zero derivative would not imply being constant). If $\lambda_1, \ldots, \lambda_n \in \mathbf{C}$ are the components of Y, we conclude that $X = \lambda_1 X_1 + \cdots + \lambda_n X_n$, and $(X_1, \ldots, X_n)$ is a generating system. The same computation shows that the coefficients $\lambda_1, \ldots, \lambda_n$ are unique, so that $(X_1, \ldots, X_n)$ is indeed a basis of $\mathcal{F}_A(U)$. $\quad\square$

Remark 7.18. If $w = 0$, then it is clear that, for all z_0 in U, the vectors $X_1(z_0), \ldots, X_n(z_0)$ are linearly dependent over $\mathbf{C}$. One can prove algebraically the much stronger statement that $X_1, \ldots, X_n$ are linearly dependent over $\mathbf{C}$ (with coefficients that do not depend on z): this is the "wronskian lemma" (Lemma 1.12 in [**vS03**]; also see exercise 3 at the end of the chapter).

Corollary 7.19. *If S_A admits a fundamental matricial solution $\mathcal{X} = [X_1, \ldots, X_n]$, then, for any basis $(Y_1, \ldots, Y_n)$ of $\mathcal{F}_A(U)$, the matrix $\mathcal{Y} := [Y_1, \ldots, Y_n]$ is a fundamental matricial solution.*

Proof. We write $Y_i = \mathcal{X}P_i$, with $P_i \in \mathbf{C}^n$ as in the theorem. Then, putting $P := [P_1, \ldots, P_n]$, one has $\mathcal{Y} = \mathcal{X}P$ with $P \in \mathrm{Mat}_n(\mathbf{C})$. Since P links two bases, $\det P \neq 0$. Thus $\det \mathcal{Y} = (\det \mathcal{X})(\det P)$ does not vanish. $\quad\square$

Corollary 7.20. *Let U be a nonempty domain. Let $f_1, \ldots, f_n \in \mathcal{F}_{\underline{a}}(U)$. Write $w(z) := (w_n(f_1, \ldots, f_n))(z)$. If there exists $z_0 \in U$ such that $w(z_0) \neq 0$, then w vanishes nowhere and $(f_1, \ldots, f_n)$ is a basis of $\mathcal{F}_A(U)$. In this case, any basis of $\mathcal{F}_{\underline{a}}(U)$ is a fundamental system of solutions of $E_{\underline{a}}$.*

Proof. This just uses the isomorphism $f \mapsto X_f$ from $\mathcal{F}_{\underline{a}}(U)$ to $\mathcal{F}_A(U)$, where $A := A_{\underline{a}}$ (Proposition 7.8). $\quad\square$

As the following result shows, the existence of a fundamental matricial solution or a fundamental system of solutions actually corresponds to an "optimal case",[5] where the solution space has dimension n.

Theorem 7.21. *For any domain $U \subset \Omega$, one has $\dim_{\mathbf{C}} \mathcal{F}_A(U) \leq n$.*

---

[5]In old literature on functional equations, authors said that they had a "full complement of solutions" when they reached the maximum reasonable number of independent solutions.

Proof. This can be deduced from the "wronskian lemma" (see Remark 7.18), but we shall do it using linear algebra over the field $K := \mathcal{M}(U)$. Since elements of $\mathcal{F}_A(U)$ belong to K^n, the maximum number of K-linearly independent elements $X_1, \ldots, X_k \in \mathcal{F}_A(U)$ is some integer $k \leq n$. If $k = 0$, then $\mathcal{F}_A(U) = \{0\}$ and the result is trivial, so assume $k \geq 1$. Choose such elements $X_1, \ldots, X_k$; we are going to prove that they form a basis of the $\mathbf{C}$-linear space $\mathcal{F}_A(U)$ and the conclusion will follow.

Since $X_1, \ldots, X_k$ are K-linearly independent elements, they are plainly $\mathbf{C}$-linearly independent elements. Now let $X \in \mathcal{F}_A(U)$. By the maximality property of $X_1, \ldots, X_k$, one can write $X = f_1 X_1 + \cdots + f_k X_k$, with $f_1, \ldots, f_k \in K = \mathcal{M}(U)$. Derivating this relation and using S_A, we find:

$$X' = f_1' X_1 + \cdots + f_k' X_k + f_1 X_1' + \cdots + f_k X_k',$$

whence:

$$
\begin{aligned}
AX &= f_1' X_1 + \cdots + f_k' X_k + f_1 A X_1 + \cdots + f_k A X_k \\
&= f_1' X_1 + \cdots + f_k' X_k + A(f_1 X_1 + \cdots + f_k X_k),
\end{aligned}
$$

whence at last:

$$f_1' X_1 + \cdots + f_k' X_k = 0 \implies f_1' = \cdots = f_k' = 0,$$

since $X_1, \ldots, X_k$ are K-linearly independent; this implies $f_1, \ldots, f_k \in \mathbf{C}$. $\square$

Corollary 7.22. *For any domain $U \subset \Omega$, one has $\dim_{\mathbf{C}} \mathcal{F}_{\underline{a}}(U) \leq n$.*

Proof. Again use the isomorphism $f \mapsto X_f$ from $\mathcal{F}_{\underline{a}}(U)$ to $\mathcal{F}_A(U)$, where $A := A_{\underline{a}}$ (Proposition 7.8). $\square$

7.3. The existence theorem of Cauchy

From now on, we shall state and prove theorems for systems S_A and leave it to the reader to translate them into results about equations $E_{\underline{a}}$. Also, we shall call S_A a differential equation, without explicitly saying "vectorial".

Lemma 7.23. (i) *Let $A \in \mathrm{Mat}_n(\mathbf{C}[[z]])$ be a matrix with coefficients in $\mathbf{C}[[z]]$. Then there is a unique $\mathcal{X} \in \mathrm{Mat}_n(\mathbf{C}[[z]])$ such that $\mathcal{X}' = A\mathcal{X}$ and $\mathcal{X}(0) = I_n$. Moreover, $\mathcal{X} \in \mathrm{GL}_n(\mathbf{C}[[z]])$.*

(ii) *Suppose that A has a strictly positive radius of convergence, i.e., $A \in \mathrm{Mat}_n(\mathbf{C}\{z\})$. Then $\mathcal{X}$ also has a strictly positive radius of convergence, i.e., $X \in \mathrm{GL}_n(\mathbf{C}\{z\})$.*

Proof. (i) Write $A = A_0 + zA_1 + \cdots$ and $\mathcal{X} = \mathcal{X}_0 + z\mathcal{X}_1 + \cdots$. Then:

$$
\begin{cases} \mathcal{X}' = A\mathcal{X}, \\ \mathcal{X}(0) = I_n, \end{cases} \iff \begin{cases} \mathcal{X}_0 = I_n, \\ \forall k \geq 0, \ (k+1)\mathcal{X}_{k+1} = A_0\mathcal{X}_k + \cdots + A_k\mathcal{X}_0, \end{cases}
$$

so that it can be recursively solved and admits a unique solution. Then $(\det \mathcal{X})(0) = 1$, so that $\det \mathcal{X}$ is invertible in $\mathbf{C}[[z]]$; as already noted (Remark 7.16), this implies $\mathcal{X} \in \mathrm{GL}_n(\mathbf{C}[[z]])$.

(ii) We use the same norm as in Section 1.4. Choose $R > 0$ and strictly smaller than the r.o.c. of A. Then the $\|A_k\| R^k$ tend to 0 as $k \to +\infty$, so that they are bounded: there exists $C > 0$ such that $\|A_k\| \leq CR^{-k}$ for all $k \geq 0$. If necessary, reduce R so that moreover $CR \leq 1$. We are going to prove that $\|\mathcal{X}_k\| \leq R^{-k}$ for all $k \geq 0$ which will yield the conclusion. The inequality is obvious for $k = 0$. Assume it for all indexes up to k. Then:

$$\|\mathcal{X}_{k+1}\| \leq \frac{1}{k+1} \sum_{i+j=k} \|A_i\| \|\mathcal{X}_j\| \leq \frac{1}{k+1} \sum_{i+j=k} (CR^{-i})(R^{-j})$$
$$= (CR)R^{-(k+1)} \leq R^{-(k+1)}.$$

$\square$

Note that the proof says nothing about the radius of convergence of $\mathcal{X}$ as compared to that of A; they may be different.

Exercise 7.24. Give an example of such a difference. (A scalar example does it.)

Theorem 7.25 (Cauchy existence theorem). *Let* $A \in \mathrm{Mat}_n(\Omega)$. *Then, for all* $z_0 \in \Omega$, *there exists a domain* $U \subset \Omega$, $U \ni z_0$ *such that the map* $X \mapsto X(z_0)$ *from* $\mathcal{F}_A(U)$ *to* $\mathbf{C}^n$ *is an isomorphism. In other words, on such a domain, the Cauchy problem* $\begin{cases} X' = AX, \\ X(z_0) = X_0 \end{cases}$ *admits a unique solution for any initial condition* $X_0 \in \mathbf{C}^n$.

Proof. Suppose for a moment that we put $B(z) := A(z_0 + z)$ and $Y(z) := X(z_0 + z)$. Then the Cauchy problem $\begin{cases} X' = AX, \\ X(z_0) = X_0 \end{cases}$ is equivalent to $\begin{cases} Y' = AY, \\ Y(0) = X_0. \end{cases}$ In other words, we may (and shall) assume from the beginning that $z_0 = 0$.

From the previous lemma, there exists $\mathcal{X} \in \mathrm{GL}_n(\mathbf{C}\{z\})$ such that $\mathcal{X}' = A\mathcal{X}$ and $\mathcal{X}(0) = I_n$. Let $U \subset \Omega$ be any nonempty domain containing 0 on which $\mathcal{X}$ converges and $\det \mathcal{X}$ does not vanish. Then we may look for X in the form $\mathcal{X}Y$. With the same computation as in Theorem 7.17, we find that our Cauchy problem is equivalent to $\begin{cases} Y' = 0, \\ Y(0) = X_0, \end{cases}$ that is, to $Y = X_0$. Therefore, its unique solution on U is $\mathcal{X}X_0$. $\square$

Corollary 7.26. *For all $z_0 \in \Omega$, there exists a domain $U \subset \Omega$, $U \ni z_0$, such that $\dim_{\mathbf{C}} \mathcal{F}_A(U) = n$.*

Remark 7.27. We can obtain from this that there are no *singular* solutions on any open subset of Ω. Indeed, suppose X is a solution in a punctured disk $\overset{\circ}{\mathrm{D}}(z_0, R) \setminus \{z_0\}$, where $\overset{\circ}{\mathrm{D}}(z_0, R) \subset \Omega$ and $R > 0$. Let $\mathcal{X}$ be a fundamental solution on $\overset{\circ}{\mathrm{D}}(z_0, R') \subset \overset{\circ}{\mathrm{D}}(z_0, R)$, $R' > 0$. Then $X = \mathcal{X}X_0$ on $\overset{\circ}{\mathrm{D}}(z_0, R')$, with $X_0 \in \mathbf{C}^n$; thus z_0 is a removable singularity of X.

Exercise 7.28. Deduce from Cauchy's theorem that $\dim_{\mathbf{C}} \mathcal{F}_A(\Omega) \leq n$ for any domain Ω. (Use the injectivity of the restriction map $\mathcal{F}_A(\Omega) \to \mathcal{F}_A(U)$.)

7.4. The sheaf of solutions

Let Ω be a nonempty domain of **S**, $a_1, \ldots, a_n \in \mathcal{O}(\Omega)$ and $A \in \mathrm{Mat}_n(\mathcal{O}(\Omega))$. Remember that we have denoted $\mathcal{F}_{\underline{a}}(U)$, resp. $\mathcal{F}_A(U)$, the complex linear space of solutions of $E_{\underline{a}}$, resp. S_A, on an open subset $U \subset \Omega$. In this section, we shall study at the same time some topological and algebraic properties of the maps $\mathcal{F}_{\underline{a}} : U \mapsto \mathcal{F}_{\underline{a}}(U)$ and $\mathcal{F}_A : U \mapsto \mathcal{F}_A(U)$. To this end, we shall indifferently write $\mathcal{F}$ either for $\mathcal{F}_{\underline{a}}$ or for $\mathcal{F}_A$.

Sheaves. The first important property is that $\mathcal{F}$ is a *sheaf*. (For the general theory, see [**God73**]; the book [**Ahl78**] of Ahlfors uses a less flexible presentation, resting on "espaces étalés".) To be precise, $\mathcal{F}$ is a *sheaf of complex linear spaces over* Ω, meaning that it associates to every open subset $U \subset \Omega$ a **C**-linear space $\mathcal{F}(U)$, with the following extra structure and conditions (which are of course completely obvious in the case of $\mathcal{F}_{\underline{a}}$ and $\mathcal{F}_A$):

(1) If $V \subset U$ are two open subsets of Ω, there is a morphism (linear map) of *restriction*:

$$\rho_V^U : \mathcal{F}(U) \to \mathcal{F}(V).$$

In general, an element $s \in \mathcal{F}(U)$ is called a "section" of $\mathcal{F}$ over U (in our case of interest, sections are solutions f or X) and we shall often write $s_{|V} := \rho_V^U(s) \in \mathcal{F}(V)$ as its restriction to V. The restriction maps must satisfy natural compatibility conditions: if $V = U$, it is the identity map of $\mathcal{F}(U)$; if $W \subset V \subset U$, then the restriction map $\mathcal{F}(U) \to \mathcal{F}(W)$ is the composite of the restriction maps $\mathcal{F}(U) \to \mathcal{F}(V)$ and $\mathcal{F}(V) \to \mathcal{F}(W)$. Stated in formulas:

$$\rho_U^U = \mathrm{Id}_{\mathcal{F}(U)} \text{ and } \rho_W^V \circ \rho_V^U = \rho_W^U.$$

(2) Given an open covering $U = \bigcup U_i$ of the open subset $U \subset \Omega$ by open subsets $U_i \subset U$, there arises a map $\mathcal{F}(U) \to \prod \mathcal{F}(U_i)$, defined as $s \mapsto (s_i)$, where $s \in \mathcal{F}(U)$ and the s_i are the restrictions $s_{|U_i} = \rho^U_{U_i}(s)$.

Then the second requirement is that this map be injective: a section is totally determined by its restrictions to an open covering.

(3) In the previous construction, it is a consequence of the compatibility condition stated before that one has for all i, j the equality $(s_i)_{|U_i \cap U_j} = (s_j)_{|U_i \cap U_j}$ on $U_i \cap U_j$; indeed, $\rho^{U_i}_{U_i \cap U_j} \circ \rho^U_{U_i} = \rho^U_{U_i \cap U_j} = \rho^{U_j}_{U_i \cap U_j} \circ \rho^U_{U_j}$.

Our last requirement is that, conversely, for every open covering $U = \bigcup U_i$ and for every family of sections $(s_i) \in \prod \mathcal{F}(U_i)$, if this family satisfies the compatibility condition that for all i, j one has the equality $(s_i)_{|U_i \cap U_j} = (s_j)_{|U_i \cap U_j}$, then there exists a section $s \in \mathcal{F}(U)$ such that, for all i, one has $s_i = s_{|U_i}$. The section s is of course unique, because of the second requirement.

Local systems. We will not give the precise and general definition of a "local system" of linear spaces here (see however Section 8.2), but rather state the properties of $\mathcal{F} = \mathcal{F}_{\underline{a}}$ or $\mathcal{F}_A$ which imply that it is indeed a local system, and then obtain the consequences, using direct arguments (taking in account that our "sections" are actually functions or vectors of functions). But most of what we are going to do until the end of this chapter is valid in a much more general form, and the reader *should* look at the beginning of the extraordinary book of Deligne [**Del70**] to see the formalism. The basic fact is the following.

Proposition 7.29. *Every $a \in \Omega$ has a neighborhood $U \subset \Omega$ on which the sheaf $\mathcal{F}$ is "constant", meaning that, for every nonempty domain $V \subset U$, the restriction map $\mathcal{F}(U) \to \mathcal{F}(V)$ is bijective.*

Proof. Choose for U the disk of convergence of a fundamental system of solutions (case $\mathcal{F} = \mathcal{F}_{\underline{a}}$) or a fundamental matricial solution (case $\mathcal{F} = \mathcal{F}_A$); if necessary, shrink the disk so that $U \subset \Omega$. Then we know that $\dim_{\mathbf{C}} \mathcal{F}(U) = n$ (Section 7.3). Now, for every nonempty domain $V \subset U$, we know that the restriction map $\mathcal{F}(U) \to \mathcal{F}(V)$ is injective (principle of analytic continuation) and that $\dim_{\mathbf{C}} \mathcal{F}(V) \leq n$ (Theorem 7.21). $\square$

Definition 7.30. We say that an open set U is *trivializing* for $\mathcal{F}$ if Proposition 7.29 applies, *i.e.*, if the restriction of $\mathcal{F}$ to U is constant in the above sense.

Germs. We defined germs of functions and germs of solutions, but germs can actually be defined for any sheaf $\mathcal{F}$ on Ω as follows. Call a "local element" of $\mathcal{F}$ a pair (U, s), where U is an open subset of Ω and $s \in \mathcal{F}(U)$. Fix a point $a \in \Omega$. We say that two local elements (U_1, s_1) and (U_2, s_2) define the same germ at a if there exists an open neighborhood V of a such that $V \subset U_1 \cap U_2$ and $(s_1)_{|V} = (s_2)_{|V}$. The germ defined by a local element (U, s) at $a \in U$ will be denoted s_a and called *the germ of s at a*. The set of all germs at a is written $\mathcal{F}_a$. The set $\mathcal{F}_a$ (do not confuse this notation with that of $\mathcal{F}_{\underline{a}}$!) is the *stalk* or *fiber* of the sheaf $\mathcal{F}$ at a. In our case (sheaf of linear spaces), germs can be added and multiplied by scalars, so that $\mathcal{F}_a$ is actually a linear space. Moreover, there is for $a \in U$ a natural map $\mathcal{F}(U) \to \mathcal{F}_a$ and it is of course a linear map. Now, since our sheaves are local systems, we get the following consequence of the previous proposition.

Corollary 7.31. *For all sufficiently small connected neighborhoods of a, the linear maps $\mathcal{F}(U) \to \mathcal{F}_a$ are bijective.*

In the particular case of $\mathcal{F}_{\underline{a}}$ and $\mathcal{F}_A$, there is an additional structure and property that do not make sense for general local systems. Indeed, there is an "initial condition" map from $\mathcal{F}(U)$ to $\mathbf{C}^n$, which sends $f \in \mathcal{F}_{\underline{a}}(U)$ to $(f(a), \dots, f^{(n-1)}(a))$ and $X \in \mathcal{F}_A(U)$ to $X(a)$. We know from Cauchy's theorem that it is bijective for all sufficiently small connected neighborhoods of a.

Corollary 7.32 (of Cauchy's theorem). *The initial condition map induces an isomorphism $\mathcal{F}_a \mapsto \mathbf{C}^n$.*

7.5. The monodromy representation

Monodromy. We shall play the usual game (that of Chapters 5 and 6) but described in a more general guise. For greater generality, see Chapter 8; for even greater generality, see the book of Deligne [**Del70**]. To begin with, we fix $a, b \in \Omega$ and a path γ from a to b in Ω. We cover the image curve by "sufficiently small" disks $D_0, \dots, D_N$ in such a way that $a \in D_0$, $b \in D_N$ and, for $i = 1, \dots, N$, $D_i \cap D_{i-1} \neq \emptyset$. Here, "sufficiently small" just means trivializing (Definition 7.30). Then there are isomorphisms (linear bijections) $\mathcal{F}_a \leftarrow \mathcal{F}(D_0)$; $\mathcal{F}(D_{i-1}) \to \mathcal{F}(D_i \cap D_{i-1}) \leftarrow \mathcal{F}(D_i)$ for $i = 1, \dots, N$; and $\mathcal{F}(D_N) \to \mathcal{F}_b$. Composing all these maps, we get an isomorphism $s \mapsto s^\gamma$ from $\mathcal{F}_a$ to $\mathcal{F}_b$. We know that this isomorphism depends only on the homotopy class of γ in $\Pi_1(\Omega; a, b)$. In case $\mathcal{F}$ is $\mathcal{F}_{\underline{a}}$ or $\mathcal{F}_A$, this follows from the principle of monodromy. For an arbitrary local system, this can be proved in exactly the same way (see the book of Deligne). To summarize, we get a map:

$$\Pi_1(\Omega; a, b) \to \mathrm{Iso}(\mathcal{F}_a, \mathcal{F}_b).$$

Last, we know that the composition of paths (or homotopy classes) gives rise to the composition of these maps: the proof is just as easy in the case of an arbitrary local system as in the case of $\mathcal{F}_{\underline{a}}$ or $\mathcal{F}_A$. Therefore, taking $a = b$, we get an anti-morphism of groups:[6]

$$\rho_{\mathcal{F},a} : \pi_1(\Omega; a) \to \mathrm{GL}(\mathcal{F}_a).$$

This is the *monodromy representation* of the local system $\mathcal{F}$ at the base point a.

Remark 7.33. Call the "opposite" of a group G with multiplication $x.y$ the group G° having the same elements but the group law $x \star y := y.x$. Then an anti-morphism of groups $G \to H$ is the same thing as a morphism $G^\circ \to H$. For this reason, it is sometimes said that the monodromy representation is:

$$\left(\pi_1(\Omega; a)\right)^\circ \to \mathrm{GL}(\mathcal{F}_a).$$

However, the distinction will have no consequence for us.

Definition 7.34. The *monodromy group* of $\mathcal{F}$ at the base point a is the image of the monodromy representation:

$$\mathrm{Mon}(\mathcal{F}, a) := \mathrm{Im}\, \rho_{\mathcal{F},a} \subset \mathrm{GL}(\mathcal{F}_a).$$

In the case of $\mathcal{F}_{\underline{a}}$ and $\mathcal{F}_A$, writing z_0 as the base point, we will denote $\rho_{\underline{a},z_0}$ or ρ_{A,z_0} as the monodromy representation and $\mathrm{Mon}(E_{\underline{a}}, z_0)$ or $\mathrm{Mon}(S_A, z_0)$ as the monodromy group.

Now we illustrate by an easy result the reason one can consider monodromy theory as a "galoisian" theory. Remember that the map $\mathcal{F}(\Omega) \mapsto \mathcal{F}_a$ is injective, so that one can identify a "global section" $s \in \mathcal{F}(\Omega)$ with its germ s_a at a: global sections are just germs which can be extended all over Ω.

Theorem 7.35. *The germ $s \in \mathcal{F}_a$ is fixed by the monodromy action if, and only if, $s \in \mathcal{F}(\Omega)$.*

Proof. The hypothesis means: $\forall g \in \pi_1(\Omega, a)$, $\rho_{\mathcal{F},a}(g)(s) = s$. In this case, this translates to: for every loop γ based at a, one has $s^\gamma = s$. This implies that, for every point $b \in \Omega$ and every path γ from a to b, the section $s^\gamma \in \mathcal{F}_b$ depends on b alone and not on the path γ. Then one can glue together all these germs to make a section $s \in \mathcal{F}(\Omega)$ whose germ at a is the given one. (We already met a similar argument in Section 5.2.) $\square$

Exercise 7.36. (i) Why is the map $\mathcal{F}(\Omega) \mapsto \mathcal{F}_a$ injective?

(ii) Prove rigorously the glueing argument in the theorem.

[6]Beware that the letter ρ here has nothing to do with that of the restriction maps ρ_V^U for a sheaf. Both uses of the letter are extremely common and we hope they will cause no confusion.

Remark 7.37. The similarity of algebraic equations to Galois theory is as follows. Suppose $P(x) = 0$ is an irreducible equation over $\mathbf{Q}$. Then a rational expression $A(x_1, \ldots, x_n)/B(x_1, \ldots, x_n)$ in the roots x_i of P is a rational number if, and only if, it is left invariant by all permutations of the roots. Here, the field K of all rational expressions of the roots plays the role of $\mathcal{F}_a$; the base field $\mathbf{Q}$ plays the role of $\mathcal{F}(\Omega)$; the symmetric group plays the role of the fundamental group; and the Galois group, which is the image of the symmetric group, plays the role of the monodromy group. One good source to pursue the analogy further is the book [**DD05**] by Adrien and Régine Douady; also see [**Ram93**].

Corollary 7.38. (i) *If the monodromy group of S_A is trivial, then S_A admits a fundamental matricial solution on Ω.*

(ii) *If the monodromy group of $E_{\underline{a}}$ is trivial, then $E_{\underline{a}}$ admits a fundamental system of solutions on Ω.*

Corollary 7.39. (i) *For every simply-connected domain U of Ω, S_A admits a fundamental matricial solution on U.*

(ii) *For every simply-connected domain U of Ω, $E_{\underline{a}}$ admits a fundamental system of solutions on U.*

Proof. Thanks to Cauchy's theorem, analytic continuation of a given fundamental matricial solution, resp. fundamental system of solutions, is possible everywhere; and the triviality of the monodromy group means that they all are mutually compatible, whence can be patched together.   □

Using the last of the "basic properties" of Section 3.1, one obtains:

Corollary 7.40. (i) *The r.o.c. of every fundamental matricial solution of S_A at z_0 is $\geq d(z_0, \partial\Omega)$.*

(ii) *The r.o.c. of every fundamental system of solutions of $E_{\underline{a}}$ at z_0 is $\geq d(z_0, \partial\Omega)$.*

Proof. Indeed, all local solutions can be extended to holomorphic functions on the corresponding disk (since this disk is simply connected).   □

Dependency on the base point a. If $a, b \in \Omega$, since Ω is arcwise connected (as any domain in $\mathbf{C}$), there is a path γ from a to b. This path induces an isomorphism $\phi_\gamma : \pi_1(\Omega; a) \to \pi_1(\Omega; b)$, $[\lambda] \mapsto [\gamma^{-1}.\lambda.\gamma]$. Actually, ϕ_γ only depends on the homotopy class $[\gamma]$ of γ, so we could as well denote it by $\phi_{[\gamma]}$, but this does not matter. Therefore, all the fundamental groups of Ω are isomorphic, but beware that the isomorphisms are not canonical.

The path γ also induces an isomorphism $u_\gamma : \mathcal{F}_a \to \mathcal{F}_b$, $s \mapsto s^\gamma$, whence, by conjugation, an isomorphism $\psi_\gamma : \mathrm{GL}(\mathcal{F}_a) \to \mathrm{GL}(\mathcal{F}_b)$, $u \mapsto u_\gamma \circ u \circ u_\gamma^{-1}$. (Again, these isomorphisms only depend on the homotopy class $[\gamma]$.) This gives a commutative diagram:

$$
\begin{array}{ccc}
\pi_1(\Omega; a) & \xrightarrow{\ \rho_a\ } & \mathrm{GL}(\mathcal{F}_a) \\[2pt]
{\scriptstyle \phi_\gamma}\Big\downarrow & & \Big\downarrow{\scriptstyle \psi_\gamma} \\[2pt]
\pi_1(\Omega; b) & \xrightarrow[\ \rho_b\ ]{} & \mathrm{GL}(\mathcal{F}_b)
\end{array}
$$

Indeed, the element $[\lambda] \in \pi_1(\Omega; a)$ goes down to $[\gamma^{-1}.\lambda.\gamma]$, then right to $u_\gamma \circ u_\lambda \circ u_\gamma^{-1}$; and it goes right to u_λ, then down to $u_\gamma \circ u_\lambda \circ u_\gamma^{-1}$.

As a consequence, ψ_γ sends $\mathrm{Mon}(\mathcal{F}, a) = \mathrm{Im}\, \rho_a$ to $\mathrm{Mon}(\mathcal{F}, b) = \mathrm{Im}\, \rho_b$; the monodromy groups of $\mathcal{F}$ at different points are all isomorphic (but the isomorphisms are not canonical).

Exercise 7.41. Let γ' be another path from a to b. Then $\phi_\gamma^{-1} \circ \phi_{\gamma'}$ is an automorphism of the group $\pi_1(\Omega; a)$ and $\phi_{\gamma'} \circ \phi_\gamma^{-1}$ is an automorphism of the group $\pi_1(\Omega; b)$. Show that these are inner automorphisms (that is, of the form $g \mapsto g_0 g g_0^{-1}$).

Matricial monodromy representation. Let $\mathcal{B}$ be a basis of $\mathcal{F}_a$, whence an isomorphism $\mathbf{C}^n \to \mathcal{F}_a$, $X_0 \mapsto \mathcal{B} X_0$. The automorphism u_λ of analytic continuation along a loop λ based at a transforms $\mathcal{B}$ into a new base $\mathcal{B}^\lambda = \mathcal{B} M_{[\lambda]}$, where $M_{[\lambda]} \in \mathrm{GL}_n(\mathbf{C})$. It transforms an element $X = \mathcal{B} X_0 \in \mathcal{F}_a$ into $X^\lambda = \mathcal{B}^\lambda X_0 = \mathcal{B}(M_{[\lambda]} X_0) \in \mathcal{F}_a$. Therefore, in the space $\mathbf{C}^n$, the automorphism of analytic continuation is represented by the linear map $X_0 \mapsto M_{[\lambda]} X_0$. In this way, our monodromy representation has been conjugated to a *matricial monodromy representation* $\pi_1(\Omega, a) \to \mathrm{GL}_n(\mathbf{C})$, $[\lambda] \mapsto M_{[\lambda]}$. Note that, as usual, this is an anti-morphism of groups. Its image is the *matricial monodromy group* of $\mathcal{F}$ at a with respect to the basis $\mathcal{B}$. So let $\mathcal{B}' = \mathcal{B} P$ be another basis, with $P \in \mathrm{GL}_n(\mathbf{C})$. The corresponding monodromy matrices are defined by the relations $\mathcal{B}'^\lambda = \mathcal{B}' M'_{[\lambda]}$, where $M'_{[\lambda]} \in \mathrm{GL}_n(\mathbf{C})$. On the other hand,

$$\mathcal{B}'^\lambda = (\mathcal{B} P)^\lambda = \mathcal{B}^\lambda P = \mathcal{B} M_{[\lambda]} P = \mathcal{B}' P^{-1} M_{[\lambda]} P \text{ so that } M'_{[\lambda]} = P^{-1} M_{[\lambda]} P.$$

Therefore, changing the base yields a conjugated representation: the matricial monodromy groups of $\mathcal{F}$ at a with respect to various bases are all conjugated as subgroups of $\mathrm{GL}_n(\mathbf{C})$.

In the case of the sheaf of solutions of a differential system or equation, the "initial condition map" $\mathcal{F}_{A_{z_0}} \to \mathbf{C}^n$, $X \mapsto X(z_0)$ or $\mathcal{F}_{\underline{a}_{z_0}} \to \mathbf{C}^n$,

$f \mapsto (f(z_0), \ldots, f^{(n-1)}(z_0))$ allows us to make a canonical choice of a basis; indeed, we can use the one whose image by the initial condition map is the canonical basis of $\mathbf{C}^n$. For instance, in the case of systems, this means that $\mathcal{B}$ has as elements the columns of the fundamental matricial solution such that $\mathcal{X}(z_0) = I_n$. However, this natural choice is not always the best to describe the matricial monodromy group:

Example 7.42. Consider the equation $z^2 f'' - z f' + f = 0 \Leftrightarrow f'' - z^{-1} f' + z^{-2} f = 0$ on $\Omega := \mathbf{C}^*$, and take the base point $z_0 := 1$ and the usual fundamental loop λ such that $I(0, \lambda) = +1$. For the obvious basis $\mathcal{B} := (z, z \log z)$, one has $\mathcal{B}^\lambda = (z, z(\log z + 2i\pi)) = \mathcal{B} M_{[\lambda]}$ with $M_{[\lambda]} = \begin{pmatrix} 1 & 2i\pi \\ 0 & 1 \end{pmatrix}$.

However, the basis which corresponds to the "canonical" initial conditions $\begin{pmatrix} 1 & 0 \\ 0 & 1 \end{pmatrix}$ is rather $\mathcal{B}' = (z - z \log z, z \log z)$, that is, BP with $P := \begin{pmatrix} 1 & 0 \\ -1 & 1 \end{pmatrix}$. The corresponding monodromy matrix is

$$M'_{[\lambda]} = P^{-1} M_{[\lambda]} P = \begin{pmatrix} 1 - 2i\pi & 2i\pi \\ -2i\pi & 1 + 2i\pi \end{pmatrix}.$$

It is clearly easier to compute the matricial monodromy group generated by $M_{[\lambda]}$ than the matricial monodromy group generated by $M'_{[\lambda]}$.

Exercise 7.43. Check that $\mathcal{B}'^\lambda = \mathcal{B}' M'_{[\lambda]}$ in the example.

7.6. Holomorphic and meromorphic equivalences of systems

Gauge transformations and equivalence of systems. The underlying idea here is that of change of unknown function. Instead of studying (or trying to solve) the system $X' = AX$, we introduce $Y := FX$, where F is an invertible matrix of functions. We then find that

$$Y' = (FX)' = F'X + FX' = F'X + FAX = (F' + FA)F^{-1}Y,$$

i.e., $Y' = BY$, where $B := F'F^{-1} + FAF^{-1}$. Conversely, since F is invertible, one can see that if $Y' = BY$, then $X := F^{-1}Y$ satisfies $X' = AX$. The matrix F as well as the map $X \mapsto FX$ is called a *gauge transformation*. Note that there is no corresponding notion for scalar equations.

Definition 7.44. We shall write $F[A] := F'F^{-1} + FAF^{-1}$. If $B = F[A]$, we shall write $F : A \simeq B$ or $F : S_A \simeq S_B$, or even[7] $F : A \to B$ or $F : S_A \to S_B$.

---

[7]This is the notation for the more general notion of "morphism" (see the remark at the end of the chapter and Chapter 8 for a more complete discussion), so in this case we would say explicitly "the equivalence $F : A \to B$" or "the isomorphism $F : A \to B$".

If $F \in \mathrm{GL}_n(\mathcal{O}(\Omega))$, we shall say that the systems S_A and S_B on Ω are *holomorphically equivalent* or *holomorphically isomorphic*, and we write $A \underset{h}{\sim} B$ or $S_A \underset{h}{\sim} S_B$.

If $F \in \mathrm{GL}_n(\mathcal{M}(\Omega))$, we shall say that the systems S_A and S_B on Ω are *meromorphically equivalent* or *meromorphically isomorphic*, and we write $A \underset{m}{\sim} B$ or $S_A \underset{m}{\sim} S_B$.

Remark 7.45. *These notions are provisional.* Indeed, if one says nothing about the behavior of the gauge transformation at the singularities or at the boundary of their domain of definition, these equivalence relations are much too weak. But they are a good intermediate step on our way to the more significant classification criteria that we shall encounter in Chapter 9.

Example 7.46. The relation $A = F[0_n]$ means that F is a fundamental matricial solution of S_A. Thus, $A \underset{h}{\sim} 0_n$, resp. $A \underset{m}{\sim} 0_n$, means that A admits a fundamental matricial solution that is holomorphic, resp. meromorphic, on Ω.

Proposition 7.47. *One has the equalities $I_n[A] = A$ and $G[F[A]] = (GF)[A]$, and the logical equivalence $B = F[A] \Leftrightarrow A = F^{-1}[B]$.*

Proof. Easy computations, left to the reader!   □

Exercise 7.48. Do these computations.

Corollary 7.49. *Holomorphic and meromorphic equivalences are indeed equivalence relations.*

Remark 7.50. One can say that the group $\mathrm{GL}_n(\mathcal{O}(\Omega))$ operates on $\mathrm{Mat}_n(\mathcal{O}(\Omega))$, but the similar statement for $\mathrm{GL}_n(\mathcal{M}(\Omega))$ is not quite correct: the operation is partial, since $F[A]$ could have poles.

The following result says that, in some sense, one does not really generalize the theory by studying systems rather than scalar differential equations.

Theorem 7.51 (Cyclic vector lemma). *Every system S_A is meromorphically equivalent to a system coming from a scalar equation $E_{\underline{a}}$, i.e., the system $S_{A_{\underline{a}}}$.*

Proof. We want to find $F \in \mathrm{GL}_n(\mathcal{M}(\Omega))$ such that $F' + FA = A_{\underline{a}}F$ for some functions $a_1, \ldots, a_n$. We shall actually find such an F with holomorphic coefficients. To ensure that $F \in \mathrm{GL}_n(\mathcal{M}(\Omega))$, it will then be sufficient to ensure that $(\det F)(z_0) \neq 0$ at one arbitrary point $z_0 \in \Omega$. For simplicity and without loss of generality, we can as well assume that $0 \in \Omega$ and choose

$z_0 := 0$. Call $L_1, \dots, L_n$ the lines of F. Then

$$F' + FA = A_{\underline{a}}F \Longleftrightarrow \begin{cases} L_1' + L_1 A = L_2, \dots, L_{n-1}' + L_{n-1}A = L_n \\ \text{and} \\ L_n' + L_n A = -(a_n L_1 + \cdots + a_1 L_n). \end{cases}$$

Since $a_1, \dots, a_n$ are not imposed, it is therefore enough to choose L_1 holomorphic on Ω and such that the sequence defined by $L_{i+1} := L_i' + L_i A$ for $i = 1, \dots, n-1$ produces a basis $(L_1, \dots, L_n)$ of $\mathcal{M}(\Omega)^n$. (Such a vector L_1 is called a *cyclic vector*, whence the name of the theorem.) To that end, we shall simply require that $\det(L_1, \dots, L_n)(0) = 1$.

Now, choose a fundamental matricial solution $\mathcal{X}$ at 0 such that $\mathcal{X}(0) = I_n$. Putting $M_i := L_i \mathcal{X}$ (recall that these are line vectors), one sees that $L_{i+1} := L_i' + L_i A \Leftrightarrow M_{i+1} = M_i'$. Thus, we must choose $M_i := M_1^{(i-1)}$ for $i = 2, \dots, n$. On the other hand, $\det(M_1, \dots, M_n)(0) = \det(L_1, \dots, L_n)(0)$ since $\det \mathcal{X}(0) = 1$. Therefore, we look for M_1 such that $\det(M_1, \dots, M_1^{(n-1)})(0) = 1$. This is very easy: calling $E_1, \dots, E_n$ the canonical basis of $\mathbf{C}^n$, we could just take $M_1 = E_1 + zE_2 + \cdots + \dfrac{z^{n-1}}{(n-1)!} E_n$. However, in that case, $L_1 := M_1 \mathcal{X}^{-1}$ would not be holomorphic on the whole of Ω. Therefore we *truncate* the vector $M_1 \mathcal{X}^{-1}$ (which is a power series at 0) in order to eliminate terms containing z^n. This does not change the condition on the first $(n-1)$ derivatives at 0 of M_1.

To summarize: L_1 is the truncation of the line vector

$$\left(E_1 + zE_2 + \cdots + \frac{z^{n-1}}{(n-1)!} E_n \right) \mathcal{X}^{-1}$$

up to degree $(n-1)$; the line vectors $L_2, \dots, L_n$ are defined by the recursive formulas $L_{i+1} := L_i' + L_i A$ for $i = 1, \dots, n-1$; the gauge transformation matrix F has lines $L_2, \dots, L_n$; then $F[A] = A_{\underline{a}}$ for some $\underline{a}$. $\quad\square$

Remark 7.52. We cannot conclude that S_A is holomorphically equivalent to $S_{A_{\underline{a}}}$, because F is holomorphic on Ω but F^{-1} might have poles if $\det F$ has zeros. A more serious drawback of the theorem is that the a_i are meromorphic on Ω and not necessarily holomorphic. There is no corresponding result that guarantees an equation with holomorphic coefficients.

The problem of finding cyclic vectors is a practical one and software dedicated to formal treatment of differential equations uses more efficient algorithms than the one shown above, which requires finding a fundamental matricial solution. Note however that taking a vector of polynomials of degree $(n-1)$ for L_1 *at random* will yield a cyclic vector with probability 1! (Suggestion: prove it.)

Example 7.53. Let $A := \begin{pmatrix} a & b \\ c & d \end{pmatrix}$, where $a, b, c, d \in \mathcal{O}(\Omega)$. Take $L_1 :=$ $(1, 0)$. Then $L_2 := L_1' + L_1 A = (a, b)$, so that (L_1, L_2) is a basis except if $b = 0$, which is obviously an exceptional condition. In the same way, $(0, 1)$ is a cyclic vector, except if $c = 0$.

Meromorphic equivalence and sheaves of solutions. We now describe more precisely the effect of a meromorphic gauge transformation on solutions. Let $F : A \simeq B$ be such a transformation. Let $X \in \mathcal{F}_A(U)$ be a solution of S_A on U. Then FX is a meromorphic solution of $\mathcal{F}_B$ on U. But we know from Remark 7.27 that then FX is analytic: $FX \in \mathcal{F}_B(U)$. Thus, $X \mapsto FX$ is a linear map $\mathcal{F}_A(U) \to \mathcal{F}_B(U)$. For the same reason, $Y \mapsto F^{-1}Y$ is a linear map $\mathcal{F}_B(U) \to \mathcal{F}_A(U)$, and it is the inverse of the previous one, so that they are isomorphisms.

Definition 7.54. An *isomorphism* $\phi : \mathcal{F} \to \mathcal{F}'$ *of sheaves of complex linear spaces* on Ω is a family (ϕ_U) indexed by the open subsets U of Ω, where each $\phi_U : \mathcal{F}(U) \to \mathcal{F}'(U)$ is an isomorphism (of complex linear spaces) and where the family is compatible with the restriction maps in the following sense: if $V \subset U$, then the diagram

$$
\begin{array}{ccc}
\mathcal{F}(U) & \xrightarrow{\phi_U} & \mathcal{F}'(U) \\
\downarrow & & \downarrow \\
\mathcal{F}(V) & \xrightarrow{\phi_V} & \mathcal{F}'(V)
\end{array}
$$

is commutative (the vertical maps are the restriction maps).

Example 7.55. If $A = A_{\underline{a}}$, the maps $\mathcal{F}_{\underline{a}}(U) \to \mathcal{F}_A(U)$ defined in Proposition 7.8 make up an isomorphism of sheaves.

Theorem 7.56. *The systems S_A and S_B are meromorphically equivalent if, and only if, the sheaves $\mathcal{F}_A$ and $\mathcal{F}_B$ are isomorphic.*

Proof. We just proved that $S_A \underset{h}{\sim} S_B$ implies $\mathcal{F}_A \simeq \mathcal{F}_B$. (The compatibility with restriction maps was obviously satisfied.) Assume conversely that (ϕ_U) is an isomorphism from $\mathcal{F}_A$ to $\mathcal{F}_B$. From the compatibility with restriction maps, it follows that (ϕ_U) induces isomorphisms $\phi_{z_0} : \mathcal{F}_{A,z_0} \to \mathcal{F}_{B,z_0}$ between the spaces of germs at an arbitrary $z_0 \in \Omega$. As a consequence, a fundamental matricial solution $\mathcal{X}$ of S_A at z_0 has as an image a fundamental matricial solution $\mathcal{Y} := \phi_{z_0}(\mathcal{X})$ of S_B at z_0.

Again from the compatibility conditions (and from the fact that $\mathcal{F}_A, \mathcal{F}_B$ are local systems), it follows that along a path γ from z_0 to z_1, endowed with the usual small disks $D_0, \ldots, D_N$, the successive continuations $\mathcal{X}_0 :=$ $\mathcal{X}, \ldots, \mathcal{X}_N$ and $Y_0 := \mathcal{Y}, \ldots, \mathcal{Y}_N$ satisfy $\phi_{D_i}(\mathcal{X}_i) = \mathcal{Y}_i$. In other words, $\phi_{z_1}(\mathcal{X}^\gamma) = \mathcal{Y}^\gamma$. It is in particular true that, for any loop λ based at z_0, one has $\phi_{z_0}(\mathcal{X}^\lambda) = \mathcal{Y}^\lambda$.

On the other hand, $\mathcal{X}^\lambda = \mathcal{X} M_{[\lambda]}$ and $\mathcal{Y}^\lambda = \mathcal{Y} N_{[\lambda]}$ (the monodromy matrices) and, by linearity of ϕ_{z_0}, one has:

$$\mathcal{Y} N_{[\lambda]} = \mathcal{Y}^\lambda = \phi_{z_0}(\mathcal{X}^\lambda) = \phi_{z_0}(\mathcal{X} M_{[\lambda]}) = \phi_{z_0}(\mathcal{X}) M_{[\lambda]} = \mathcal{Y} M_{[\lambda]},$$

i.e., $M_{[\lambda]} = N_{[\lambda]}$.

Now, if we set $F := \mathcal{Y}\mathcal{X}^{-1} \in \mathrm{GL}_n(\mathcal{M}_{z_0})$ (the meromorphic germs at z_0), we see that

$$F^\lambda = \mathcal{Y}^\lambda (\mathcal{X}^\lambda)^{-1} = (\mathcal{Y} M_{[\lambda]})(\mathcal{X} M_{[\lambda]})^{-1} = \mathcal{Y}\mathcal{X}^{-1} = F.$$

As in the "galoisian" Theorem 7.35, we conclude that $F \in \mathrm{GL}_n(\mathcal{M}(\Omega))$.

Last, from the facts that $\mathcal{X}' = A\mathcal{X}$ and $(F\mathcal{X})' = B(F\mathcal{X})$, it follows that $F' + FA = BF$ (we use the fact that $\mathcal{X}$ is invertible). □

Exercise 7.57. Write down the computation that leads from $\mathcal{X}' = A\mathcal{X}$ and $(F\mathcal{X})' = B(F\mathcal{X})$ to $F' + FA = BF$.

Note that essential use was made of monodromy considerations. The information carried by the local systems $\mathcal{F}_A$, $\mathcal{F}_B$ is clearly topological in nature.

Meromorphic equivalence and monodromy representations. Again consider a meromorphic isomorphism $F : A \to B$. Fix $z_0 \in \Omega$ and fundamental matricial solutions $\mathcal{X}$, $\mathcal{Y}$ of S_A, S_B at z_0. (We do not require that $\mathcal{Y} = F\mathcal{X}$.) The monodromy representations $\pi_1(\Omega, z_0) \to \mathrm{GL}(\mathcal{F}_{A,z_0})$, $[\lambda] \mapsto M_{[\lambda]}$ and $\pi_1(\Omega, z_0) \to \mathrm{GL}(\mathcal{F}_{B,z_0})$, $[\lambda] \mapsto N_{[\lambda]}$ are characterized by the relations $\mathcal{X}^\lambda = \mathcal{X} M_{[\lambda]}$ and $\mathcal{Y}^\lambda = \mathcal{Y} N_{[\lambda]}$.

Since $F\mathcal{X}$ is an invertible matrix and a matricial solution of S_B at z_0 (because of the relation $F' = BF - FA$), it can be written $F\mathcal{X} = \mathcal{Y}P$, where $P \in \mathrm{GL}_n(\mathbf{C})$.

Exercise 7.58. Write down the computation underlying the previous deduction.

Since F is globally defined on Ω, it is fixed by the monodromy $F^\lambda = F$ for any loop λ based at z_0. The effect of λ on the equality $F\mathcal{X} = \mathcal{Y}P$ is $(F\mathcal{X})^\lambda = (\mathcal{Y}P)^\lambda$, whence

$$F^\lambda \mathcal{X}^\lambda = \mathcal{Y}^\lambda P^\lambda \Longrightarrow F\mathcal{X}^\lambda = \mathcal{Y}^\lambda P \Longrightarrow F\mathcal{X} M_{[\lambda]} = \mathcal{Y} N_{[\lambda]} P$$
$$\Longrightarrow \mathcal{Y}P M_{[\lambda]} = \mathcal{Y} N_{[\lambda]} P \Longrightarrow P M_{[\lambda]} = N_{[\lambda]} P.$$

This property is an instance of the following definition.

Definition 7.59. Two linear representations $G \xrightarrow{\rho} \mathrm{GL}(E)$ and $G \xrightarrow{\rho'} \mathrm{GL}(E')$ are said to be *equivalent* or *conjugate* or *isomorphic* if there exists a linear isomorphism $p : E \to E'$ such that:

$$\forall g \in G, \ \rho'(g) \circ p = p \circ \rho(g), \quad i.e., \quad \rho'(g) = p \circ \rho(g) \circ p^{-1}.$$

Example 7.60. All the monodromy representations attached to S_A at the point z_0 are isomorphic. This includes the intrinsic representation $\pi_1(\Omega, z_0) \to \mathrm{GL}(\mathcal{F}_{A,z_0})$ as well as all the matricial representations arising from the choice of a basis of $\mathcal{F}_{A,z_0}$.

Theorem 7.61. *Two systems are meromorphically equivalent if, and only if, their monodromy representations at some arbitrary point are isomorphic.*

Proof. We have already proved that if $S_A \underset{m}{\sim} S_B$, then their monodromy representations at z_0 are conjugate.

So we assume conversely that these representations are conjugate, *i.e.*, there exists $P \in \mathrm{GL}_n(\mathbf{C})$ such that, for all loops λ based at z_0, one has $PM_{[\lambda]} = N_{[\lambda]}P$ (where these monodromy matrices come from $\mathcal{X}$, $\mathcal{Y}$ and z_0 chosen as before).

We then put $F := \mathcal{Y}P\mathcal{X}^{-1}$. This is a meromorphic germ of the invertible matrix and, since $F\mathcal{X} = \mathcal{Y}P$ is a matricial solution of S_B, the usual computation implies $F' + FA = BF$. On the other hand, the effect of monodromy on F is as follows:

$$F^\lambda = (\mathcal{Y}P\mathcal{X}^{-1})^\lambda = \mathcal{Y}^\lambda P^\lambda (\mathcal{X}^\lambda)^{-1} = \mathcal{Y}N_{[\lambda]}PM_{[\lambda]}^{-1}\mathcal{X}^{-1} = \mathcal{Y}P\mathcal{X}^{-1} = F.$$

Thus, by the usual "galoisian" argument, F is global: $F \in \mathrm{GL}_n(\mathcal{M}(\Omega))$ and $S_A \underset{m}{\sim} S_B$. $\square$

Corollary 7.62. *The monodromy group* $\mathrm{Mon}(\mathcal{F}_A, z_0)$ *is trivial if, and only if, the system* $\mathcal{F}_A$ *admits a meromorphic fundamental matricial solution.*

Proof. Indeed, the second statement is equivalent to being meromorphically equivalent to 0_n (see the last exercise at the end of the chapter). $\square$

Remark 7.63. In all three definitions of isomorphism (meromorphic equivalence of differential systems, isomorphism of sheaves, conjugacy of representations), one can relax the requirement of bijectivity. One thereby obtains the notion of *morphism* (of differential systems, of sheaves, of representations) and one proves more generally that (meromorphic) morphisms from S_A to S_B, and morphisms from $\mathcal{F}_A$ to $\mathcal{F}_B$ and morphisms from the monodromy representation $\pi_1(\Omega, z_0) \to \mathrm{GL}(\mathcal{F}_{A,z_0})$ to the monodromy representation $\pi_1(\Omega, z_0) \to \mathrm{GL}(\mathcal{F}_{B,z_0})$ correspond bijectively to each other. This is the functorial point of view; see the next chapter and also the book of Deligne [**Del70**].

Exercises

(1) Study the rational function $f_k(z) := \dfrac{z^k}{z^2 - 1}$ on **S** for $k \in \mathbf{N}$: what are its zeros, its poles and their orders? Whenever possible, do it using both coordinates z and w.

(2) How can equation $zf'' + f' = 0$ be expressed at infinity?

(3) Prove the wronskian lemma mentioned in Remark 7.18 and deduce from it Theorem 7.21. (Let $\mathcal{X}$ have columns $X_1, \dots, X_n$ and be such $\mathcal{X}' = A\mathcal{X}$. If $\det \mathcal{X} = 0$, there exists a nontrivial column vector C of functions such that $\mathcal{X}C = 0$. An easy calculation yields $\mathcal{X}C' = 0$. Choosing C with as many zero components as possible and eliminating one nonzero component between the two relations $\mathcal{X}C = \mathcal{X}C' = 0$ implies $C' = fC$ for some $f \neq 0$ and then the existence of a nontrivial column vector C_0 of complex numbers such that $\mathcal{X}C_0 = 0$.)

(4) Check that $U \mapsto \mathcal{O}(U)$ is a sheaf of **C**-algebras, and that the **C**-algebra of germs $\mathcal{O}_a$ is $\mathbf{C}\{z-a\}$. (However, beware that $\mathcal{O}$ is not a local system.)

(5) Give a necessary and sufficient condition for two rank one systems $x' = ax$ and $y' = by$ to be holomorphically, resp. meromorphically, equivalent.

(6) Starting from an arbitrary system of rank 2, find an equivalent scalar equation. (Method: it is not very difficult by brute force.)

(7) Starting from an arbitrary system, prove rigorously that a cyclic vector can be chosen at random.

(8) In the proof of Theorem 7.56, give the details in the definition of the isomorphisms ϕ_{z_0}.

(9) The construction of the monodromy representation really depends only on the local system $\mathcal{F}$ (not the fact that it comes from a differential system). Prove that two local systems are isomorphic (as sheaves) if, and only if, the attached monodromy representations are equivalent.

(10) Prove rigorously the following fact: a system is meromorphically equivalent to the trivial system with matrix 0_n if, and only if, it admits a meromorphic fundamental matricial solution.

A functorial point of view on analytic continuation: Local systems

We fix a domain, *i.e.*, a nonempty connected open set $\Omega \subset \mathbf{C}$.

8.1. The category of differential systems on Ω

As usual, we really only consider complex analytic linear differential systems.

We now introduce a *category* $\mathfrak{Ds}$. Its *objects* are systems $S_A : X' = AX$ with $A \in \mathrm{Mat}_n(\mathcal{O}(\Omega))$. Its *morphisms* from S_A to S_B are matrices[1] $F \in \mathrm{Mat}_{p,n}(\mathcal{O}(\Omega))$ such that $F' = BF - FA$. We then write $F : S_A \to S_B$ and say that it is a morphism or an *arrow* in $\mathfrak{Ds}$ with *source* S_A and *target* S_B.

The definition of morphisms is set to ensure the following property: if X is a solution of S_A, then $Y := FX$ is a solution of S_B. Indeed:

$$Y' = F'X + FX' = F'X + FAX = (F' + FA)X = BFX = BY.$$

___
[1] By abuse of language, matrices $F \in \mathrm{Mat}_{p,n}(\mathcal{M}(\Omega))$, resp. $F \in \mathrm{Mat}_{p,n}(\mathbf{C}((z)))$, such that $F' = BF - FA$ will be called "meromorphic morphisms", resp. "formal morphisms", although we shall not associate them with a category.

We note the following easy facts:

(1) The identity matrix $I_n \in \mathrm{Mat}_n(\mathcal{O}(\Omega))$ is a morphism of S_A into itself. Indeed $I_n' = 0_n = AI_n - I_nA$. In the general notation of category theory (see below), it would be noted Id_{S_A}.

(2) The composition of $F : S_A \to S_B$ and $G : S_B \to S_C$ is a morphism $GF : S_A \to S_C$. Indeed:

$$(GF)' = G'F + GF' = (CG - GB)F + G(BF - FA)$$
$$= CGF - GBF + GBF - GFA = C(GF) - (GF)A.$$

In the general notation of category theory, it would be noted $G \circ F$.

(3) The identity morphisms are "neutral": if $F : S_A \to S_B$ is a morphism in $\mathfrak{Ds}$, then $\mathrm{Id}_{S_B} \circ F = F = F \circ \mathrm{Id}_{S_A}$.

(4) The composition, when defined, is associative: if $G \circ F$ and $H \circ G$ are defined, then so are $H \circ (G \circ F)$ and $(H \circ G) \circ F$ and they are equal: $H \circ (G \circ F) = (H \circ G) \circ F$. Of course, we write it $H \circ G \circ F$.

As we are going to see, objects and morphisms as well as identities and compositions belong to the structure being defined, while neutrality and associativity belong to its axioms.

General definitions of category theory. We stay at a very elementary level. For details, see the famous book by Saunders Mac Lane "Categories for the working mathematician" [**Mac98**].

A *category* $\mathcal{C}$ comprises:

(1) A "class"[2] $\mathrm{Ob}(\mathcal{C})$ of *objects*.

(2) For any two objects $X, Y \in \mathrm{Ob}(\mathcal{C})$, a set $\mathrm{Mor}_{\mathcal{C}}(X, Y)$ of *morphisms* or *arrows*. If $f \in \mathrm{Mor}_{\mathcal{C}}(X, Y)$, we write $f : X \to Y$ and say that X is the *source* and Y the *target* of X. So it is implicitly part of the axioms that if $X' \neq X$ or $Y' \neq Y$, then $\mathrm{Mor}_{\mathcal{C}}(X, Y)$ and $\mathrm{Mor}_{\mathcal{C}}(X', Y')$ are disjoint.[3]

(3) For every object $X \in \mathrm{Ob}(\mathcal{C})$, a particular *endomorphism* $\mathrm{Id}_X \in \mathrm{Mor}_{\mathcal{C}}(X, X)$ is called the *identity morphism of X* or for short the *identity of X*.

---

[2]We shall not be formal about this word, which belongs to sophisticated set theory. A class somehow looks like a set, but one is entitled to speak of the category of sets, whose objects form the class of sets, although the set of sets does not exist.

[3]Since a same matrix $F \in \mathrm{Mat}_{p,n}(\mathcal{O}(\Omega))$ can be such that $F' = B_1 F - FA_1$ and $F' = B_2 F - FA_2$, where $A_1 \neq A_2$ or $B_1 \neq B_2$, our previous definition of morphism formally contradicts this implicit axiom. This can be repaired by saying that a morphism from A to B really is a triple (A, F, B) or a diagram $A \xrightarrow{F} B$ such that the above conditions hold. But this kind of formal rigor is not really important here.

(4) For any three objects $X, Y, Z \in \mathrm{Ob}(\mathcal{C})$, a composition law $(f, g) \mapsto g \circ f$ from $\mathrm{Mor}_{\mathcal{C}}(X, Y) \times \mathrm{Mor}_{\mathcal{C}}(X, Y)$ to $\mathrm{Mor}_{\mathcal{C}}(X, Y)$, so that we can consider that the composition $g \circ f$ is a partially defined operation, restricted to pairs (f, g) such that the source of g is the target of f.

These four data, which constitute the structure, are subjected to the following two axioms:

(1) Identity morphisms are neutral: if $f : X \to Y$ is a morphism, then $\mathrm{Id}_Y \circ f = f = f \circ \mathrm{Id}_X$.

(2) Composition is associative: if $f : X \to Y$, $g : Y \to Z$ and $h : Z \to T$ are morphisms with compatible sources and targets, then $h \circ (g \circ f) = (h \circ g) \circ f$. We shall of course write $h \circ g \circ f : X \to T$.

Isomorphisms in $\mathcal{C}$ are *invertible* arrows: $f : X \to Y$ is an isomorphism if there exists $g : Y \to X$ such that $g \circ f = \mathrm{Id}_X$ and $f \circ g = \mathrm{Id}_Y$. Such a g is then unique (easy!) and we write it f^{-1} and call it the *inverse* of f. Clearly f^{-1} is then invertible and $(f^{-1})^{-1} = f$. Likewise, if $f : X \to Y$ and $g : Y \to Z$ are isomorphisms, then so is $g \circ f$ and $(g \circ f)^{-1} = f^{-1} \circ g^{-1}$ (also easy!). Two objects X, Y are *isomorphic* if there is an isomorphism $f : X \to Y$. Isomorphy is plainly an equivalence relation on $\mathrm{Ob}(\mathcal{C})$.

The usual terminology applies: *endomorphisms* have the same source and target, *automorphisms* are at the same time endomorphisms and isomorphisms.

Exercise 8.1. (i) Prove in all generality the properties of isomorphisms quoted above.

(ii) Prove in all generality that automorphisms of an object X form a group $\mathrm{Aut}(X)$.

8.2. The category $\mathfrak{Ls}$ of local systems on Ω

We want to define the category $\mathfrak{Ls}$ of *local systems* on Ω. First we shall define the category $\mathfrak{Sh}$ of sheaves of complex vector spaces on Ω and then we shall define $\mathfrak{Ls}$ as a *full subcategory* of $\mathfrak{Sh}$, meaning that $\mathrm{Ob}(\mathfrak{Ls}) \subset \mathrm{Ob}(\mathfrak{Sh})$ and that, for $X, Y \in \mathrm{Ob}(\mathfrak{Ls})$, one has $\mathrm{Mor}_{\mathfrak{Ls}}(X, Y) = \mathrm{Mor}_{\mathfrak{Sh}}(X, Y)$.

The category $\mathfrak{Sh}$. The objects of $\mathfrak{Sh}$ are sheaves of complex vector spaces. Such an $\mathcal{F} \in \mathrm{Ob}(\mathfrak{Sh})$ associates to each open subset $U \subset \Omega$ a complex vector space $\mathcal{F}(U)$ and to each pair $V \subset U$ a linear map $\rho_V^U : \mathcal{F}(U) \to \mathcal{F}(V)$, called the restriction map, all of these subjected to the rules described in Section 7.4.

A morphism ϕ from $\mathcal{F}$ to $\mathcal{F}'$ in $\mathfrak{Sh}$ is given by a family of linear maps $\phi_U : \mathcal{F}(U) \to \mathcal{F}'(U)$ compatible with the restriction maps in the following sense: $\phi_V \circ \rho_V^U = \rho'^U_V \circ \phi_U$. In other words, there are commutative diagrams:

$$
\begin{array}{ccc}
\mathcal{F}(U) & \xrightarrow{\phi_U} & \mathcal{F}'(U) \\
\rho_V^U \downarrow & & \downarrow \rho'^U_V \\
\mathcal{F}(V) & \xrightarrow{\phi_V} & \mathcal{F}'(V)
\end{array}
$$

Obviously, the family of all $\mathrm{Id}_{\mathcal{F}(U)}$ is an endomorphism of $\mathcal{F}$, and if $\phi : \mathcal{F} \to \mathcal{F}'$, $\phi' : \mathcal{F}' \to \mathcal{F}''$ are morphisms in $\mathfrak{Sh}$, the family $\phi' \circ \phi$ of all $\phi'_U \circ \phi_U : \mathcal{F}(U) \to \mathcal{F}''(U)$ is a morphism $\mathcal{F} \to \mathcal{F}''$ in $\mathfrak{Sh}$. Quite as obviously, the axioms of categories (neutrality, associativity) are satisfied.

Local systems on Ω. Here, we are not going to use the most general possible definitions (see for instance [**Del70**]) but some that are equivalent in our situation and also very convenient.

For every sheaf on $\mathcal{F} \in \mathrm{Ob}(\mathfrak{Sh})$, the restriction of $\mathcal{F}$ to an open subset D of Ω is the sheaf $\mathcal{F}_{|D}$ on D defined by $\mathcal{F}_D(U) := \mathcal{F}(U)$ for every open subset U of D; restriction maps are defined the obvious way.

Given a complex vector space E, the constant sheaf $\underline{E}$ on D sends each open subset U of D to the space $\underline{E}(U)$ of locally constant maps $U \to E$. If U is connected, $\underline{E}(U)$ has a natural identification with E. If U has r connected components (which are by necessity open), $\underline{E}(U)$ has a natural identification with E^r.

Exercise 8.2. (i) Check that locally constant maps $U \to E$ are the same as continuous maps with E equipped with the discrete topology.

(ii) Check that $\underline{E}$ is indeed a sheaf.

We say that a sheaf $\mathcal{F}$ in $\mathfrak{Sh}$ is a *local system*[4] *on* Ω if Ω can be covered by connected open subsets D such that the restriction sheaf $\mathcal{F}_D$ is isomorphic to a constant sheaf $\underline{E}$ for some finite-dimensional vector space E. Such a neighborhood is said to trivialize $\mathcal{F}$. One can then prove that every connected simply-connected open subset is trivializing (this will actually follow from the relation with $\pi_1(\Omega)$ to be described in Section 8.4).

---

[4]This is not the most general definition (see for instance [**Del70**]) but it is equivalent in our situation and it is very convenient.

Exercise 8.3. (i) Show the equivalence of this definition with the one we used previously (Section 7.4).

(ii) Show that all the stalks $\mathcal{F}_a$ have the same dimension, which is that of any of the spaces E appearing in trivializations.

8.3. A functor from differential systems to local systems

Let $S_A : X' = AX$ be an object of $\mathfrak{Ds}$. Then, by Cauchy's theorem, setting $\mathcal{F}_A(U) := \{$solutions of S_A over $U\}$ defines a local system (disks are trivializing). So we have defined a mapping $\Phi : S_A \rightsquigarrow \mathcal{F}_A$ from $\mathrm{Ob}(\mathfrak{Ds})$ to $\mathrm{Ob}(\mathfrak{Ls})$. We write $\Phi(S_A) = \mathcal{F}_A$. The use of the strange arrow $\rightsquigarrow$ instead of the more familiar $\mapsto$ is motivated by the fact that a functor is not exactly the same thing as a plain map between sets.

Now let $F : S_A \to S_B$ be a morphism in $\mathfrak{Ds}$. As observed above, if X is a solution of S_A, then FX is a solution of S_B. More precisely, if X is holomorphic over some open subset $U \subset \Omega$, then FX is holomorphic over U, so that $\phi_U : X \mapsto FX$ is a linear map from $\mathcal{F}_A(U)$ to $\mathcal{F}_B(U)$. Clearly, the family of all the ϕ_U is a morphism $\phi : \mathcal{F}_A \to \mathcal{F}_B$ in $\mathfrak{Ls}$. We write it $\Phi(F)$. (So we use the same notation for the image $\Phi(S_A)$ of an object S_A or the image $\Phi(F)$ of a morphism F.) It is easy to check that all those mappings $\mathrm{Mor}_{\mathfrak{Ds}}(S_A, S_B) \to \mathrm{Mor}_{\mathfrak{Ls}}(\mathcal{F}_A, \mathcal{F}_B)$ have the following algebraic properties: $\Phi(\mathrm{Id}_{S_A}) = \mathrm{Id}_{\mathcal{F}_A}$ and, whenever $G \circ F$ is defined, $\Phi(G \circ F) = \Phi(G) \circ \Phi(F)$.

Definition 8.4. Let $\mathcal{C}$, $\mathcal{C}'$ be two categories and let Φ be a mapping from $\mathrm{Ob}(\mathcal{C})$ to $\mathrm{Ob}(\mathcal{C}')$ and also, using the same letter Φ, a collection of mappings $\mathrm{Mor}_{\mathcal{C}}(X, Y) \to \mathrm{Mor}_{\mathcal{C}'}(\Phi(X), \Phi(Y))$ for all $X, Y \in \mathrm{Ob}(\mathcal{C})$. We say that Φ is a *(covariant) functor* from $\mathcal{C}$ to $\mathcal{C}'$ if $\Phi(\mathrm{Id}_X) = \mathrm{Id}_{\Phi(X)}$ for all $X \in \mathrm{Ob}(\mathcal{C})$ and if $\Phi(G \circ F) = \Phi(G) \circ \Phi(F)$ for all $X, Y, Z \in \mathrm{Ob}(\mathcal{C})$ and $F \in \mathrm{Mor}_{\mathcal{C}}(X, Y)$ and $G \in \mathrm{Mor}_{\mathcal{C}}(Y, Z)$.

Examples 8.5. (i) We just defined a covariant functor Φ from $\mathfrak{Ds}$ to $\mathfrak{Ls}$.

(ii) Taking on objects $(X, x_0) \rightsquigarrow \pi_1(X, x_0)$, one can (and you should!) define a covariant functor from the category of pointed topological spaces (X is a topological space, $x_0 \in X$ and the morphisms are the continuous maps respecting the marked point) to the category of groups.

(iii) Sending a vector space V to its dual V^* and a linear map $f : V \to W$ to its transpose $f^* : W^* \to V^*$ defines a *contravariant* functor from the category of vector spaces to itself. Formally, with the notation of the previous definition, Φ maps $\mathrm{Mor}_{\mathcal{C}}(X, Y)$ to $\mathrm{Mor}_{\mathcal{C}'}(\Phi(Y), \Phi(X))$ and the last rule is to be replaced by $\Phi(G \circ F) = \Phi(F) \circ \Phi(G)$.

Every (covariant or contravariant) functor $\Phi : \mathcal{C} \to \mathcal{C}'$ has the following important property: if f is an isomorphism in $\mathcal{C}$, then $\Phi(f)$ is an isomorphism in $\mathcal{C}'$; and, more precisely, $\Phi(f^{-1}) = (\Phi(f))^{-1}$.

We are now going to express the fact that the functor we have defined from $\mathfrak{D}\mathfrak{s}$ to $\mathfrak{L}\mathfrak{s}$ is an "equivalence" of categories. Indeed, "isomorphisms" of categories are not really interesting (one never meets any!) and the weaker notion of equivalence is as near as we can reasonably hope to come. To explain what follows in an example, let us consider the functor $(X, x_0) \rightsquigarrow \pi_1(X, x_0)$, which is in no way an equivalence of categories, but which is "essentially surjective". Indeed, for any group G, we can find a pointed topological space (X, x_0) such that $\pi_1(X, x_0) \simeq G$. But can we reasonably hope for equality? Can we believe that there is a simply-connected space whose fundamental group is $\{0\}$, where $0 \in \mathbf{Z}$? Is this number zero the homotopy equivalence class of all loops at x_0 in some space X?

Definition 8.6. (i) We say that the (covariant) functor $\Phi : \mathcal{C} \to \mathcal{C}'$ is *essentially surjective* if every object of $\mathcal{C}'$ is isomorphic to $\Phi(X)$ for some $X \in \mathrm{Ob}(\mathcal{C})$.

(ii) We say that Φ is *faithful*, resp. *full*, resp. *fully faithful*, if, for any $X, Y \in \mathrm{Ob}(\mathcal{C})$, the map $f \mapsto \Phi(f)$, $\mathrm{Mor}_{\mathcal{C}}(X, Y) \to \mathrm{Mor}_{\mathcal{C}'}(\Phi(X), \Phi(Y))$, is injective, resp. surjective, resp. bijective.

(iii) We say that Φ is an *equivalence of categories* if it is essentially surjective and fully faithful.

Conditions (ii) can obviously be adapted to the case of a contravariant functor; then in (iii) one speaks of *anti-equivalence of categories*.

Exercise 8.7. Prove that the category of finite-dimensional complex vector spaces with linear maps is equivalent to the category whose objects are the $\mathbf{C}^n$, $n \in \mathbf{N}$, and whose morphisms $\mathbf{C}^n \to \mathbf{C}^p$ are the matrices in $\mathrm{Mat}_{p,n}(\mathbf{C})$.

One can prove that then there is a functor $G : \mathcal{C}' \to \mathcal{C}$ such that the obviously defined endofunctors $G \circ F$ and $F \circ G$ are, in some natural sense, "isomorphic" to the identity functors of $\mathcal{C}$ and $\mathcal{C}'$ (see the exercises at the end of the chapter).

Theorem 8.8. *The functor $\Phi : S_A \rightsquigarrow \mathcal{F}_A$ is an equivalence of the category $\mathfrak{D}\mathfrak{s}$ with the category $\mathfrak{L}\mathfrak{s}$.*

Proof. (i) Φ is faithful. Let $A \in \mathrm{Mat}_n(\mathcal{O}(\Omega))$ and $B \in \mathrm{Mat}_p(\mathcal{O}(\Omega))$. Let F_1, F_2 be two morphisms $S_A \to S_B$ such that $\Phi(F_1) = \Phi(F_2)$. Here, $\mathrm{Mor}_{\mathfrak{D}\mathfrak{s}}(S_A, S_B)$ and $\mathrm{Mor}_{\mathfrak{L}\mathfrak{s}}(\mathcal{F}_A, \mathcal{F}_B)$ are groups (indeed, complex vector spaces) and $F \mapsto \Phi(F)$ is a group morphism (indeed, $\mathbf{C}$-linear), and so we can suppose that $F := F_1 - F_2 : S_A \to S_B$ is such that $\Phi(F) = 0$. This means that $FX = 0 \in \mathcal{F}_B(U)$ for all $U \subset \Omega$ and $X \in \mathcal{F}_A(U)$. So take an open disk D and a fundamental matricial solution $\mathcal{X}$ on D: applying the assumption to the columns of $\mathcal{X}$, we see that $F\mathcal{X} = 0$ and thus $F = 0$ on D. By the principle of analytic continuation or because disks cover Ω, we conclude that $F = F_1 - F_2 = 0$.

(ii) Φ is full. With A and B as above, let $\phi : \mathcal{F}_A \to \mathcal{F}_B$ be a morphism of sheaves between these two local systems. Choose an open disk D and a fundamental matricial solution $\mathcal{X} = [X_1, \ldots, X_n]$ (columns) on D. Let $\phi_D(\mathcal{X}) := [\phi_D(X_1), \ldots, \phi_D(X_n)]$, by assumption a matricial solution of S_B on D. The matrix $F_D := \phi_D(\mathcal{X})\mathcal{X}^{-1}$ is made of holomorphic functions. Any other fundamental solution writes $\mathcal{X}' = \mathcal{X}C$, $C \in \mathrm{GL}_n(\mathbf{C})$, so that $\phi_D(\mathcal{X}') = \phi_D(\mathcal{X})C$ by linearity and $F_D := \phi_D(\mathcal{X}')\mathcal{X}'^{-1}$, so F_D does not depend on our particular choice. Therefore, if $D' \subset D$ is another disk, $F_{D'}$ is the restriction of F_D (because one can choose as a fundamental matricial solution on D' the restriction of $\mathcal{X}$). By the usual argument of continuation and connectedness, all the F_D patch into a unique $F \in \mathrm{Mat}_{p,n}(\mathcal{O}(\Omega))$. Coming back to the disk D and writing $\mathcal{Y} := \phi_D(\mathcal{X}) = F_D\mathcal{X}$, since $\mathcal{Y}$ is a matricial solution of S_B, starting from $\mathcal{Y}' = B\mathcal{Y}$ and $\mathcal{Y} = F_D\mathcal{X}$, we compute:

$$F_D\mathcal{X}' + F_D'\mathcal{X} = BF_D\mathcal{X} \implies F_D A\mathcal{X} + F_D'\mathcal{X} = BF_D\mathcal{X} \implies F_D A + F_D' = BF_D$$

since $\mathcal{X}$ is invertible. This being true on all D, one gets that $FA + F' = BF$, i.e., $F \in \mathrm{Mor}_{\mathfrak{D}_s}(S_A, S_B)$ and it is plainly an antecedent of $\phi \in \mathrm{Mor}_{\mathfrak{L}_s}(\mathcal{F}_A, \mathcal{F}_B)$ by the functor Φ.

(iii) Φ is essentially surjective. This is a much more difficult statement and we shall omit its proof. We shall prove a particular case (yet in some sense stronger) in Chapter 12. The courageous reader can look at [**Del70**], Theorem 2.17, or [**For81**], §§30 and 31, or [**vS03**], §6.2.   □

8.4. From local systems to representations of the fundamental group

We shall prove here a result in algebraic topology; since our goal is to apply it to differential equations, we do not state it in all possible generality; the reader should look at [**Del70, vS03**] or also the talk by Grothendieck at the Bourbaki seminar (exposé 141).

Let $\mathcal{F}$ be a local system on Ω and $a \in \Omega$. From the axioms defining local systems, one proves the following facts through the same arguments in our concrete example of Section 7.4 (sheaves of solutions):

(1) If $D \ni a$ is an open disk (more generally a simply-connected domain) in Ω, then $\mathcal{F}(D) \to \mathcal{F}_a$ is an isomorphism.

(2) If D, D' are two intersecting open disks in Ω (more generally two intersecting simply-connected domains such that $D \cap D'$ is connected), then $\mathcal{F}(D) \to \mathcal{F}(D \cap D') \leftarrow \mathcal{F}(D')$ are isomorphisms.

(3) If γ is a path in Ω with origin a and extremity b, and if we cover it by a chain of disks $D_0 \ni a, \dots, D_n \ni b$ such that D_i intersects D_{i-1} for $i = 1, \dots, n$, then the composition isomorphism $\mathcal{F}_a \simeq \mathcal{F}_b$ obtained from the chain

$$\mathcal{F}_a \leftarrow \mathcal{F}(D_0) \to \mathcal{F}(D_0 \cap D_1) \leftarrow \mathcal{F}(D_1)$$
$$\to \cdots \leftarrow \mathcal{F}(D_{n-1}) \to \mathcal{F}(D_{n-1} \cap D_n) \leftarrow \mathcal{F}(D_n) \to \mathcal{F}_b$$

only depends on the homotopy class $[\gamma] \in \Pi_1(\Omega; a, b)$.

(4) If one composes the path γ from a to b with the path γ' from b to c (both paths in Ω), then the morphism $\mathcal{F}_a \to \mathcal{F}_c$ induced by the composite path $\gamma.\gamma'$ is the composition of the morphism $\mathcal{F}_a \to \mathcal{F}_b$ induced by the path γ with the morphism $\mathcal{F}_b \to \mathcal{F}_c$ induced by the path γ'.

Note that the order of composition of the morphisms reverses the order of composition of the paths. From this, one gets an anti-representation

$$\pi_1(\Omega; a) \to \mathrm{GL}(\mathcal{F}_a).$$

We shall prefer to use the representation

$$\rho_\mathcal{F} : \pi_1(\Omega; a)^\circ \to \mathrm{GL}(\mathcal{F}_a),$$

where we write G° as the *opposite* of a group G; it has the same underlying set but with reversed law defined by $a \star b := b.a$. (We already met this notion in Remark 7.33.) Note that G and G° share the same unit and the same inversion map, and they are isomorphic through the isomorphism $x \mapsto x^{-1}$.

Now let $\mathcal{F}'$ be another local system and let $\phi : \mathcal{F} \to \mathcal{F}'$ be a morphism of sheaves. All diagrams

$$
\begin{array}{ccc}
\mathcal{F}(D) & \xrightarrow{\rho^D_{D \cap D'}} & \mathcal{F}(D \cap D') \\
{\scriptstyle \phi_D}\big\downarrow & & \big\downarrow{\scriptstyle \phi_{D \cap D'}} \\
\mathcal{F}'(D) & \xrightarrow[\rho'^D_{D \cap D'}]{} & \mathcal{F}'(D \cap D')
\end{array}
\qquad \text{and} \qquad
\begin{array}{ccc}
\mathcal{F}(D) & \xrightarrow{\rho^D_a} & \mathcal{F}_a \\
{\scriptstyle \phi_D}\big\downarrow & & \big\downarrow{\scriptstyle \phi_a} \\
\mathcal{F}'(D) & \xrightarrow[\rho'^D_a]{} & \mathcal{F}'_a
\end{array}
$$

are commutative. The chain above and the corresponding one for $\mathcal{F}'$ thus give rise to a commutative diagram

$$
\begin{array}{ccc}
\mathcal{F}_a & \longrightarrow & \mathcal{F}_b \\
\phi_a \downarrow & & \downarrow \phi_b \\
\mathcal{F}'_a & \longrightarrow & \mathcal{F}'_b
\end{array}
$$

and, specializing to $a = b$ and replacing the path γ by a loop λ:

$$
\begin{array}{ccc}
\mathcal{F}_a & \xrightarrow{\rho_{\mathcal{F}}([\lambda])} & \mathcal{F}_a \\
\phi_a \downarrow & & \downarrow \phi_a \\
\mathcal{F}'_a & \xrightarrow[\rho_{\mathcal{F}'}([\lambda])]{} & \mathcal{F}'_a
\end{array}
$$

One says that ϕ_a is a *morphism of representations* from $\rho_{\mathcal{F}}$ to $\rho_{\mathcal{F}'}$.

Definition 8.9. Let $\rho : G \to \mathrm{GL}(V)$ and $\rho' : G \to \mathrm{GL}(V')$ be two complex linear representations of the same group G. A *morphism* $f : \rho \to \rho'$ is a **C**-linear map $f : V \to V'$ which *intertwines* ρ and ρ' in the following sense:

$$
\forall g \in G, \ \rho'(g) \circ f = f \circ \rho(g).
$$

This defines a category $\mathfrak{Rep}_{\mathbf{C}}(G)$ of complex linear representations of G; its full subcategory of finite-dimensional representations is written $\mathfrak{Rep}^f_{\mathbf{C}}(G)$.

With the short notation for representations, *i.e.*, $g.v := \rho(g)(v)$, the intertwining condition reads:

$$
\forall g \in G, \ \forall v \in V, \ f(g.v) = g.f(v).
$$

Theorem 8.10. *Define a functor* Ψ *from the category* $\mathfrak{Ls}$ *of local systems to the category* $\mathfrak{Rep}^f_{\mathbf{C}} \left(\pi_1(\Omega; a)^\circ \right)$ *of finite-dimensional complex linear representations of* $\pi_1(\Omega; a)^\circ$ *which associates to the local system* $\mathcal{F}$ *the representation* $\rho_{\mathcal{F}}$ *and to the morphism* $\phi : \mathcal{F} \to \mathcal{F}'$ *in* $\mathfrak{Ls}$ *the morphism* ϕ_a *in* $\mathfrak{Rep}^f_{\mathbf{C}} \left(\pi_1(\Omega; a)^\circ \right)$. *Then* Ψ *is an equivalence of categories.*

Proof. We just saw that ϕ_a is indeed a morphism in $\mathfrak{Rep}^f_{\mathbf{C}} \left(\pi_1(\Omega; a)^\circ \right)$ and we leave to the reader the pleasure of checking that the above defines a covariant functor.

(i) Ψ **is faithful.** As in the previous case, sets of morphisms are groups here (indeed complex vector spaces) and maps induced by the functor are group morphisms (indeed **C**-linear maps), so we can consider $\phi : \mathcal{F} \to \mathcal{F}'$ such that $\phi_a = 0$. Let $D \ni a$ be an open disk. From the commutative diagram

$$\begin{array}{ccc} \mathcal{F}(D) & \xrightarrow{\ \rho_a^D\ } & \mathcal{F}_a \\ {\scriptstyle\phi_D}\downarrow & & \downarrow{\scriptstyle\phi_a} \\ \mathcal{F}'(D) & \xrightarrow[{\rho'}_a^D]{} & \mathcal{F}'_a, \end{array}$$

where vertical maps are isomorphisms, we conclude that $\phi_D = 0$. From the diagrams

$$\begin{array}{ccc} \mathcal{F}(D) & \xrightarrow{\ \rho_{D\cap D'}^D\ } & \mathcal{F}(D \cap D') \\ {\scriptstyle\phi_D}\downarrow & & \downarrow{\scriptstyle\phi_{D\cap D'}} \\ \mathcal{F}'(D) & \xrightarrow[{\rho'}_{D\cap D'}^D]{} & \mathcal{F}'(D \cap D') \end{array} \quad \text{and} \quad \begin{array}{ccc} \mathcal{F}(D') & \xrightarrow{\ \rho_{D\cap D'}^D\ } & \mathcal{F}(D \cap D') \\ {\scriptstyle\phi_{D'}}\downarrow & & \downarrow{\scriptstyle\phi_{D\cap D'}} \\ \mathcal{F}'(D') & \xrightarrow[{\rho'}_{D\cap D'}^{D'}]{} & \mathcal{F}'(D \cap D'), \end{array}$$

where horizontal maps are isomorphisms, we conclude that $\phi_{D'} = 0$ if D' intersects D. By connectedness of Ω, this extends to all open disks and so $\phi = 0$.

(ii) Ψ is full. Let $\phi_{(a)} : \mathcal{F}_a \to \mathcal{F}'_a$ be a linear map which is a morphism in $\mathfrak{Rep}_{\mathbf{C}}^f (\pi_1(\Omega; a)^\circ)$; we write it $\phi_{(a)}$ because we are to prove that $\phi_{(a)} = \phi_a$ for some $\phi : \mathcal{F} \to \mathcal{F}'$. So, for any loop λ in Ω based at a, writing $f_{[\lambda]} : \mathcal{F}_a \to \mathcal{F}_a$ and $f'_{[\lambda]} : \mathcal{F}'_a \to \mathcal{F}'_a$ as the isomorphisms of analytic continuation, we have:

$$f'_{[\lambda]} \circ \phi_{(a)} = \phi_{(a)} \circ f_{[\lambda]}.$$

Let $b \in \Omega$ be arbitrary and let γ_1, γ_2 be paths from a to b in Ω. Taking $\lambda := \gamma_1 \gamma_2^{-1}$, a loop at a in Ω, we have (with obvious notation):

$$f_{[\lambda]} = f_{[\gamma_2]}^{-1} \circ f_{[\gamma_1]} \text{ and } f'_{[\lambda]} = {f'_{[\gamma_2]}}^{-1} \circ f'_{[\gamma_1]},$$

so that the intertwining relation above reads:

$$\begin{aligned} {f'_{[\gamma_2]}}^{-1} \circ f'_{[\gamma_1]} \circ \phi_{(a)} &= \phi_{(a)} \circ f_{[\gamma_2]}^{-1} \circ f_{[\gamma_1]} \\ \implies f'_{[\gamma_1]} \circ \phi_{(a)} \circ f_{[\gamma_1]}^{-1} &= f'_{[\gamma_2]} \circ \phi_{(a)} \circ f_{[\gamma_2]}^{-1} \circ f_{[\gamma_1]}, \end{aligned}$$

thus defining a linear map $\phi_{(b)} : \mathcal{F}_b \to \mathcal{F}'_b$ independently of the chosen path γ of analytic continuation from a to b. Each germ $\phi_{(b)}$ extends to a disk $D_b \ni b$ and the collection of all $\phi_{(D_b)} : \mathcal{F}_{D_b} \to \mathcal{F}'_{D_b}$ is compatible by construction (this is tedious to check algebraically[5] but almost obvious with a little drawing). So they patch into a morphism $\phi : \mathcal{F} \to \mathcal{F}'$ such that each $\phi_{(D_b)} : \mathcal{F}_{D_b} \to \mathcal{F}'_{D_b}$ is ϕ_{D_b} and in particular $\phi_{(a)} : \mathcal{F}_a \to \mathcal{F}'_a$ is ϕ_a.

---

[5]All these proofs are much easier to write if one uses the formalism in [**Del70**], which should be an encouragement to learn it.

(iii) Ψ is essentially surjective. I know of no "concrete" proof of this fact. The group $G := \pi_1(\Omega; a)^\circ$ acts on the universal covering:

$$\tilde{\Omega} \xrightarrow{p} \Omega = \tilde{\Omega}/G.$$

Starting with a representation $G \to \mathrm{GL}(V)$, we define on $\tilde{\Omega}$ the constant sheaf $\tilde{\mathcal{F}} := \underline{V}$, a trivial local system (Section 8.2). Then its *direct image* $\mathcal{G} := p_* \tilde{\mathcal{F}}$ is the sheaf on Ω defined by $U \mapsto \mathcal{G}(U) := \tilde{\mathcal{F}}\left(p^{-1}(U)\right)$. The stalks are direct sums: $\mathcal{G}_a = \bigoplus_{p(x)=a} \tilde{\mathcal{F}}_x$. The group G operates linearly on each stalk $\mathcal{G}_a$ because it operates topologically on each fiber $p^{-1}(a)$. Then the invariant subspaces $\mathcal{G}_a^G$ patch into a local system $\mathcal{F} := \mathcal{G}^G$, which is the one we are looking for. ☐

Exercise 8.11. Fill in the details for the last step; you may look for help in Deligne or Grothendieck.

Let us stress again that this equivalence is only the formal part of the Riemann-Hilbert correspondence we are aiming at. "True R.-H." involves conditions on the singularities.

Exercises

(1) Let F, G be covariant functors from $\mathcal{C}$ to $\mathcal{C}'$. Define a morphism from F to G as a family (ϕ_X) indexed by objects of $\mathcal{C}$ such that each ϕ_X is a morphism from $F(X)$ to $G(X)$ in $\mathcal{C}'$ and, moreover, for each arrow $f : X \to Y$ in $\mathcal{C}$, one has $\phi_Y \circ F(f) = G(f) \circ \phi_X$. Show that this allows us to define a category of functors from $\mathcal{C}$ to $\mathcal{C}'$.

(2) Show that the covariant functor F from $\mathcal{C}$ to $\mathcal{C}'$ is an equivalence of category if, and only if, there exists a covariant functor F' from $\mathcal{C}'$ to $\mathcal{C}$ such that $F' \circ F$ and $F \circ F'$ are respectively equivalent to the identity functors of $\mathcal{C}$ and $\mathcal{C}'$.

(3) Assume that Φ is fully faithful. Prove that $f : X \to Y$ is an isomorphism if, and only if, $\Phi(f)$ is. Find an example that shows that this fails if Φ is only faithful. (Hint: consider the "forgetful" functor which associates to a group its underlying set; this works the same for any functor whose work is to forget a structure or a part of a structure, for instance from topological groups to groups, etc.)

(4) (i) Check that $\mathfrak{Rep}_\mathbf{C}(G)$ is indeed a category and that the forgetful functor $\rho \rightsquigarrow V$, $f \rightsquigarrow f$ from $\mathfrak{Rep}_\mathbf{C}(G)$ to the category $\mathcal{V}ect_\mathbf{C}$ of complex vector spaces is essentially surjective and faithful.

    (ii) Give an example of $f : \rho \to \rho'$ which is not an isomorphism while $f : V \to V'$ is.

The Riemann-Hilbert Correspondence

Chapter 9

Regular singular points and the local Riemann-Hilbert correspondence

References for this chapter are Chapter 8.3 of the book by Ahlfors [**Ahl78**] for a gentle introduction to the problem; and the following, mostly about technical information on scalar differential equations: [**BR78**, Chap. 9], [**CL55**, Chap. 4] (mainly §§4.5 to 4.8), [**Hil97**, Chap. 5] (mainly §§5.1 to 5.3) and [**Inc44**, Chap. 16]. The book [**Inc44**] by Ince, although somewhat old-fashioned, is a great classic and rich in information.

It is traditional that the two main points of view — order n scalar equations *vs* rank n vectorial equations — although essentially equivalent (because of the cyclic vector lemma, which we proved in Theorem 7.51) are studied in parallel but separately. We faithfully followed this tradition.

It is however important to understand that there is a big difference between the case of scalar equations and that of systems. In both cases, we intend to define a notion of "moderate" singularities. There are essentially two ways to do that. One is to put some condition on the solutions of the equation or system; the condition will be of "moderate" growth in the neighborhood of the singularity. The terminology is that of "regular singular point" or equation or system. However, one also wishes to have a way to test that property directly on the equation or system. The corresponding notion is that of singularity, or equation, or system "of the first kind"; this is an old terminology, and an equivalent but more modern one is that of "fuchsian" singularity, or equation, or system. We shall rather use the older terminology here to emphasize the down-to-earth approach (without vector bundles, etc.) followed in this book.

Now the big difference is this: for scalar equations, the two notions ("regular singular" or "of the first kind") are equivalent; this is Fuchs criterion (see Theorem 9.30). But for systems, the condition "of the first kind" is strictly stronger (see Exercise 9.16 in Section 9.3). Actually, it is not really intrinsic (it is not conserved by gauge transformations; see the same exercise) but the more intrinsic condition "regular singular" is not easy to test: there is essentially no other way to test it than to use the cyclic vector lemma (Theorem 7.51) to reduce the problem to the case of a scalar equation.

9.1. Introduction and motivation

Many special functions discovered in mathematics and in physics since the eighteenth century were found to be solutions of linear differential equations with polynomial coefficients. This includes for instance the exponential and logarithm functions, the functions z^α and the hypergeometric series, which we shall study in detail in Chapter 11; but also many others, see the books quoted above. Note however that this does *not* include the most famous Gamma, Zeta and Theta functions: but these do satisfy other kinds of "functional equations" that also proved useful in their study.

We want to make a *global* study of such functions. It turned out (as experience in the domain was accumulated) that their global behavior is extremely dependent on the singularities of the equation, that is, the singularities of the coefficients a_i when the equation is written in the form E_a. So it is worth starting with a *local* study near the singularities. The most important features of a solution near a singularity are:

(1) Its monodromy. This says how far the solution is from being uniform. Solutions "ramify" and for this reason singularities are sometimes called "branch points".

(2) Its rate of growth (or decay), which may be moderate or have the physical character of an explosion.

That monodromy alone is not a sufficient feature to characterize solutions near a singularity is shown by the following examples: $zf' - f = 0$, $zf' + f = 0$ and $z^2 f' + f = 0$. The first one has z as a basis of solutions: the singularity is only apparent. The second one has $1/z$ as a basis of solutions: the solutions have a simple pole. The third one has $e^{1/z}$ as a basis of solutions: the solutions may have exponential growth or decay, or even spiraling, according to the direction in approaching the singularity 0. Yet, in all three cases, the monodromy is trivial.

So the first step of the study (beyond ordinary points where Cauchy's theorem applies) is the case of a singularity where all solutions have moderate growth. For uniform solutions near 0, this excludes $e^{1/z}$ (and actually all solutions having an essential singularity at 0) but includes all meromorphic solutions, since the condition $f(z) = O(z^{-N})$ (see the corollary of Theorem 3.10) means that they have polynomial growth as functions of $1/z$. However, for "multivalued" solutions (*i.e.*, those with nontrivial monodromy), the definition has to be adapted:

Example 9.1. The analytic continuation L of the log function along the infinite path $\gamma(t) := e^{-t+ie^{t^2}}$, $t \in \mathbf{R}_+$, takes the value $L(\gamma(t)) = -t+ie^{t^2}$, so that $|L(\gamma(t))| \geq e^{t^2}$, while $|\gamma(t)| = e^{-t}$: clearly the condition $L(z) = O(z^{-N})$ is satisfied for no N along this path, which approaches 0 while spiraling as $t \to +\infty$. (This example will be detailed in Section 9.2.)

Exercise 9.2. Find a similar example with a function z^α. (Use a path that spirals very fast, while approaching 0 very slowly.)

So we shall give in Section 9.2 a reasonable definition of "moderate growth" (one which would not exclude the logarithm and z^α functions). We shall then find that equations all of whose solutions have moderate growth can be classified by their monodromy: this is the local Riemann-Hilbert correspondence.[1] Of course, the equivalence relation used (through gauge transformations) must be adapted so as to take into account what goes on at 0: there will be *a new definition of meromorphic equivalence*

Conventions for this chapter. All systems and equations considered will have coefficients which are holomorphic in some punctured disk $\overset{\bullet}{\mathrm{D}} := \overset{\circ}{\mathrm{D}}(0,R) \setminus \{0\}$, and which are meromorphic at 0; in other words, they will be elements of the field $\mathcal{M}_0 = \mathbf{C}(\{z\})$ (meromorphic germs at 0). This means that we consider only the singularity 0: other points will be treated in examples. (One can always change variables so as to reduce the problem to this case.)

Definition 9.3. Let $A, B \in \mathrm{Mat}_n(\mathbf{C}(\{z\}))$. We say that A and B, or S_A and S_B, are *meromorphically equivalent at* 0 if there exists $F \in \mathrm{GL}_n(\mathbf{C}(\{z\}))$ such that $F[A] = B$, *i.e.*, $F' = BF - FA$. We then write $A \sim B$ or $S_A \sim S_B$.

---

[1]The global Riemann-Hilbert correspondence will be studied in Chapters 11 and 12.

Figure 9.1. Thickened logarithmic spiral

Note that since the zeros and poles of a nontrivial meromorphic function are isolated, A and B are then holomorphically equivalent in some punctured neighborhood of 0. However, the above condition is much stronger than holomorphic equivalence away from the poles. We shall find out that the fact that F has at most poles establishes a strong link between A and B.

9.2. The condition of moderate growth in sectors

To understand the condition we are going to introduce, let us examine more closely the example of the logarithm in the previous section.

Example 9.4. Let $\gamma(t) := e^{-t+\mathrm{i}e^{t^2}}$, for $t \geq 0$. Then $\gamma(t) \to 0$ as $t \to +\infty$. This infinite path starts at $\gamma(0) = e^{\mathrm{i}} \in \mathbf{C} \setminus \mathbf{R}_-$, the domain of the principal determination of the logarithm, and $\log \gamma(t) = -t + \mathrm{i}e^{t^2}$ as long as the path does not leave the domain, that is, $t < \sqrt{\ln \pi}$.

To give a proper meaning to analytic continuation, we "thicken" the image curve (which is a spiral) into a simply-connected domain $U \supset \operatorname{Im} \gamma$ such that $0 \in \overline{U}$. Let

$$V := \{(t, u) \in \mathbf{R}^2 \mid t \geq 0 \text{ and } |u| \leq \operatorname{argsh}(\pi e^{-t^2})\}.$$

The latter condition implies that:

$$(t, u_1), (t, u_2) \in V \Rightarrow \left| e^{u_2 + t^2} - e^{u_1 + t^2} \right| < 2\pi$$

and allows us to deduce that $(t, u) \mapsto e^{-t+\mathrm{i}e^{t^2 + u}}$ is a homeomorphism from V to a "thick spiral" $U \supset \operatorname{Im} \gamma$ (the latter curve being the image of the subset $u = 0$ of V). Since V is simply connected, so is U. By a (now) standard continuity argument, the determination of the logarithm on the simply-connected domain U which coincides with log on their common domain is given by the formula $L\left(e^{-t+\mathrm{i}e^{t^2 + u}}\right) = -t + \mathrm{i}e^{t^2 + u}$; in particular, for $t \in \mathbf{R}_+$, it takes the value $L(\gamma(t)) = -t + \mathrm{i}e^{t^2}$, so that (as we already saw)

$|L(\gamma(t))| \geq e^{t^2}$, while $|\gamma(t)| = e^{-t}$. Clearly the condition $L(z) = O(z^{-N})$ is not satisfied in U for any N.

We now consider a punctured disk $\overset{\bullet}{D} := \overset{\circ}{D}(0, R) \setminus \{0\}$ and an analytic germ f at some point $a \in \overset{\bullet}{D}$ such that f admits analytic continuation along all paths in $\overset{\bullet}{D}$ originating in a. By Cauchy's Theorem 7.25, this is in particular the case for solutions of linear differential equations. The collection F of all germs obtained from f in this way is called a *multivalued function* on $\overset{\bullet}{D}$. For any open subset $U \subset \overset{\bullet}{D}$, an analytic function on U, all of whose germs belong to the collection F, is called a *determination of F on U*. (Such a determination is therefore an element of $\mathcal{O}(U)$.) For a simply-connected domain U such determinations always exist. For an arbitrary domain U, determinations subject to initial conditions, if they exist, are unique, *i.e.*, two determinations of F on U which take the same value w_0 at some $z_0 \in U$ are equal. Clearly, one can linearly combine, multiply and derive multivalued functions: they form a differential algebra (see Section 5.2) containing $\mathcal{O}(\overset{\bullet}{D})$.

In the following definition, we shall use the following notation for open angular sectors with vertex at 0; if $0 < b - a < 2\pi$, then:

$$S_{a,b} := \{re^{i\theta} \mid r > 0 \text{ and } a < \theta < b\}.$$

Definition 9.5. We say that a multivalued function F on $\overset{\bullet}{D}$ has *moderate growth in sectors* if, for any $a, b \in \mathbf{R}$ such that $0 < b - a < 2\pi$ and for any determination f of F on the simply-connected domain $U := \overset{\bullet}{D} \cap S_{a,b}$, there exists $N \in \mathbf{N}$ such that $f(z) = O(z^{-N})$ as $z \to 0$ in U. Beware however that the exponent N may depend on the sector and on the determination.

Note that restricting to a smaller punctured disk (with radius $R' < R$) does not affect the condition (or its negation), so, in practice, we expect it to be true (or false) for some unspecified R and do not usually mention the punctured disk $\overset{\bullet}{D}$.

Example 9.6. Any determination of the logarithm on $\overset{\bullet}{D} \cap S_{a,b}$ is such that $L(re^{i\theta}) = \ln r + i(\theta + 2k_0\pi)$ for $r > 0$ and $a < \theta < b$, where k_0 is fixed. Thus $|L(z)| \leq |\ln r| + C$ for some fixed C, so that $L(z) = O(1/z)$ for $z \to 0$ in U: the (multivalued) logarithm function has moderate growth in sectors.

Exercise 9.7. Prove that z^α has moderate growth in sectors.

Basic and obvious properties. For the following, the proof is left as an (easy) exercise:

- Multivalued functions with moderate growth in sectors are closed under linear combinations and multiplication: they form a **C**-algebra.
- The matricial function $z^A = \exp(A \log z)$, where $A \in \mathrm{Mat}_n(\mathbf{C})$, has moderate growth in sectors, meaning that all its coefficients have moderate growth in sectors; this follows from the example and the exercise above, combined with the previous property.
- For a uniform function on $\overset{\bullet}{\mathrm{D}}$, the condition of moderate growth in sectors is equivalent to being meromorphic after the corollary of Theorem 3.10. (Hint: use the compactness of the circle of directions $S^1 := \mathbf{R}/2\pi\mathbf{Z}$.)

A nonobvious property is that multivalued functions with moderate growth in sectors actually form a *differential* **C**-algebra (the definition was given in Section 3.4).

Lemma 9.8. *If g is analytic and bounded in $U' := \overset{\bullet}{\mathrm{D}} \cap S_{a',b'}$, then, for any $a,b \in \mathbf{R}$ such that $a' < a < b < b'$, the function zg' is bounded in $U := \overset{\bullet}{\mathrm{D}} \cap S_{a,b}$.*

Proof. By elementary geometry, there exists a constant $C > 0$ such that:

$$\forall z \in S_{a,b} , \ d(z, \partial S_{a',b'}) \geq C |z| .$$

Now, assuming $|g| \leq M$ on U and using *Cauchy estimates*[2] [**Ahl78**, Chap. 4, §2.3, p. 122], one gets:

$$|g'(z)| \leq \frac{M}{d(z, \partial S_{a',b'})} \implies |zg'(z)| \leq \frac{M}{C}.$$

$\square$

Exercise 9.9. Use trigonometry to compute the constant C in the lemma.

Theorem 9.10. *If f has moderate growth in sectors on $\overset{\bullet}{\mathrm{D}}$, so has f'.*

Proof. For a given sector $S_{a,b}$ with $0 < b - a < 2\pi$, fix a slightly bigger one $S_{a',b'}$ with $a' < a < b < b'$ such that $b' - a' < 2\pi$. Choose N such that $g := z^N f$ is bounded on $\overset{\bullet}{\mathrm{D}} \cap S_{a',b'}$. Then $zg' - Ng$ is bounded on $\overset{\bullet}{\mathrm{D}} \cap S_{a,b}$ after the lemma, so $f' = (zg' - Ng)z^{-N-1} = O(z^{-N-1})$ on $\overset{\bullet}{\mathrm{D}} \cap S_{a,b}$. $\square$

---

[2]Cauchy estimates are easy enough to prove starting from Cauchy formulas given in Corollary 3.25.

9.3. Moderate growth condition for solutions of a system

Let $A \in \mathrm{Mat}_n(\mathbf{C}(\{z\}))$ be holomorphic on the punctured disk $\overset{\bullet}{\mathrm{D}} := \overset{\circ}{\mathrm{D}}(0,R) \setminus \{0\}$. Let $\mathcal{X}$ be a fundamental matricial solution of the system S_A at $a \in \overset{\bullet}{\mathrm{D}}$. Then, if $\mathcal{X}$ has moderate growth in sectors, the same is true for all fundamental matricial solutions of S_A at any point of $\overset{\bullet}{\mathrm{D}}$ (because they can be obtained from $\mathcal{X}$ by analytic continuation and multiplication by a constant matrix), and therefore also for all (vector) solutions.

Definition 9.11. We say that S_A *has a regular singular point at* 0, or that S_A *is regular singular at* 0, if it has a fundamental matricial solution at some point with moderate growth in sectors.

The following properties are then obvious:

(1) The system S_A is regular singular at 0 if, and only if, all its (vector) solutions at some point have moderate growth in sectors.

(2) If S_A has a regular singular point at 0 and if $A \sim B$ (meromorphic equivalence at 0), then S_B is regular singular at 0. Indeed, if $B = F[A]$ with $F \in \mathrm{GL}_n(\mathbf{C}(\{z\}))$ and if $\mathcal{X}$ is a fundamental matricial solution of S_A, then $F\mathcal{X}$ is a fundamental matricial solution of S_B and it has moderate growth in sectors.

Example 9.12. If $A = z^{-1}C$ with $C \in \mathrm{Mat}_n(\mathbf{C})$, then z^C is a fundamental matricial solution and it has moderate growth in sectors, so 0 is a regular singular point for S_A.

This example is in some sense "generic":

Theorem 9.13. *If the system* $X' = AX$ *is regular singular at* 0, *then there is a matrix* $C \in \mathrm{Mat}_n(\mathbf{C})$ *such that* $A \sim z^{-1}C$, *i.e.*, $A = F[z^{-1}C]$ *for some* $F \in \mathrm{GL}_n(\mathbf{C}(\{z\}))$.

Proof. Let $\mathcal{X}$ be a fundamental matricial solution at some point $a \in \overset{\bullet}{\mathrm{D}}$. The result of analytic continuation along the fundamental loop λ is $\mathcal{X}^\lambda = \mathcal{X}M$ for some invertible monodromy matrix $M \in \mathrm{GL}_n(\mathbf{C})$. From Section 4.4, we know that there exists a matrix $C \in \mathrm{Mat}_n(\mathbf{C})$ such that $e^{2\mathrm{i}\pi C} = M$. Let $F := \mathcal{X}z^{-C}$. Then $F^\lambda = \mathcal{X}^\lambda(z^{-C})^\lambda = \mathcal{X}Me^{-2\mathrm{i}\pi C}z^{-C} = \mathcal{X}MM^{-1}z^{-C} = \mathcal{X}z^{-C} = F$, that is, F is uniform; but since $\mathcal{X}$ and z^{-C} also have moderate growth, so has F. Therefore, $F \in \mathrm{GL}_n(\mathbf{C}(\{z\}))$, and of course $F[z^{-1}C] = A$. $\square$

Definition 9.14. We say that S_A *has a singularity of the first kind at* 0, or that 0 *is a singularity of the first kind for* S_A, if A has at most a simple pole at 0, *i.e.*, $zA \in \mathrm{Mat}_n(\mathbf{C}\{z\})$. We also sometimes say improperly for short that S_A *is of the first kind* (*at* 0).

A more modern terminology, totally equivalent to the above one, is that of *fuchsian singularity*, but we shall stick to the older one.

Examples 9.15. (1) The rank one system $zf' = \alpha f$ is of the first kind.

(2) If $a_1, \ldots, a_n$ all have at most a simple pole at 0 and if $A := A_{\underline{a}}$, then S_A has a singularity of the first kind at 0.

(3) Suppose p and q are holomorphic at 0. We vectorialize the equation $f'' + (p/z)f' + (q/z^2)f = 0$ by putting $X := \begin{pmatrix} f \\ zf' \end{pmatrix}$ so that $X' = AX$ with $A = z^{-1}\begin{pmatrix} 0 & 1 \\ -q & 1-p \end{pmatrix}$: this is a system of the first kind.

Exercise 9.16. Under the same assumptions, vectorialize the equation $f'' + (p/z)f' + (q/z^2)f = 0$ by putting $Y := \begin{pmatrix} f \\ f' \end{pmatrix}$ so that $Y' = BY$ with $B = \begin{pmatrix} 0 & 1 \\ -q/z^2 & -p/z \end{pmatrix}$. If $q(0) \neq 0$, this is not a system of the first kind. It is equivalent to the system above (which is of the first kind) through the gauge transform $X = FY$, $A = F[B]$, where $F := \mathrm{Diag}(1, z)$.

Theorem 9.17. *The system S_A is regular singular at 0 if, and only if, it is meromorphically equivalent to a system having a singularity of the first kind at 0.*

Proof. The previous Theorem 9.13 shows a stronger version of one implication. For the converse implication, it will be enough to construct a fundamental matricial solution for a system having a singularity of the first kind at 0, and to prove that this solution has moderate growth in sectors. This will be done in the next section (Theorem 9.18 below).   □

The main difficulty in using this theorem is that it is not easy to see if a given system is meromorphically equivalent to one of the first kind (it is easy for *scalar* equations; see Theorem 9.30).

9.4. Resolution of systems of the first kind and monodromy of regular singular systems

The following result and method of resolution are, in essence, due to Fuchs and Frobenius.

Theorem 9.18. *The system $X' = z^{-1}AX$, $A \in \mathrm{Mat}_n(\mathbf{C}\{z\})$, has a fundamental matricial solution of the form $\mathcal{X} = Fz^C$, $F \in \mathrm{GL}_n(\mathbf{C}(\{z\}))$ and $C \in \mathrm{Mat}_n(\mathbf{C})$.*

Proof. This amounts to saying that F is a meromorphic gauge transformation from $z^{-1}C$ to $z^{-1}A$, *i.e.*, $zF' = AF - FC$. This "simplification" of A into a constant matrix $C = F^{-1}AF - zF^{-1}F'$ will be achieved in two main steps.

First step: Elimination of resonances. *Resonancies*[3] here are occurrences of pairs of distinct eigenvalues $\lambda \neq \mu \in \mathrm{Sp}(A(0))$ such that $\mu - \lambda \in \mathbf{N}$. The second step of the resolution of our system (using a Birkhoff gauge transformation) will require that there be no resonancies, so we begin by eliminating them. To do this, we alternate constant gauge transformations (with matrix in $\mathrm{GL}_n(\mathbf{C})$) and "shearing" gauge transformations: these are transformations with diagonal matrices $S_{k,l} := \mathrm{Diag}(z, \ldots, z, 1, \ldots, 1)$ (k times z and l times 1).

We begin by triangularizing $A(0) \in \mathrm{Mat}_n(\mathbf{C})$. Note that if $P \in \mathrm{GL}_n(\mathbf{C})$, then $P' = 0$ and the gauge transformation $P[A] = PAP^{-1}$ is just a conjugation; also, we then have $P[A](0) = PA(0)P^{-1}$, which we can choose to be upper triangular. Assume that there are resonancies and choose $\lambda \neq \mu \in \mathrm{Sp}(A(0))$ such that $m := \mu - \lambda \in \mathbf{N}^*$ is as big as possible. The sum of all such m for all pairs of eigenvalues will serve as a "counter" for our algorithm; it strictly decreases at each step. We can choose P such that the eigenvalue μ is concentrated in the upper left block of $PA(0)P^{-1}$:

$$A' := PAP^{-1} = \begin{pmatrix} a & b \\ c & d \end{pmatrix}, \text{ where } a(0) \in \mathrm{Mat}_k(\mathbf{C}), \mathrm{Sp}(a(0)) = \{\mu\};$$

$$b(0) = 0_{k,l}; \quad c(0) = 0_{l,k}; \quad d(0) \in \mathrm{Mat}_l(\mathbf{C}), \mu \notin \mathrm{Sp}(d(0)).$$

We now apply the shearing gauge transformation $S_{k,l}$ to A' and get:

$$A'' := S_{k,l}[A'] = \begin{pmatrix} a - I_k & z^{-1}b \\ zc & d \end{pmatrix},$$

so that $A''(0)$ has the same eigenvalues as $A(0)$, except that all μ have been transformed to $\mu - 1 = \lambda + m - 1$: the total quantity of resonancies has strictly decreased. Iterating the process, we get rid of all resonancies.

Second step: Birkhoff gauge transformation. A *Birkhoff gauge transformation* has matrix $F = I_n + zF_1 + \cdots$ (it can be formal or convergent).

Proposition 9.19. *If $A \in \mathrm{Mat}_n(\mathbf{C}[[z]])$ has no resonancies, then there is a unique formal Birkhoff gauge transformation F such that $F^{-1}AF - zF^{-1}F' = A(0)$.*

---

[3]The name is due to a similar situation in the study of real differential equations with periodic coefficients; it has a physical meaning related to exceptional increase of amplitude or energy of some periodic phenomena.

Proof. The condition $F = I_n + zF_1 + \cdots$, i.e., $F(0) = I_n$, implies that F is invertible, so that the relation $F^{-1}AF - zF^{-1}F' = A(0)$ is equivalent to $zF' = AF - FA(0)$. Writing $A = A_0 + zA_1 + \cdots$, this is equivalent to: $F_0 = I_n$ and, for $k \geq 1$,

$$kF_k = A_0F_k + \cdots + A_kF_0 - F_kA_0$$
$$\Longleftrightarrow F_k(A_0 + kI_n) - A_0F_k = A_1F_{k-1} + \cdots + A_kF_0.$$

Note that, by the assumption of nonresonancy, the matrices $A_0 + kI_n$ and A_0 have no common eigenvalue for $k \geq 1$. Using the following lemma, we conclude that the F_k are unique and can be recursively calculated using the following formula:

$$F_k := \Phi_{A_0,A_0+kI_n}^{-1}(A_1F_{k-1} + \cdots + A_kF_0). \qquad \square$$

Lemma 9.20. *Let $P \in \mathrm{Mat}_n(\mathbf{C})$ and $Q \in \mathrm{Mat}_p(\mathbf{C})$ and define the linear map $\Phi_{P,Q}(X) := XQ - PX$ from $\mathrm{Mat}_{n,p}(\mathbf{C})$ into itself. Then, the eigenvalues of $\Phi_{P,Q}$ are the $\mu - \lambda$, where $\lambda \in \mathrm{Sp}(P)$ and $\mu \in \mathrm{Sp}(Q)$. In particular, if P and Q have no common eigenvalue, then $\Phi_{P,Q}$ is bijective.*

Proof. If P and Q are each in triangular form, then $\Phi_{P,Q}$ is triangular in the canonical basis of $\mathrm{Mat}_{n,p}(\mathbf{C})$ put in the right order. In general, if P and P' are conjugate and Q and Q' are conjugate, then $\Phi_{P,Q}$ and $\Phi_{P',Q'}$ are conjugate. $\qquad \square$

Exercise 9.21. Give details for the above arguments: which order of the basis makes $\Phi_{P,Q}$ triangular ? which automorphism of $\mathrm{Mat}_{n,p}(\mathbf{C})$ conjugates $\Phi_{P,Q}$ and $\Phi_{P',Q'}$?

Proposition 9.22. *In the previous proposition, if A moreover converges, $A \in \mathrm{Mat}_n(\mathbf{C}\{z\})$, then the Birkhoff gauge transformation also converges, $F \in \mathrm{GL}_n(\mathbf{C}\{z\})$.*

Proof. We use very basic functional analysis (normed vector spaces). When k grows, $\Phi_{A_0,A_0+kI_n} \sim k\,\mathrm{Id}$ in $\mathrm{End}(\mathrm{Mat}_{n,p}(\mathbf{C}))$, so for an adequate norm: $\left\|\Phi_{A_0,A_0+kI_n}^{-1}\right\| \sim 1/k$. Therefore, there is a constant $D > 0$ such that

$$\|F_k\| \leq (D/k)\sum_{i=0}^{k-1}\|A_{k-i}\|\,\|F_i\|.$$

Also, by the hypothesis of convergence, there exist $C, R > 0$ such that $\|A_j\| \leq CR^{-j}$. Putting $g_k := R^k\|F_k\|$, one sees that $g_k \leq (CD/k)\sum_{i=0}^{k-1}g_i$. Now, increasing C or D, we can clearly assume that $CD \geq 1$ and an easy induction gives $g_k \leq (CD)^k$, whence $\|F_k\| \leq (CD/R)^k$. $\qquad \square$

This ends the proof of Theorem 9.18. $\qquad \square$

Corollary 9.23. *If we choose as a fundamental matricial solution* $\mathcal{X} := Fz^C$ *as constructed in the theorem, we find the monodromy matrix:*

$$M_{[\lambda]} = \mathcal{X}^{-1}\mathcal{X}^\lambda = z^{-C}F^{-1}Fz^C e^{2i\pi C} = e^{2i\pi C}.$$

The local Riemann-Hilbert correspondence. Remember that we explained in Definition 7.59 when two linear representations are equivalent. In the case of the group $\pi_1(\mathbf{C}^*, a)$, a linear representation is completely characterized by the image $M \in \mathrm{GL}_n(\mathbf{C})$ of the standard generator (the homotopy class of the fundamental loop λ), and the representations characterized by $M, N \in \mathrm{GL}_n(\mathbf{C})$ are equivalent if, and only if, the matrices M and N are conjugate, *i.e.*, $N = PMP^{-1}$, $P \in \mathrm{GL}_n(\mathbf{C})$. When we apply these general facts to the monodromy representation of a particular system, we identify $\pi_1(\mathbf{C}^*, a)$ with $\pi_1(\overset{\bullet}{\mathrm{D}}, a)$.

Theorem 9.24. *Associating to a regular singular system $X' = z^{-1}AX$, $A \in \mathrm{Mat}_n(\mathbf{C}\{z\})$, one of its monodromy representations $\rho : \pi_1(\mathbf{C}^*, a) \to \mathrm{GL}(\mathcal{F}_a)$ yields a bijection between meromorphic equivalence classes of regular singular systems on the one hand and isomorphism classes of linear representations of the fundamental group on the other hand.*

Proof. Suppose $A, B \in \mathrm{Mat}_n(\mathbf{C}\{z\})$ define equivalent systems: $z^{-1}A \sim z^{-1}B$, and let F be the corresponding gauge transformation. Then, if $\mathcal{X}$, $\mathcal{Y}$ respectively are fundamental matricial solutions for these two systems, one has $F\mathcal{X} = \mathcal{Y}P$ for some $P \in \mathrm{GL}_n(\mathbf{C})$ and their respective monodromy matrices $M := \mathcal{X}^{-1}\mathcal{X}^\lambda$ and $N := \mathcal{Y}^{-1}\mathcal{Y}^\lambda$ satisfy the relation

$$N = (F\mathcal{X}P^{-1})^{-1}(F\mathcal{X}P^{-1})^\lambda = P\mathcal{X}^{-1}F^{-1}(F\mathcal{X}^\lambda P^{-1}) = PMP^{-1},$$

that is, according to Definition 7.59, we have equivalent representations. Therefore, the mapping from classes of systems to classes of representations mentioned in the theorem is well defined.

Conversely, if we are given A and B, the fundamental matricial solutions $\mathcal{X}$, $\mathcal{Y}$, the monodromy matrices $M := \mathcal{X}^{-1}\mathcal{X}^\lambda$ and $N := \mathcal{Y}^{-1}\mathcal{Y}^\lambda$ and a conjugating matrix $P \in \mathrm{GL}_n(\mathbf{C})$ such that $N = PMP^{-1}$, then, setting $F := \mathcal{Y}P\mathcal{X}^{-1}$, we see first that

$$F^\lambda = \mathcal{Y}^\lambda P(\mathcal{X}^\lambda)^{-1} = \mathcal{Y}NPM^{-1}\mathcal{X}^{-1} = \mathcal{Y}P\mathcal{X}^{-1} - F,$$

that is, F is uniform; and since the systems are regular singular, so that $\mathcal{X}$, $\mathcal{Y}$ have moderate growth in sectors, so has F which is therefore a meromorphic gauge transformation relating the two systems. Thus the mapping from classes of systems to classes of representations mentioned in the theorem is injective.

Last, since any $M \in \mathrm{GL}_n(\mathbf{C})$ can be written $e^{2i\pi C}$, we know from the corollary to the previous theorem that the above mapping from classes of systems to classes of representations is surjective. $\qquad\square$

Exercise 9.25. In the proof of injectivity above, rigorously check that the meromorphic gauge transformation F indeed relates the two systems and that F has moderate growth in sectors.

9.5. Moderate growth condition for solutions of an equation

We now look for a condition on the functions $a_i(z)$ ensuring that the scalar equation $E_{\underline{a}}$ has a basis of solutions at some point $a \in \overset{\bullet}{\mathrm{D}}$ having moderate growth in sectors. Of course, in this case, every basis at every point has moderate growth, and all solutions have moderate growth.

Definition 9.26. We say that $E_{\underline{a}}$ *has a regular singular point at* 0, or that $E_{\underline{a}}$ *is regular singular at* 0, if it has a basis of solutions at some point with moderate growth in sectors.

Examples 9.27. (i) If all a_i have a simple pole at 0, then the system with matrix $A_{\underline{a}}$ is of the first kind, hence it is regular singular, and $E_{\underline{a}}$ is too.

(ii) If p has a simple pole and q a double pole, then vectorializing with $\begin{pmatrix} f \\ zf' \end{pmatrix}$ yields a system of the first kind, so that $f'' + pf' + qf = 0$ is regular singular at 0.

The last example suggests that we should use the differential operator $\delta := z\dfrac{d}{dz}$ (which is sometimes called the "Euler differential operator") instead of the differential operator $D := \dfrac{d}{dz}$. Both are "derivations", which means that they are $\mathbf{C}$-linear and satisfy "Leibniz' rule":

$$D(fg) = fD(g) + D(f)g,$$
$$\delta(fg) = f\delta(g) + \delta(f)g.$$

Both operate at will on $\mathbf{C}(z)$, on $\mathbf{C}\{z\}$, on $\mathbf{C}(\{z\})$ and even on $\mathbf{C}[[z]]$ and $\mathbf{C}((z))$. We are going to do some elementary differential algebra with them.

The operator $\delta = zD$ is the composition of operator $D : f \mapsto Df$ and of operator $z : f \mapsto zf$. This composition is not commutative: the operator Dz sends f to $D(zf) = f + zD(f)$, so that one can write $Dz = zD + 1$, where 1 denotes the identity operator $f \mapsto 1.f = f$.

Lemma 9.28. (i) *One has the equality* $z^k D^k = \delta(\delta - 1) \cdots (\delta - k + 1)$.

(ii) *For* $k \geq 2$, *the operator* δ^k *is equal to* $z^k D^k +$ *a linear combination of* $zD, \ldots, z^{k-1} D^{k-1}$.

Proof. (i) can be proved easily by induction, but it is simpler to look at the effect of both sides acting on z^m, $m \in \mathbf{Z}$. Clearly, $\delta(z^m) = mz^m$, so that, for any polynomial $P \in \mathbf{C}[X]$, one has $P(\delta)(z^m) = P(m)z^m$ (these are the classical rules about polynomials in endomorphisms and eigenvectors). On the other hand, we know that $D^k(z^m) = m(m-1) \cdots (m - k + 1)z^{m-k}$, so that $z^k D^k(z^m) = m(m-1) \cdots (m - k + 1)z^m = \delta(\delta - 1) \cdots (\delta - k + 1)z^m$.

(ii) We now have a triangular system of relations that we can solve recursively:

$$\begin{cases} zD = \delta, \\ z^2 D^2 = \delta^2 - \delta, \\ z^3 D^3 = \delta^3 - 3\delta^2 + 2\delta, \\ \cdots\cdots\cdots\cdots\cdots \end{cases} \implies \begin{cases} \delta = zD, \\ \delta^2 = z^2 D^2 + zD, \\ \delta^3 = z^3 D^3 + 3z^2 D^2 + zD, \\ \cdots\cdots\cdots\cdots\cdots\cdots \end{cases}$$

$\square$

Using these relations, we can transform a differential equation $f^{(n)} + b_1 f^{(n-1)} + \cdots + b_n f = 0$ into one involving δ in the following way: rewrite the equation in the more symbolic form $(D^n + b_1 D^{n-1} + \cdots + b_n)f = 0$; then multiply by z^n on the left and replace $z^n D^i$ by $z^{n-i}(z^i D^i) = z^{n-i}\delta \cdots (\delta - i + 1)$. The process can be reversed.

Proposition 9.29. *Assume* $z^n(D^n + b_1 D^{n-1} + \cdots + b_n) = \delta^n + a_1 \delta^{n-1} + \cdots + a_n$. *Then:*

$$\left(v_0(b_1) \geq -1, \ldots, v_0(b_n) \geq -n\right) \iff a_1, \ldots, a_n \in \mathbf{C}\{z\}.$$

Proof. Call T the triangular matrix such that $(1, zD, \ldots, z^n D^n) = (1, \delta, \ldots, \delta^n)T$. According to the lemma above, its coefficients are in $\mathbf{C}$ and its diagonal coefficients are all equal to 1. Then, setting $a_0, b_0 := 1$,

$$\begin{pmatrix} a_n \\ \vdots \\ a_0 \end{pmatrix} = T \begin{pmatrix} b_n z^n \\ \vdots \\ b_0 \end{pmatrix},$$

which shows that the a_i are linear combinations with constant coefficients of the $b_i z^i$, and conversely. $\square$

The following criterion shows that, contrary to the case of systems, it is very easy to check if all the solutions of an equation have moderate growth at 0. This is a justification for the utility of cyclic vectors.

Theorem 9.30 (Fuchs criterion). *The equivalent equations $f^{(n)} + b_1 f^{(n-1)} + \cdots + b_n f = 0$ and $\delta^n f + a_1 \delta^{n-1} f + \cdots + a_n f = 0$ are regular singular at 0 if, and only if, $v_0(b_i) \geq -i$ for $i = 1, \ldots, n$; or equivalently $a_1, \ldots, a_n \in \mathbf{C}\{z\}$.*

In essence, the theorem says that, for scalar equations, a regular singular point is the same thing as a *singularity of the first kind* or a *fuchsian singularity*.

Proof. It follows from the proposition above that the two criteria stated are indeed equivalent. Suppose they are verified. Then the system obtained by the vectorialization $X := \begin{pmatrix} f \\ \delta f \\ \vdots \\ \delta^{n-1} f \end{pmatrix}$ has the matrix $z^{-1} A_{\underline{a}}$, so it is of the first kind and thus regular singular; so the equation is also regular singular. Conversely assume that the equation is regular singular. If $n = 1$, the equation can be seen as a system $f' = -b_1 f$ of rank 1 and we apply the criterion for systems: there exists a meromorphic nonzero u such that $u[-b_1] = -b_1 + u'/u$ has a simple pole at 0. But u'/u itself has a simple pole at 0, so b_1 also has, which is the desired conclusion. The (nontrivial) proof for the general case $n \geq 2$ will be by induction and it will require several steps.

Step 1. So we suppose that $n \geq 2$, that $a_1, \ldots, a_n \in \mathbf{C}(\{z\})$ and that all the solutions of the equation $\delta^n f + a_1 \delta^{n-1} f + \cdots + a_n f = 0$ have moderate growth.

Lemma 9.31. *The equation has at least one solution of the form $f = u z^{\alpha}$, where $\alpha \in \mathbf{C}$ and $u = 1 + u_1 z + \cdots \in \mathbf{C}\{z\}$.*

Proof. Let $(f_1, \ldots, f_n)$ be a fundamental system of solutions at some point and let $M \in \mathrm{GL}_n(\mathbf{C})$ be the matrix of its monodromy along the fundamental loop λ, i.e., $(f_1^{\lambda}, \ldots, f_n^{\lambda}) = (f_1, \ldots, f_n)M$. We triangularize M: there is $P \in \mathrm{GL}_n(\mathbf{C})$ such that $M = PTP^{-1}$ and T is upper triangular; call β its first diagonal coefficient. Since M is invertible, so is T and $\beta \neq 0$, so that we can write $\beta = e^{2i\pi\alpha}$ for some $\alpha \in \mathbf{C}$. Then $(g_1, \ldots, g_n) := (f_1, \ldots, f_n)P$ is a fundamental system of solutions and

$$(g_1^{\lambda}, \ldots, g_n^{\lambda}) = (f_1^{\lambda}, \ldots, f_n^{\lambda})P = (f_1, \ldots, f_n)MP$$
$$= (g_1, \ldots, g_n)P^{-1}MP = (g_1, \ldots, g_n)T,$$

which implies in particular that $g_1^{\lambda} = \beta g_1$. Thus, $g_1 z^{-\alpha}$ is uniform and has moderate growth (by the assumption of regularity), so it is a meromorphic function $cz^m u$ with u as indicated in the statement of the lemma. Then, changing α to $\alpha + m$, we find that g_1 is a solution of the form required. $\square$

Step 2. We thus have a particular solution of the form $f_0 = uz^\alpha$ as above. We now look for the equation the solutions of which are the f/z^α, where f is a solution of $\delta^n f + a_1 \delta^{n-1} f + \cdots + a_n f = 0$. This is a change of unknown function. So we put $f = z^\alpha g$, and, noticing that $\delta.z^\alpha = z^\alpha.(\delta + \alpha)$, we obtain:

$$\delta^n(z^\alpha g) + a_1 \delta^{n-1}(z^\alpha g) + \cdots + a_n(z^\alpha g)$$
$$= z^\alpha \big((\delta + \alpha)^n g + a_1 (\delta + \alpha)^{n-1} g + \cdots + a_n g\big),$$

whence a new equation $\delta^n g + b_1 \delta^{n-1} g + \cdots + b_n g = 0$, where the a_i, b_i are related by the formula:

$$X^n + b_1 X^{n-1} + \cdots + b_n = (X + \alpha)^n + a_1(X + \alpha)^{n-1} + \cdots + a_n.$$

In particular, the a_i are holomorphic if, and only if, the b_i are. And of course, since $f = z^\alpha g$ has moderate growth if, and only if, g has moderate growth, we see that the equation $\delta^n f + a_1 \delta^{n-1} f + \cdots + a_n f = 0$ is regular singular if, and only if, the equation $\delta^n g + b_1 \delta^{n-1} g + \cdots + b_n g = 0$ is. Therefore, we are led to prove the theorem for the latter equation. But we know that this one has a particular solution $u = 1 + u_1 z + \cdots \in \mathbf{C}\{z\}$.

Step 3. We are now going to operate a euclidean division of polynomials, but in a noncommutative setting!

Lemma 9.32. *Let $v \in \mathbf{C}(\{z\})$. Then every differential operator $\delta^m + p_1 \delta^{m-1} + \cdots + p_m$ with $p_1, \ldots, p_m \in \mathbf{C}(\{z\})$ can be written in the form $(\delta^{m-1} + q_1 \delta^{m-2} + \cdots + q_{m-1})(\delta - v) + w$, where $w, q_1, \ldots, q_{m-1} \in \mathbf{C}(\{z\})$.*

Proof. We note first that $\delta^{i-1} v = \delta^i - \delta^{i-1}(\delta - v)$ is a sum of terms $r_j \delta^j$, $j = 0, \ldots, i-1$, with all $r_j \in \mathbf{C}(\{z\})$. From this we deduce by induction that the theorem is true for each differential operator δ^i. Then we get the conclusion by linear combination. $\qquad\square$

Now we apply the lemma with $v := \delta(u)/u$ and get:

$$\delta^n + b_1 \delta^{n-1} + \cdots + b_n = (\delta^{n-1} + c_1 \delta^{n-2} + \cdots + c_{n-1})(\delta - v) + w.$$

Applying this to u and noting that $(\delta - v)u = \delta(u) - vu = 0$, we get $wu = 0$ so that $w = 0$. Therefore, we have the equality:

$$\delta^n + b_1 \delta^{n-1} + \cdots + b_n = (\delta^{n-1} + c_1 \delta^{n-2} + \cdots + c_{n-1})(\delta - v).$$

Step 4. Now we will apply the induction hypothesis to the new operator. We note that, if $g_1, \ldots, g_n$ are a basis of solutions of the equation $\delta^n g + b_1 \delta^{n-1} g + \cdots + b_n g = 0$, then the $h_i := (\delta - v)g_i$ are solutions of $\delta^{n-1} h + c_1 \delta^{n-2} h + \cdots + c_{n-1} h = 0$, and, of course, they have moderate growth (since the g_i and v have). We choose the basis such that $g_n = u$. Then $h_1, \ldots, h_{n-1}$ are linearly independent. Indeed, if $\lambda_1 h_1 + \cdots + \lambda_{n-1} h_{n-1} = 0$, then the function $g := \lambda_1 g_1 + \cdots + \lambda_{n-1} g_{n-1}$ satisfies $\delta(g) = vg$; but since $v = \delta(u)/u$, this implies $\delta(g/u) = 0$, whence $\lambda_1 g_1 + \cdots + \lambda_{n-1} g_{n-1} = \lambda g_n$, which is only possible if all $\lambda_i = 0$. Therefore, $h_1, \ldots, h_{n-1}$ are a basis of solutions of $\delta^{n-1} h + c_1 \delta^{n-2} h + \cdots + c_{n-1} h = 0$ and, since they have moderate growth, this is a regular singular equation. By the inductive hypothesis of the theorem, all $c_i \in \mathbf{C}\{z\}$. But then, from $\delta^n + b_1 \delta^{n-1} + \cdots + b_n = (\delta^{n-1} + c_1 \delta^{n-2} + \cdots + c_{n-1})(\delta - v)$ and the fact that $v \in \mathbf{C}\{z\}$, we conclude that all $b_i \in \mathbf{C}\{z\}$, which ends the induction step. $\square$

9.6. Resolution and monodromy of regular singular equations

We shall only present the basic cases and examples. The general case is rather complicated because of resonancies, which anyway are exceptional. For the complete algorithms (mostly due to Fuchs and Frobenius), see the references given at the beginning of the chapter.

We start with the equation $\delta^n f + a_1 \delta^{n-1} f + \cdots + a_n f = 0$, which we assume to be regular singular at 0, $i.e.$, $a_1, \ldots, a_n \in \mathbf{C}\{z\}$. As we saw in the proof of Theorem 9.30, there certainly is a solution of the form $f = z^\alpha u$, $\alpha \in \mathbf{C}$, $u = 1 + u_1 z + \cdots \in \mathbf{C}\{z\}$. Actually, it even follows from the argument given there that, if the monodromy is semi-simple, there is a whole basis of solutions of this form. From the already proven relation $\delta z^\alpha = z^\alpha(\delta + \alpha)$, we obtain the equation:

$$(\delta + \alpha)^n u + a_1(\delta + \alpha)^{n-1} u + \cdots + a_n u = \delta^n u + b_1 \delta^{n-1} u + \cdots + b_n u = 0,$$

where, setting $P(X) := X^n + a_1 X^{n-1} + \cdots + a_n$ and $Q(X) := X^n + b_1 X^{n-1} + \cdots + b_n$, we have $Q(X) = P(X + \alpha)$. Now in the equation $Q(\delta)(u) = 0$, all terms are multiples of z and thus vanish at 0, except maybe $b_n u$, so that $b_n(0)u(0) = 0$. But $u(0) = 1$, so that $b_n(0) = 0$, whence the $indicial$ $equation$:

$$\alpha^n + a_1(0)\alpha^{n-1} + \cdots + a_n(0) = 0.$$

This is a $necessary$ condition for α to be a possible exponent. It can be proved that, if α is a nonresonant root of this equation, $i.e.$, no $\alpha + k$, $k \in \mathbf{N}^*$, is a root, then the condition is also sufficient. We only prove a slightly weaker result.

Theorem 9.33. *If the indicial equation has n distinct and nonresonant roots $\alpha_1, \ldots, \alpha_n$ (so that $\alpha_i - \alpha_j \notin \mathbf{Z}$ for $i \neq j$), then there is a fundamental system of solutions of the form $f_i = z_i^\alpha u_i$, $u_i \in \mathbf{C}\{z\}$, $u_i(0) = 1$.*

Proof. The matrix of the corresponding system is $A := A_{\underline{a}}$ and the characteristic polynomial of the companion matrix $A(0)$ is the one defining the indicial equation: therefore, by assumption, $A(0)$ is nonresonant in the sense of Section 9.4. So the system has a fundamental matricial solution $Fz^{A(0)}$, where F is a Birkhoff matrix. Since $A(0)$ is semi-simple, we can write $A(0) = P\mathrm{Diag}(\alpha_1, \ldots, \alpha_n)P^{-1}$ and conclude that $FP\mathrm{Diag}(z^{\alpha_1}, \ldots, z^{\alpha_n})$ is another fundamental matricial solution. Its first line gives the fundamental system of solutions of the required form. □

From here on, we only study examples!

Example 9.34. We consider $\delta^2 f - \delta f = 0$. The indicial equation is $\alpha^2 - \alpha = 0$, whence two roots 0 and 1. In this case both roots yield solutions, because 1 and z are indeed solutions (and a fundamental system of solutions since their wronskian is 1).

Actually, the system obtained by vectorialization has constant matrix $A := \begin{pmatrix} 0 & 1 \\ 0 & 1 \end{pmatrix}$ and $A^2 = A$, so that we easily compute $z^A = \begin{pmatrix} 1 & z-1 \\ 0 & z \end{pmatrix}$, whence the fundamental system of solutions $(1, z-1)$.

Example 9.35. We consider $\delta^2 f - \delta f - zf = 0$. The indicial equation is again $\alpha^2 - \alpha = 0$, whence two roots 0 and 1. In this case there is a problem. Indeed, looking for a power series solution $\sum_{n \geq 0} f_n z^n$, we find the equivalent relations $(n^2 - n)f_n - f_{n-1} = 0$. This implies that $f_0 = 0$, f_1 is free and can be taken equal to 1, and the other coefficients are recursively computed: $f_n = \dfrac{1}{n!(n-1)!}$ for $n \geq 1$. This has the required form for $\alpha = 1$, a nonresonant root, but the resonant root 0 gave nothing.

To see why, we vectorialize the equation: here, $A := \begin{pmatrix} 0 & 1 \\ -z & 1 \end{pmatrix}$. But $A(0)$ has roots 0 and 1, it is resonant and we must transform it before solving (since it is not constant, computing z^A would not make any sense). The algorithm was explained in Section 9.4. The shearing transform $\begin{pmatrix} 1 & 0 \\ 0 & z \end{pmatrix}$ gives $B := F[A] = \begin{pmatrix} 0 & z \\ -1 & 0 \end{pmatrix}$. But now, $B(0)$ is nonresonant but nilpotent and $z^{B(0)}$ contains a logarithm, so the fundamental matricial solutions of B and A also do.

Example 9.36. Consider the divergent series

$$f := \sum_{n \geq 1} (n-1)! z^n$$

(Euler series[4]). It satisfies the nonhomogeneous first-order equation $f = z + z\delta f$, that is, $(1 - z\delta)f = z$. Since $z = \delta z$, the series f is a solution of

---

[4]For a very interesting account of the importance of this series, see the monograph [**Ram93**].

$(1-\delta)(1-z\delta)f = 0$, that is, $z\delta^2 f - \delta f + f = 0$, or $\delta^2 f - (1/z)\delta f + (1/z)f = 0$, an *irregular* equation.

Example 9.37. The Bessel equation is $z^2 f'' + z f' + (z^2 - \alpha^2)f$. It is regular (*i.e.*, it has only ordinary points) in $\mathbf{C}^*$ and it has a regular singular point at 0. With the Euler differential operator, the equation becomes $\delta^2 f + (z^2 - \alpha^2)f = 0$. The indicial equation is (using by necessity the letter x for the unknown) $x^2 - \alpha^2 = 0$, which has roots $\pm\alpha$. We assume that $2\alpha \notin \mathbf{Z}$, so that both exponents give rise to a solution. For instance, putting $f = z^\alpha g$ gives rise to the equation $(\delta+\alpha)^2 g + (z^2 - \alpha^2)g = 0$, *i.e.*, $\delta^2 g + 2\alpha\delta g + z^2 g = 0$. We look for g in the form $g = g_0 + g_1 z + \cdots$, with $g_0 = 1$. This gives for all $n \geq 0$ the relation $(n^2 + 2\alpha n)g_n + g_{n-2} = 0$ (with the natural convention that $g_{-1} = g_{-2} = 0$) so that g_1, and then all g_n with odd index n, are 0. For even indexes, setting $h_n := g_{2n}$ we find $h_0 = 1$ and $h_n = \dfrac{-1}{4n(n+\alpha)}h_{n-1}$ for $n \geq 1$, whence $h_n = \dfrac{(-1/4)^n}{n!(\alpha+1)\cdots(\alpha+n)}$. In the end, we get the solution:

$$f(z) = z^\alpha \sum_{n\geq 0} \frac{(-1/4)^n}{n!(\alpha+1)\cdots(\alpha+n)}z^{2n}.$$

It is customary to consider a constant multiple of this solution, the *Bessel function*:

$$J_\alpha(z) := (z/2)^\alpha \sum_{n\geq 0} \frac{(-1)^n}{n!\Gamma(n+\alpha+1)}(z/2)^{2n}.$$

Here, the Gamma function[5] Γ is an analytic function on $\mathbf{C} \setminus (-\mathbf{N})$, which satisfies the functional equation $\Gamma(z+1) = z\Gamma(z)$, from which one obtains immediately that $\Gamma(n+\alpha+1) = (\alpha+1)\cdots(\alpha+n)\Gamma(\alpha+1)$ and then that $f(z) = 2^\alpha \Gamma(\alpha+1)J_\alpha(z)$.

Example 9.38. The Airy function is defined, for real x, by $Ai(x) := \dfrac{1}{\pi}\int_0^{+\infty} \cos(t^3/3 + xt)\,dt$. It is not difficult to prove that it is well defined and satisfies the differential equation $Ai''(x) = x\,Ai(x)$. So we decide to study the complex differential equation $f'' = xf$. It is regular on $\mathbf{C}$, from which one deduces that it has a fundamental system of uniform solutions in $\mathbf{C}$, so that the Airy function can be extended to an entire function. To study it at infinity, we set $w := 1/z$ and $g(w) := f(z) = f(1/w)$, so that $g'(w) = -w^{-2}f'(1/w)$ and $g''(w) = w^{-4}f''(1/w) + 2w^{-3}f'(1/w)$. We end with the differential equation $g'' + 2w^{-1}g' - w^{-5}g = 0$, which is irregular at $w = 0$. Actually, the asymptotic behavior of the Airy function at infinity was the origin of the discovery by Stokes of the so-called "Stokes phenomenon" (see the book of Ramis already quoted [**Ram93**]).

---

[5]More will be said on the Gamma function in the next chapter.

Example 9.39. Define the *Pochhammer symbols* (using the Gamma function introduced above) by:

$$(\alpha)_n = \alpha(\alpha + 1) \cdots (\alpha + n - 1) = \frac{\Gamma(\alpha + n)}{\Gamma(\alpha)}.$$

For instance, $(1)_n = n!$. The *hypergeometric series of Euler-Gauss* is defined as:

$$F(\alpha, \beta, \gamma; z) := \sum_{n \geq 0} \frac{(\alpha)_n (\beta)_n}{n!(\gamma)_n} z^n.$$

We consider α, β, γ as parameters and z as the variable, so we will write $F(z)$ for short here instead of $F(\alpha, \beta, \gamma; z)$. We must assume that $\gamma \notin -\mathbf{N}$ for this series to be defined. We also assume that $\alpha, \beta \notin -\mathbf{N}$, so that it is not a polynomial. Then the radius of convergence is 1. The coefficients $f_n := \frac{(\alpha)_n (\beta)_n}{n!(\gamma)_n}$ satisfy the recursive relation $(n+1)(n+\gamma)f_{n+1} = (n+\alpha)(n+\beta)f_n$. We multiply both sides by z^{n+1} and sum for all $n \geq 0$. Recalling that $\delta(\sum f_n z^n) = \sum n f_n z^n$ and $\delta^2(\sum f_n z^n) = \sum n^2 f_n z^n$, we obtain the equality:

$$\delta(\delta + \gamma - 1)F = z(\delta + \alpha)(\delta + \beta)F,$$

from which follows the differential equation:

$$(1 - z)\delta^2 F + (\gamma - 1 - (\alpha + \beta)z)\delta F - \alpha\beta z F = 0.$$

This equation is regular singular at 0. The indicial equation is $x^2 + (\gamma - 1)x = 0$. If we assume that $\gamma \notin \mathbf{Z}$, we find that there is a fundamental basis made up of a power series with constant term 1 (this is $F(z)$) and of a solution $z^{1-\gamma}G(z)$, where G is a power series with constant term 1. The study of the hypergeometric equation will be continued in the next chapter.

Exercise 9.40. Show that G is itself a hypergeometric series with different parameters.

Exercises

(1) Prove that $1/\log$ is not the solution of a linear differential equation with polynomial coefficients.

(2) Are the equations $zf' - f = 0$, $zf' + f = 0$ and $z^2 f' + f = 0$ meromorphically equivalent at 0?

(3) Prove that $e^{1/z}$ does not have moderate growth in sectors.

(4) Show that for all $k, l \in \mathbf{N}$, $D^k z^l$ is a linear combination of operators $z^i D^j$.

(5) Compute the triangular matrix T such that $(1, zD, z^2D^2, z^3D^3, z^4D^4) = (1, \delta, \delta^2, \delta^3, \delta^4)T$.

(6) At the end of Step 2 of the proof of Fuchs criterion (Theorem 9.30), is $f = z^\alpha g$ a gauge transformation in the sense of Section 7.6?

(7) What can be said about the Bessel equation when $2\alpha \in \mathbf{Z}$? In all cases, how does it behave at infinity?

Chapter 10

Local Riemann-Hilbert correspondence as an equivalence of categories

Many problems which were initially formulated as classification problems, *i.e.*, finding a one-to-one correspondence between two types of objects (or between sets of equivalence classes of interesting objects) were translated in the second half of the twentieth century into the language of categories. One significant while very elementary example is linear algebra. Mathematicians of the nineteenth century were usually content to find that sets admitting linear operations (in geometry, analysis, etc.) depended on a certain number of constants: this is what we call nowadays the *dimension* of a linear space, indeed the one and only isomorphism invariant. However nowadays we are interested in linear maps that are not injective (and want to understand their kernel) or not surjective (and want to understand their cokernel). Linear algebra is more interesting with all linear maps and not only isomorphisms.

So we are going to formulate the Riemann-Hilbert correspondence as an equivalence of categories; remember that this concept was defined in Section 8.3.

10.1. The category of singular regular differential systems at 0

We shall now define a category $\mathcal{E}_f^{(0)}$. First we define the bigger category $\mathcal{E}^{(0)}$. We need $\mathcal{M}_0 = \mathbf{C}(\{z\})$, the field of meromorphic germs at 0, *i.e.*, the quotient field of $\mathbf{C}\{z\}$. It is equal to the ring $\mathbf{C}\{z\}[1/z]$ of convergent Laurent series (not generalized, only a finite number of terms with negative exponents is allowed).

The objects of $\mathcal{E}^{(0)}$ are differential systems $S_A : X' = AX$, where $A \in \mathrm{Mat}_n(\mathcal{M}_0)$; the superscript $-^{(0)}$ stands for "local at 0". If $A \in \mathrm{Mat}_n(\mathcal{M}_0)$ and $B \in \mathrm{Mat}_p(\mathcal{M}_0)$, morphisms from A to B are matrices $F \in \mathrm{Mat}_{p,n}(\mathcal{M}_0)$ such that $F' = BF - FA$. As usual, this just means that $X' = AX$ implies $Y' = BY$, where $Y := FX$.

Now we define the category $\mathcal{E}_f^{(0)}$ as the full subcategory of $\mathcal{E}^{(0)}$ whose objects are the systems S_A such that 0 is a regular singular point; the subscript $-_f$ stands for "fuchsian" (recall this word was defined in Chapter 9).

For every $R > 0$ we set $\overset{\bullet}{\mathrm{D}}_R := \overset{\circ}{\mathrm{D}}(0, R) \setminus \{0\}$, the punctured open disk; thus, $\overset{\bullet}{\mathrm{D}}_\infty = \mathbf{C}^*$. We define $\mathcal{E}_{f,R}^{(0)}$ as the full subcategory of $\mathcal{E}_f^{(0)}$ whose objects are the systems S_A such that A is meromorphic at 0 and holomorphic over $\overset{\bullet}{\mathrm{D}}_R$, *i.e.*, $A \in \mathrm{Mat}_n(\mathcal{M}_0 \cap \mathcal{O}(\overset{\bullet}{\mathrm{D}}_R))$. We note that, if $F : S_A \to S_B$ is a morphism in $\mathcal{E}_{f,R}^{(0)}$, then F itself has coefficients in $\mathcal{M}_0 \cap \mathcal{O}(\overset{\bullet}{\mathrm{D}}_R)$. Indeed, letting $a \in \overset{\bullet}{\mathrm{D}}_R$ and $\mathcal{X}, \mathcal{Y}$ be fundamental matricial solutions of S_A, S_B at a, we must have $F\mathcal{X} = \mathcal{Y}C$ with $C \in \mathrm{Mat}_{p,n}(\mathbf{C})$, so $F = \mathcal{Y}C\mathcal{X}^{-1}$ is holomorphic at a and extends to a multivalued function on the whole of $\overset{\bullet}{\mathrm{D}}_R$ by Chapter 7; but we know in advance it is uniform.

Clearly, $\mathcal{E}_f^{(0)} = \bigcup_{R>0} \mathcal{E}_{f,R}^{(0)}$ (ascending family while R decreases to 0). It is besides more correct to say that the category $\mathcal{E}_f^{(0)}$, which is not a set, is the "direct limit" of the categories $\mathcal{E}_{f,R}^{(0)}$ (see [**Mac98**] for details about this concept which will not be needed in this course[1]).

Lemma 10.1. *Let $R > 0$. Then $\mathcal{E}_{f,R}^{(0)} \to \mathcal{E}_f^{(0)}$ is an equivalence of categories.*

---

[1] The dual concept of "inverse limit" is explored in Section 16.3. Note however that germs, as defined in exercise 1 to Chapter 2 (or, more generally, in Section 7.4), are an example of direct limit.

Proof. Since $\mathcal{E}_{f,R}^{(0)}$ is defined as a full subcategory of $\mathcal{E}_f^{(0)}$, this amounts to proving that it is *essential*, *i.e.*, that every object of $\mathcal{E}_f^{(0)}$ is isomorphic (in $\mathcal{E}_f^{(0)}$) to an object of $\mathcal{E}_{f,R}^{(0)}$. Indeed, the proof of Theorem 9.18 shows that S_A is equivalent to a system S_B with $B = z^{-1}C$ and $C \in \mathrm{Mat}_n(\mathbf{C})$. $\qquad\square$

We want to build a category equivalence from $\mathcal{E}_f^{(0)}$ to $\mathfrak{Rep}_{\mathbf{C}}^f(\pi_1^\circ)$ for some realization of the fundamental group π_1 to be made precise (categories $\mathfrak{Rep}_{\mathbf{C}}^f(G)$ were introduced in Definition 8.9). We shall be able to build an explicit category equivalence from $\mathcal{E}_{f,\infty}^{(0)}$ to $\mathfrak{Rep}_{\mathbf{C}}^f(\pi_1^\circ)$ (Section 10.3) and we shall want to compose it with an equivalence from $\mathcal{E}_f^{(0)}$ to $\mathcal{E}_{f,\infty}^{(0)}$. To do this, we must in some sense invert the equivalence from $\mathcal{E}_{f,\infty}^{(0)}$ to $\mathcal{E}_f^{(0)}$ of the lemma. The (very formal) process will now be explained (in Section 10.2). The reader might notice the strong relationship of what follows with the exercises in Section 8.4.

10.2. About equivalences and isomorphisms of categories

Admitting the existence of a category $\mathcal{C}at$ of all categories, the objects, resp. morphisms, of which are categories, resp. functors, it is clear that an isomorphism in $\mathcal{C}at$ is a functor F which is bijective on objects and fully faithful. This is exactly equivalent to the existence of an inverse functor (which would be the precise translation of the isomorphy condition).

However, looking at the arguments presented in Section 8.3, you can guess why this never happens in practice. The best we can get instead is an equivalence of categories, which is something weaker than an isomorphism. It is obvious that the identity functor is an equivalence of categories and is rather easier to check than composing equivalences of categories yields an equivalence of categories. What about inverting?

Let $F : \mathcal{C} \to \mathcal{D}$ be an equivalence of categories. For each object Y in $\mathcal{D}$, arbitrarily choose an object X in $\mathcal{C}$ and an isomorphism $u_Y : F(X) \to Y$. This is in principle possible from the fact that F is essentially surjective. However, this seems to use some very general axiom of choice, but we shall not delve into this aspect here. We are going to make the association $Y \rightsquigarrow X$ into a functor. So set $G(Y) := X$ (the particular object chosen above). Note however that, in order to define G as a functor, we shall need not only the choice of X but also the choice of u_Y.

Now let Y' be another object in $\mathcal{D}$, $X' = G(Y')$ the selected object of $\mathcal{C}$ and $u_{Y'} : F(X') \to Y'$ the corresponding isomorphism in $\mathcal{D}$. For every morphism $g : Y \to Y'$ in $\mathcal{D}$, we have a morphism $u_{Y'}^{-1} \circ g \circ u_Y : F(X) \to F(X')$ in $\mathcal{D}$. Thus, since F is fully faithful, we have a well-defined morphism $F^{-1}\left(u_{Y'}^{-1} \circ g \circ u_Y\right) : X \to X'$ in $\mathcal{C}$. We call it $G(g)$; it is a morphism $X \to X'$, i.e., $G(Y) \to G(Y')$. It is characterized by $F(G(g)) = u_{Y'}^{-1} \circ g \circ u_Y$.

Clearly, if $Y' = Y$ (so that $X' = G(Y') = G(Y) = X$) and $g = \mathrm{Id}_Y$, we find that $G(g) = \mathrm{Id}_X$. On the other hand, if $g : Y \to Y'$ is arbitrary and if moreover we have $g' : Y' \to Y''$ and $X'' := G(Y'')$, then

$$F\left(G(g') \circ G(g)\right) = F(G(g')) \circ F(G(g)) = u_{Y''}^{-1} \circ g' \circ u_{Y'} \circ u_{Y'}^{-1} \circ g \circ u_Y$$
$$= u_{Y''}^{-1} \circ (g' \circ \circ g) \circ u_Y = F\left(G(g' \circ g)\right),$$

whence $G(g') \circ G(g) = G(g' \circ g)$ by faithfulness of F. So we know that G is a functor $\mathcal{D} \to \mathcal{C}$.

To express the relation of F with G, we have to understand in which sense $G \circ F$ and $F \circ G$ are respectively some kinds of approximations of the identity functors $\mathrm{Id}_\mathcal{C}$ and $\mathrm{Id}_\mathcal{D}$. We are going to exhibit:

(1) A family $(\phi_X)_{X \in \mathrm{Ob}(\mathcal{C})}$ of morphisms $\phi_X : G(F(X)) \to X$, such that for every morphism $f : X \to X'$, one has a commutative diagram:

$$
\begin{array}{ccc}
G \circ F(X) & \xrightarrow{\ \phi_X\ } & X \\
{\scriptstyle G \circ F(f)} \downarrow & & \downarrow {\scriptstyle f} \\
G \circ F(X') & \xrightarrow[\ \phi_{X'}\ ]{} & X'
\end{array}
$$

This is called a *natural transformation* from the functor $G \circ F$ to the identity functor $\mathrm{Id}_\mathcal{C}$. (Actually, it is a morphism in the category of functors from $\mathcal{C}$ to itself, an endofunctor of that category.)

(2) A family $(\psi_Y)_{Y \in \mathrm{Ob}(\mathcal{D})}$ of morphisms $\psi_Y : F(G(Y)) \to Y$, such that for every morphism $g : Y \to Y'$, one has a commutative diagram:

$$
\begin{array}{ccc}
F \circ G(Y) & \xrightarrow{\ \psi_Y\ } & Y \\
{\scriptstyle F \circ G(g)} \downarrow & & \downarrow {\scriptstyle g} \\
F \circ G(Y') & \xrightarrow[\ \psi_{Y'}\ ]{} & Y'
\end{array}
$$

This is again a natural transformation, from the functor $F \circ G$ to the identity functor $\mathrm{Id}_\mathcal{D}$.

We leave it to the reader to check that, setting $\psi_Y := u_Y$ and $\phi_X := F^{-1}(u_{F(X)})$, we indeed meet these constraints. Moreover, all the ϕ_X are isomorphisms in $\mathcal{C}$ and all the ψ_Y are isomorphisms in $\mathcal{D}$, so that our natural transformations are actually isomorphisms $G \circ F \simeq \mathrm{Id}_{\mathcal{C}}$ and $F \circ G \simeq \mathrm{Id}_{\mathcal{D}}$. We then say that F and G are *quasi-inverses* of each other.

We have proved that every equivalence of categories $F : \mathcal{C} \to \mathcal{D}$ admits a quasi-inverse $G : \mathcal{D} \to \mathcal{C}$. Conversely, every functor admitting a quasi-inverse is an equivalence of categories (see the exercises at the end of the chapter), so G itself is then an equivalence of categories from $\mathcal{D}$ to $\mathcal{C}$.

10.3. Equivalence with the category of representations of the local fundamental group

We now define the functor from $\mathcal{E}_{f,\infty}^{(0)}$ to $\mathfrak{Rep}_{\mathbf{C}}^f(\pi_1^\circ)$. To a system $S_A : X' = AX$, where $A \in \mathrm{Mat}_n(\mathcal{M}_0 \cap \mathcal{O}(\overset{\bullet}{\mathrm{D}}_\infty))$, writing $\mathcal{F}_{A,a}$ as the $\mathbf{C}$-vector space of its germs of solutions at a, we associate the monodromy representation $\rho_A : \pi_1(\mathbf{C}^*, a)^\circ \to \mathrm{GL}(\mathcal{F}_{A,a})$. Recall that if $[\lambda] \in \pi_1(\mathbf{C}^*, a)$ is the homotopy class of a loop λ in $\mathbf{C}^*$ based at a, then $\rho_A([\lambda])$ is the analytic continuation automorphism $X \mapsto X^\lambda$ of $\mathcal{F}_{A,a}$. Our functor also sends a morphism $F : S_A \to S_B$ to the linear map $X \mapsto FX$ from $\mathcal{F}_{A,a}$ to $\mathcal{F}_{B,a}$, and this linear map is indeed a morphism from ρ_A to ρ_B (it intertwines the representations).

Theorem 10.2. *The functor thus defined is an equivalence of categories from* $\mathcal{E}_{f,\infty}^{(0)}$ *to* $\mathfrak{Rep}_{\mathbf{C}}^f(\pi_1(\mathbf{C}^*, a)^\circ)$.

Proof. We first note (as we should have long ago!) that in each of our two categories $\mathcal{C} := \mathcal{E}_{f,\infty}^{(0)}$, $\mathcal{C} := \mathfrak{Rep}_{\mathbf{C}}^f(\pi_1(\mathbf{C}^*, a)^\circ)$, all the sets $\mathrm{Mor}_{\mathcal{C}}(X, Y)$ are complex vector spaces, and that the functor maps $\mathrm{Mor}_{\mathcal{E}_{f,\infty}^{(0)}}(S_A, S_B) \to \mathrm{Mor}_{\mathfrak{Rep}_{\mathbf{C}}^f(\pi_1^\circ)}(\rho_A, \rho_B)$ are $\mathbf{C}$-linear. Actually, we have "$\mathbf{C}$-linear abelian categories" and a "$\mathbf{C}$-linear additive functor"; see Appendix D. So to prove faithfulness it is enough to look at the kernels of these functor maps. If $F : S_A \to S_B$ is such that the linear map $X \mapsto FX$ is trivial, taking for X all the columns of a fundamental matricial solution of S_A at a immediately shows that $F = 0$: the functor is indeed faithful.

For fullness, consider $\phi : \mathcal{F}_{A,a} \to \mathcal{F}_{B,a}$, a $\mathbf{C}$-linear map that intertwines ρ_A and ρ_B. Let $\mathcal{X}$ be a fundamental matricial solution of S_A at a and $F := \phi(\mathcal{X})\mathcal{X}^{-1}$ (we compute $\phi(\mathcal{X})$ columnwise). Thus $F \in \mathrm{Mat}_{p,n}(\mathcal{O}_a)$. Analytic continuation of the equality $\phi(\mathcal{X}) = F\mathcal{X}$ along a loop λ in $\mathbf{C}^*$

based at a yields:

$$\rho_B([\lambda]) \circ \phi(\mathcal{X}) = F^\lambda \rho_A([\lambda])(\mathcal{X})$$

$$\Longrightarrow \phi \circ \rho_A([\lambda])(\mathcal{X}) = F^\lambda \rho_A([\lambda])(\mathcal{X}) \Longrightarrow \phi = F^\lambda,$$

so that F is uniform on $\mathbf{C}^*$. We used the equalities $\rho_A([\lambda])(\mathcal{X}) = \mathcal{X} M_\lambda$, $M_\lambda \in \mathrm{GL}_n(\mathbf{C})$ and $\phi(\mathcal{X} M_\lambda) = F^\lambda \mathcal{X} M_\lambda$.

From the fact that S_A and S_B are regular singular at 0, one obtains that $\mathcal{X}$ and $\phi(\mathcal{X})$ have moderate growth at 0, and so has $F := \phi(\mathcal{X})\mathcal{X}^{-1}$. Therefore F is meromorphic at 0 and it is a morphism in $\mathcal{E}_{f,\infty}^{(0)}$, the looked for antecedent of ϕ: our functor is indeed full.

To show that it is essentially surjective, it is much more convenient to look at matricial realizations of the monodromy representations. So let $\rho : \pi_1(\mathbf{C}^*, a)^\circ \to \mathrm{GL}_n(\mathbf{C})$ be a representation. Let $M \in \mathrm{GL}_n(\mathbf{C})$ be the image by ρ of the fundamental loop class at a and $C \in \mathrm{Mat}_n(\mathbf{C})$ such that $e^{2i\pi C} = M$. Then taking $A := z^{-1}C$ yields a system S_A in $\mathcal{E}_{f,\infty}^{(0)}$ whose image by our functor is isomorphic to ρ. $\qquad\square$

10.4. Matricial representation

Starting from an object S_A of $\mathcal{E}_{f,\infty}^{(0)}$, one can directly produce a matricial representation as follows. Let $\mathcal{X}$ be the unique fundamental matricial solution at a such that $\mathcal{X}(a) = I_n$. For every homotopy class $[\lambda] \in \pi_1(\mathbf{C}^*, a)$, let $\mathcal{X}^\lambda$ be the result of the analytic continuation along λ and let $M_\lambda \in \mathrm{GL}_n(\mathbf{C})$ be such that $\mathcal{X}^\lambda = \mathcal{X} M_\lambda$. Then $\rho'_A : [\lambda] \mapsto M_\lambda$ is a representation $\pi_1(\mathbf{C}^*, a)^\circ \to \mathrm{GL}_n(\mathbf{C})$.

The representations ρ'_A and ρ_A are actually isomorphic. Indeed, the linear isomorphism $\phi_A : C \mapsto \mathcal{X} C$ from $\mathbf{C}^n$ to $\mathcal{F}_{A,a}$ intertwines ρ'_A and ρ_A.

Proposition 10.3. *The natural transformation $S_A \rightsquigarrow \phi_A$ is an isomorphism between the functors $\rho' : S_A \rightsquigarrow \rho'_A$ and $\rho : S_A \rightsquigarrow \rho_A$ from $\mathcal{E}_{f,\infty}^{(0)}$ to $\mathfrak{Rep}_\mathbf{C}^f(\pi_1(\mathbf{C}^*, a)^\circ)$.*

Proof. Let $F : S_A \to S_B$ be a morphism in $\mathcal{E}_{f,\infty}^{(0)}$. This gives rise to a square diagram which we would like to be commutative:

$$
\begin{array}{ccc}
\rho'_A & \xrightarrow{\phi_A} & \rho_A \\
{\scriptstyle\rho'(F)}\downarrow & & \downarrow{\scriptstyle\rho(F)} \\
\rho'_B & \xrightarrow[\phi_B]{} & \rho_B
\end{array}
$$

More concretely, we want the following square to commute:

$$
\begin{array}{ccc}
\mathbf{C}^n & \xrightarrow{\ C\mapsto \mathcal{X}C\ } & \mathcal{F}_{A,a} \\
{\scriptstyle P}\downarrow & & \downarrow {\scriptstyle X\mapsto FX} \\
\mathbf{C}^p & \xrightarrow[\ D\mapsto \mathcal{Y}D\ ]{} & \mathcal{F}_{B,a}
\end{array}
$$

Here, $\mathcal{X}$ and $\mathcal{Y}$ are the fundamental matricial solutions with initial conditions $a \mapsto I_n$, $a \mapsto I_p$, and $P \in \mathrm{Mat}_{p,n}(\mathbf{C})$ is such that $F\mathcal{X} = \mathcal{Y}P$. The commutation is obvious. So we have indeed a natural transformation. Since all the maps ϕ_A are bijective, it is an isomorphism of functors. $\qquad\square$

Note that the functor ρ' arrives in the full essential subcategory of $\mathfrak{Rep}_{\mathbf{C}}^f(\pi_1(\mathbf{C}^*, a)^\circ)$ consisting of matricial representations.

Practical recipe. We take $a := 1$. We are going to describe a nondeterministic recipe that produces the matricial representations and is nevertheless functorial.

Let $S_A : X' = AX$ be arbitrary in $\mathcal{E}_f^{(0)}$. A priori, A is not defined at 1. So using the algorithm of Section 9.4, we write $A = F[z^{-1}C]$, $F \in \mathrm{GL}_n(\mathcal{M}_0)$, $C \in \mathrm{Mat}_n(\mathbf{C})$. Note that the quoted algorithm is not deterministic and the possible results F, C are not unique. So we choose one such result F, C (i.e., an execution sequence of the algorithm). Then we define the matricial representation ρ'_A as the one which sends the fundamental loop in $\pi_1(\mathbf{C}^*, a)$ to $e^{2\mathrm{i}\pi C} \in \mathrm{GL}_n(\mathbf{C})$.

Let $F : A_1 = F_1[z^{-1}C_1] \to A_2 = F_2[z^{-1}C_2]$ be a morphism in $\mathcal{E}_f^{(0)}$. Then $G := F_2^{-1} \circ F \circ F_1 : C_1 \to C_2$ is a morphism in $\mathcal{E}_f^{(0)}$, so $Gz^{C_1} = z^{C_2}P$ for some $P \in \mathrm{Mat}_{n_2,n_1}(\mathbf{C})$. The map $P : \mathbf{C}^{n_1} \to \mathbf{C}^{n_1}$ intertwines the representations; it is the morphism associated to F by our functor. Note the similarity with the construction of the quasi-inverse.

Exercise 10.4. In the recipe, describe a deterministic way of choosing F, C and prove that it yields a functor.

Exercises

(1) Prove that, if the covariant functor F admits a quasi-inverse, then it is an equivalence of categories.

(2) Define a category $\mathcal{C}$ of systems with constant coefficients as a full subcategory of $\mathcal{E}_f^{(0)}$. Show that $\mathcal{C}$ is equivalent to $\mathcal{E}_f^{(0)}$. Then describe *explicitly* an equivalence of categories of $\mathcal{C}$ with $\mathfrak{Rep}_{\mathbf{C}}^f(\pi_1(\mathbf{C}^*, a)^\circ)$.

Chapter 11

Hypergeometric series and equations

For complements to the material in this chapter, the following books are warmly recommended: [**AB94**, **IKSY91**, **WW27**, **Yos87**].

11.1. Fuchsian equations and systems

Fuchsian systems.

Definition 11.1. A *fuchsian system* is a meromorphic differential system on **S** which has only singularities of the first kind.

Recall that singularities of the first kind (or fuchsian singularities) for a system were introduced in Definition 9.14 and that this is a strictly stronger property than regular singularity (Chapter 9).

First we look at the meromorphy condition. The system $X' = AX$ is meromorphic on **C** if $A \in \mathrm{Mat}_n(\mathcal{M}(\mathbf{C}))$. At infinity, we put $w := 1/z$ and $Y(w) := X(z)$ so that $Y'(w) = B(w)Y(w)$, where $B(w) = -w^{-2}A(w^{-1})$. Thus, $X' = AX$ is meromorphic at infinity if $-w^{-2}A(w^{-1})$ is meromorphic at $w = 0$, that is, if A is meromorphic at infinity. The condition is therefore $A \in \mathrm{Mat}_n(\mathcal{M}(\mathbf{S}))$, *i.e.*, $A \in \mathrm{Mat}_n(\mathbf{C}(z))$ according to Theorem 7.5 (ii).

Next we look at the condition on singularities. The rational matrix A has a finite number of poles on **C**, and they must be simple poles. Call them $a_1, \ldots, a_m$ (of course it is possible that $m = 0$). Then by standard properties of rational functions, one can write $A = \sum_{k=1}^{m} \frac{1}{z - a_k} A_k + C$, where

C has polynomial coefficients and the A_k are constant complex matrices. Now, we want $B(w) = -w^{-2}A(w^{-1})$ to have a simple pole at $w = 0$. But $B(w) = \sum_{k=1}^{m} \dfrac{-1}{w(1-a_k w)}A_k - w^{-2}C(w^{-1})$, and so we want $w^{-2}C(w^{-1})$ to have a simple pole at $w = 0$, which is only possible if $C = 0$. We have proved:

Proposition 11.2. *Fuchsian systems have the form* $X' = \left(\sum_{k=1}^{m} \dfrac{A_k}{z - a_k} \right) X$, *where the* $A_k \in \mathrm{Mat}_n(\mathbf{C})$. $\hfill\square$

Exercise 11.3. Prove the converse.

The monodromy group of a fuchsian system. Using the local coordinate $v := z - a_k$, the system above has the form

$$v\, dX/dv = (A_k + \text{multiple of } v)X.$$

If we suppose that A_k is nonresonant, there is a fundamental matricial solution $\mathcal{X}_k = F_k(z-a_k)^{A_k}$, where F_k is a Birkhoff matrix as obtained in Section 9.4. The local monodromy (*i.e.*, calculated along a small positive fundamental loop around a_k) relative to this basis therefore has matrix $e^{2i\pi A_k}$. In the same way, at infinity, using the coordinate $w := 1/z$ and the fact that $\dfrac{z}{z - a_k} = \dfrac{1}{1 - a_k w} = 1 + \dfrac{a_k w}{1 - a_k w}$, we find that $wB(w) = -zA(z) = -\sum_{k=1}^{m} A_k + \text{multiple of } w$, so, if $\sum_{k=1}^{m} A_k$ is nonresonant, the local monodromy relative to that basis has matrix $e^{-2i\pi \sum\limits_{k=1}^{m} A_k}$.

However, it is difficult to describe the global monodromy from these local data, because all these matrices are relative to different bases (the ones obtained by Fuchs-Frobenius method near each singularity). If one fixes a basis, then one will have to use conjugates of the monodromy matrices computed above; the conjugating matrices are not easy to find.

Example 11.4. If $m = 2$, assuming nonresonancy, we have matrices A_1, A_2 and $A_\infty = -A_1 - A_2$. If one fixes a base point a and three small loops $\lambda_1, \lambda_2, \lambda_\infty$ based at a and each turning once around one of the singular points, then there is a homotopy relation, for instance $\lambda_\infty \sim \lambda_1 \lambda_2$ (the homotopy relation denoted $\sim$ was introduced in Section 5.2). If one moreover fixes a basis $\mathcal{B}$ of solutions at a, then there are monodromy matrices M_1, M_2, M_∞ such that $\mathcal{B}^{\lambda_1} = \mathcal{B}M_1$, $\mathcal{B}^{\lambda_2} = \mathcal{B}M_2$ and $\mathcal{B}^{\lambda_\infty} = \mathcal{B}M_\infty$. Then, on the one hand, $M_\infty = M_2 M_1$. On the other hand, M_1 is conjugate to $e^{2i\pi A_1}$, M_2 is conjugate to $e^{2i\pi A_2}$ and M_∞ is conjugate to $e^{2i\pi A_\infty}$: these are not equalities and the conjugating matrices are different.

Exception: The abelian cases. If the monodromy group is abelian, the problem of conjugacy disappears. There are a few cases where abelianity is guaranteed. First, if $m = 0$, the system is $X' = 0$ which has trivial monodromy. If $m = 1$, since the fundamental group of $\mathbf{S} \setminus \{a_1, \infty\} = \mathbf{C} \setminus \{a_1\}$ is isomorphic to $\mathbf{Z}$, the monodromy group is generated by $e^{2i\pi A_1}$ (local monodromy at a_1) or, alternatively, by $e^{-2i\pi A_1}$. In the same way, if $m = 2$ but there is no singularity at infinity, then $A_1 + A_2 = 0$, the local monodromies at a_1 and a_2 are respectively generated by $e^{2i\pi A_1}$ and $e^{2i\pi A_2} = e^{-2i\pi A_1}$ and the global monodromy is generated by either matrix.

Exercise 11.5. Explain topologically why we obtain inverse monodromy matrices in the last two cases.

If $m = 2$ and there is a singularity at infinity or if $m \geq 3$, the fundamental group of $\mathbf{S} \setminus \{a_1, \ldots, a_m, \infty\}$ is far from being commutative (it is the so-called "free group on m generators"). But, if $n = 1$, the linear group $\mathrm{GL}_1(\mathbf{C}) = \mathbf{C}^*$ is commutative, so the monodromy group is also commutative. This is the case of a scalar equation $f' = \left(\sum_{k=1}^{m} \dfrac{\alpha_k}{z - a_k} \right) f$. Then, without any resonancy condition, the global monodromy group is generated by the $e^{2i\pi\alpha_k}$, $k = 1, \ldots, m$ (thus a subgroup of $\mathbf{C}^*$).

In summary, the first nontrivial (nonabelian) case will be for $m = 2$, with a singularity at infinity, and $n = 2$. We shall describe it at the end of this section.

Fuchsian equations. For $n = 2$, instead of rank 2 systems, we shall rather study the monodromy representation for scalar equations of order 2.

Definition 11.6. A scalar differential equation is said to be *fuchsian* if it is meromorphic on $\mathbf{S}$ and has only regular singularities.

Recall from Chapter 9, in particular Theorem 9.30 (Fuchs criterion), that, for a scalar equation, regular singularities are the same as singularities of the first kind (*i.e.*, fuchsian singularities).

A similar argument to that given for systems shows that a meromorphic equation on $\mathbf{S}$ must have rational coefficients. However, the condition to be fuchsian is a bit more complicated. We shall just need the case $n = 2$ (see exercise (1) at the end of this chapter); the general case is described in the references given at the beginning of Chapter 9.

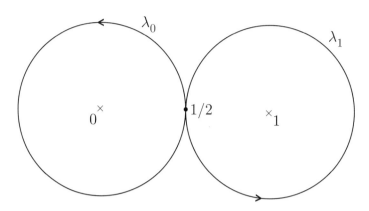

Figure 11.1. Fundamental loops of $\mathbf{S} \setminus \{0, 1, \infty\}$

It can be proved moreover that, when $n = 2$, $m = 2$ and there is a singularity at infinity, then all fuchsian equations are reducible to the "hypergeometric equations" that we are going to study in this chapter; see [**IKSY91**].

The first nonabelian case. In summary, the first nontrivial (nonabelian) case will be for $m = 2$, with a singularity at infinity, and $n = 2$. As is well known (see for instance exercise 2 to this chapter), we can take the two points a_1, a_2 to be $0, 1$. Therefore, the monodromy representation will be an anti-morphism of groups from $\pi_1(\mathbf{S} \setminus \{0, 1, \infty\}, a)$ to $\mathrm{GL}_2(\mathbf{C})$.

Here, the fundamental group is a free group on two generators. Once we choose a base point $a \in \mathbf{S} \setminus \{0, 1, \infty\}$, a small loop λ_0 turning positively once around 0 and a small loop λ_1 turning positively once around 1 (both based at a), the group is freely generated by the homotopy classes of these loops. We shall take $a := 1/2$ and the loops $\lambda_0(t) := \dfrac{1}{2}e^{2i\pi t}$, $\lambda_1(t) := 1 - \dfrac{1}{2}e^{2i\pi t}$, $t \in [0, 1]$.

A representation of $\pi_1(\mathbf{S} \setminus \{0, 1, \infty\}, 1/2)$ in $\mathrm{GL}_2(\mathbf{C})$ is therefore entirely characterized by the matrices $M_0, M_1 \in \mathrm{GL}_2(\mathbf{C})$ which are the respective images of the homotopy classes $[\lambda_0], [\lambda_1]$. Moreover, since the generators $[\lambda_0], [\lambda_1]$ are free, the matrices M_0, M_1 may be chosen at will.

The description of the monodromy representation requires the choice of a basis $\mathcal{B}$ of solutions at $1/2$. Then, the effect of monodromy is encoded in matrices $M_0, M_1, M_\infty \in \mathrm{GL}_2(\mathbf{C})$ such that $\mathcal{B}^{\lambda_0} = \mathcal{B}M_0$, $\mathcal{B}^{\lambda_1} = \mathcal{B}M_1$ and $\mathcal{B}^{\lambda_\infty} = \mathcal{B}M_\infty$, where some loop λ_∞ has been chosen. However, there must be a relation between the homotopy classes $[\lambda_0]$, $[\lambda_1]$, $[\lambda_\infty]$ and therefore a corresponding relation between the monodromy matrices M_0, M_1, M_∞. We shall take $\lambda_\infty := (\lambda_0.\lambda_1)^{-1}$, whence the relation $M_\infty M_1 M_0 = I_2$ (remember we have an anti-morphism).

11.2. The hypergeometric series

We already met hypergeometric series in Example 9.39; now we go for a more systematic study.

Definition 11.7. The *Pochhammer symbols* are defined, for $\alpha \in \mathbf{C}$ and $n \in \mathbf{N}$, by the formula:

$$(\alpha)_0 := 1 \text{ and, if } n \geq 1 \,, \, (\alpha)_n := \alpha(\alpha+1)\cdots(\alpha+n-1).$$

The *hypergeometric series* of Euler-Gauss with parameters $\alpha, \beta, \gamma \in \mathbf{C}$ is the power series:

$$F(\alpha, \beta, \gamma; z) := \sum_{n \geq 0} \frac{(\alpha)_n (\beta)_n}{n!(\gamma)_n} z^n = \sum_{n \geq 0} \frac{(\alpha)_n (\beta)_n}{(1)_n (\gamma)_n} z^n.$$

We must of course require that $\gamma \notin -\mathbf{N}$ so that the denominators do not vanish. We shall also require that $\alpha, \beta \notin -\mathbf{N}$ so that the series is not a polynomial. Last, for reasons that will appear in the next section (to avoid resonancies), we shall require that $\gamma, \alpha - \beta, \gamma - \alpha - \beta \notin \mathbf{Z}$. (The study is possible in these degenerate cases; see [**IKSY91, WW27**].)

The coefficients $f_n := \dfrac{(\alpha)_n (\beta)_n}{n!(\gamma)_n}$ satisfy the relation $f_{n+1}/f_n = (n+\alpha)(n+\beta)/(n+1)(n+\gamma)$. Since the right-hand side of this equality tends to 1 when $n \to +\infty$, the radius of convergence of the hypergeometric series is 1.

Example 11.8. From the obvious formula $(\alpha)_n = (-1)^n n! \binom{-\alpha}{n}$, we obtain that $F(\alpha, \beta, \beta; z) = (1-z)^{-\alpha}$ (the generalized binomial series).

About the Gamma function. The hypergeometric series is related in many ways to Euler's Gamma function. (For more on the Gamma function, see [**WW27**], or most books on complex analysis.) For $\Re z > 0$, one can show that the integral

$$\Gamma(z) := \int_0^{+\infty} e^{-t} t^{z-1} \, dt$$

is well defined and that the function Γ is analytic on the right half-plane. Moreover, integration by parts gives the functional equation $\Gamma(z+1) = z\Gamma(z)$. This allows one to extend Γ to the whole complex plane by putting

$\Gamma(z) := \dfrac{1}{(z)_n}\Gamma(z+n)$, where $n \in \mathbf{N}$ is chosen big enough to have $\Re(z+n) > 0$.
The extended function is holomorphic on $\mathbf{C} \setminus (-\mathbf{N})$ and has simple poles
on $-\mathbf{N}$. It still satisfies the functional equation $\Gamma(z + 1) = z\Gamma(z)$, whence
$\Gamma(z + n) = (z)_n\Gamma(z)$.

Replacing $(\alpha)_n$ by $\dfrac{\Gamma(n + \alpha)}{\Gamma(\alpha)}$, etc., the hypergeometric series can be written:

$$F(\alpha, \beta, \gamma; z) := \frac{\Gamma(\gamma)}{\Gamma(\alpha)\Gamma(\beta)} \sum_{n\geq 0} \frac{\Gamma(n + \alpha)\Gamma(n + \beta)}{\Gamma(n + 1)\Gamma(n + \gamma)} z^n.$$

Here are some special values related to the Gamma function:

$$\Gamma(1) = \int_0^{+\infty} e^{-t}\, dt = 1,$$

$$\Gamma(n) = (n - 1)! \text{ for } n \in \mathbf{N}^*,$$

$$\Gamma(1/2) = \int_{-\infty}^{+\infty} e^{-t^2}\, dt = \sqrt{\pi},$$

$$\Gamma'(1) = -\gamma,$$

where $\gamma := \lim_{n\to+\infty} (1 + 1/2 + \cdots + 1/n - \ln n)$ is the "Euler-Mascheroni
constant", a very mysterious number. We shall use other formulas related
to the Gamma function in Section 11.5.

11.3. The hypergeometric equation

We saw in the previous section that the coefficients $f_n = \dfrac{(\alpha)_n(\beta)_n}{n!(\gamma)_n}$ of
$F(\alpha, \beta, \gamma; z)$ satisfy the relation $f_{n+1}/f_n = (n + \alpha)(n + \beta)/(n + 1)(n + \gamma)$.
From this, recalling that $\delta(\sum f_n z^n) = \sum n f_n z^n$ and that $\delta^2(\sum f_n z^n) = \sum n^2 f_n z^n$, we start with $(n + 1)(n + \gamma)f_{n+1} = (n + \alpha)(n + \beta)f_n$ and calculate

$$\sum_{n\geq 0}(n + 1)(n + \gamma)f_{n+1}z^{n+1} = \sum_{n\geq 0}(n + \alpha)(n + \beta)f_n z^{n+1}.$$

Hence,

$$\sum_{n\geq 0}(n + 1)^2 f_{n+1}z^{n+1} + \sum_{n\geq 0}(\gamma - 1)(n + 1)f_{n+1}z^{n+1}$$

$$= z\left(\sum_{n\geq 0} n^2 f_n z^n + \sum_{n\geq 0}(\alpha + \beta)n f_n z^n + \sum_{n\geq 0}\alpha\beta f_n z^n\right),$$

which we rewrite as

$$(\delta^2 + (\gamma - 1)\delta)F(\alpha, \beta, \gamma; z) = z(\delta^2 + (\alpha + \beta)\delta + \alpha\beta)F(\alpha, \beta, \gamma; z),$$

that is, the hypergeometric series $F(\alpha, \beta, \gamma; z)$ is the solution of the *hyper-geometric differential equation with parameters* α, β, γ:

$$HG_{\alpha,\beta,\gamma}: \quad (1-z)\delta^2 F + ((\gamma - 1) - (\alpha + \beta)z)\delta F - \alpha\beta z F = 0.$$

For the global study, we shall also need the other form of the equality using the standard differential operator $D := d/dz$ instead of δ. Replacing δ by zD and δ^2 by $z^2 D^2 + zD$ and then dividing by z, we get the other form of the equation:

$$HG'_{\alpha,\beta,\gamma}: \quad z(1-z)D^2 F + (\gamma - (\alpha + \beta + 1)z)DF - \alpha\beta F = 0.$$

According to exercise 1 to this chapter, it should be fuchsian. It is actually obvious that equation $HG'_{\alpha,\beta,\gamma}$ is meromorphic on $\mathbf{S}$, that its only singularities in $\mathbf{C}$ are 0 and 1 and that they are regular singularities. We shall verify that ∞ is also a regular singularity. We shall also look for the local solutions at singularities, applying the method of Fuchs-Frobenius which was described in Section 9.6. In order to describe the local monodromies, we shall use the base point $1/2$ and the loops described at the end of Section 11.1.

Study at 0. We use the first form $HG_{\alpha,\beta,\gamma}$. The indicial equation is $x^2 + (\gamma - 1)x = 0$ and the two exponents $0, 1 - \gamma$ are nonresonant since we assumed that $\gamma \notin \mathbf{Z}$. Therefore, there is a unique power series solution with constant term 1 — it is clearly the hypergeometric series $F(\alpha, \beta, \gamma; z)$ itself — and a unique solution $z^{1-\gamma}G$, where G is a power series solution with constant term 1. To find G, remember that $\delta.z^{1-\gamma} = z^{1-\gamma}.(\delta + 1 - \gamma)$, so G is a solution of the equation:

$$(1-z)(\delta + 1 - \gamma)^2 G + ((\gamma - 1) - (\alpha + \beta)z)(\delta + 1 - \gamma)G - \alpha\beta z G = 0.$$

Expanding and simplifying, we find:

$$(1-z)\delta^2 G + ((1 - \gamma) - (\alpha + \beta + 2 - 2\gamma)z)\delta G - (\alpha + 1 - \gamma)(\beta + 1 - \gamma)zG = 0.$$

This is just the hypergeometric equation $HG_{\alpha+1-\gamma,\beta+1-\gamma,2-\gamma}$. Its parameters $\alpha + 1 - \gamma, \beta + 1 - \gamma, 2 - \gamma$ satisfy the same nonresonancy condition as α, β, γ, so that $G = F(\alpha + 1 - \gamma, \beta + 1 - \gamma, 2 - \gamma; z)$.

Proposition 11.9. *A basis of solutions near* 0 *is:*

$$\mathcal{B}_0 := \left(F(\alpha, \beta, \gamma; z), z^{1-\gamma} F(\alpha + 1 - \gamma, \beta + 1 - \gamma, 2 - \gamma; z) \right).$$

$\square$

Corollary 11.10. *The monodromy matrix along the loop* λ_0 *relative to the basis* $\mathcal{B}_0$ *is* $M_0 := \begin{pmatrix} 1 & 0 \\ 0 & e^{-2i\pi\gamma} \end{pmatrix}$.

Exercise 11.11. Describe the two solutions when $\beta = \gamma$.

Study at 1. We must use a local coordinate that vanishes at $z = 1$. We take $v := 1 - z$. Therefore, if $F(z) = G(v)$, then $F'(z) = -G'(v)$ and $F''(z) = G''(v)$. In symbolic notation, $D_v = -D_z$ and $D_v^2 = D_z^2$. The second form $HG'_{\alpha,\beta,\gamma}$ gives the following equation for $G(v)$:

$$v(1 - v)D_v^2 G + ((\alpha + \beta + 1 - \gamma) - (\alpha + \beta + 1)v)D_v G - \alpha\beta G = 0.$$

We recognize the hypergeometric equation $HG'_{\alpha,\beta,\alpha+\beta+1-\gamma}$ with parameters $\alpha, \beta, \alpha + \beta + 1 - \gamma$. Again for these new parameters, the nonresonancy conditions are met. We conclude:

Proposition 11.12. *A basis of solutions near 1 is:*

$$\mathcal{B}_1 := (F(\alpha, \beta, \alpha + \beta + 1 - \gamma; 1 - z),$$
$$(1 - z)^{\gamma - \alpha - \beta} F(\gamma - \beta, \gamma - \alpha, \gamma + 1 - \alpha - \beta; 1 - z)\big).$$

$\square$

Corollary 11.13. *The monodromy matrix along the loop* λ_1 *relative to the basis* $\mathcal{B}_1$ *is* $M_1 := \begin{pmatrix} 1 & 0 \\ 0 & e^{2i\pi(\gamma - \alpha - \beta)} \end{pmatrix}$.

Study at ∞. We use the coordinate $w = 1/z$. If $F(z) = G(w)$, then $zF'(z) = -wG'(w)$ which we write symbolically as $\delta_w = -\delta_z$. Likewise, $\delta_w^2 = \delta_z^2$. The equation $HG_{\alpha,\beta,\gamma}$ gives the following equation for $G(w)$:

$$(1 - \frac{1}{w})\delta_w^2 G - ((\gamma - 1) - \frac{\alpha + \beta}{w})\delta_w G - \frac{\alpha\beta}{w}G = 0,$$

or, equivalently:

$$(1 - w)\delta_w^2 G - ((\alpha + \beta) - (\gamma - 1)w)\delta_w G + \alpha\beta G = 0.$$

This is not a hypergeometric equation (miracles are not permanent!) but it is regular singular, with indicial equation $x^2 - (\alpha + \beta)x + \alpha\beta = 0$. The exponents are α and β and they are nonresonant. Therefore there are solutions of the form $w^\alpha H_1$ and $w^\beta H_2$, with H_1 and H_2 two power series with constant term 1, and they form a basis. To compute a solution $w^\alpha H$, we apply the rule $\delta_w . w^\alpha = w^\alpha(\delta_w + \alpha)$, whence the equation:

$$(1 - w)(\delta_w + \alpha)^2 H - ((\alpha + \beta) - (\gamma - 1)w)(\delta_w + \alpha)H + \alpha\beta H = 0$$
$$\Longleftrightarrow (1 - w)\delta_w^2 H + ((\alpha - \beta) - (2\alpha - \gamma + 1)w)\delta_w H - \alpha(\alpha - \gamma + 1)wH = 0.$$

We recognize the hypergeometric equation $HG_{\alpha,\alpha-\gamma+1,\alpha-\beta+1}$ with coefficients $\alpha, \alpha - \gamma + 1, \alpha - \beta + 1$ and no resonancy, so

$$H = F(\alpha, \alpha - \gamma + 1, \alpha - \beta + 1; w).$$

The calculation for a solution of the form $z^\beta H$ is symmetric and we conclude:

Proposition 11.14. *A basis of solutions near ∞ is:*

$$\mathcal{B}_\infty := ((1/z)^\alpha F(\alpha, \alpha - \gamma + 1, \alpha - \beta + 1; 1/z),$$

$$(1/z)^\beta F(\beta, \beta - \gamma + 1, \beta - \alpha + 1; 1/z)\Big).$$

$\square$

To define the local monodromy at ∞, we consider the loop $\lambda_\infty := (\lambda_0.\lambda_1)^{-1}$.

Corollary 11.15. *The monodromy matrix along the loop λ_∞ relative to the basis $\mathcal{B}_\infty$ is $M_\infty := \begin{pmatrix} e^{2i\pi\alpha} & 0 \\ 0 & e^{2i\pi\beta} \end{pmatrix}$.*

11.4. Global monodromy according to Riemann

By elementary but inspired considerations, Riemann (see the references given in the footnote on page 67) succeeded in finding *explicit generators* of the monodromy group of the hypergeometric equation, but relative to a *nonexplicit basis*. Starting from the bases $\mathcal{B}_0$, $\mathcal{B}_1$ and $\mathcal{B}_\infty$ found above, he considered transformed bases $\mathcal{C}_0$, $\mathcal{C}_1$ and $\mathcal{C}_\infty$ whose elements are constant multiples of those of $\mathcal{B}_0$, $\mathcal{B}_1$ and $\mathcal{B}_\infty$ but with unspecified coefficients. For instance:

$$\mathcal{C}_0 = \big(p_0 F(\alpha, \beta, \gamma; z), q_0 z^{1-\gamma} F(\alpha + 1 - \gamma, \beta + 1 - \gamma, 2 - \gamma; z)\big) = \mathcal{B}_0 D_0,$$

where $D_0 = \begin{pmatrix} p_0 & 0 \\ 0 & q_0 \end{pmatrix}$, $p_0, q_0 \in \mathbf{C}^*$ being unspecified, and similarly for $\mathcal{C}_1$ and $\mathcal{C}_\infty$. We are going to follow Riemann's arguments.

We shall consider all functions as defined in the cut plane:

$$\Omega := \mathbf{S} \setminus ([\infty, 0] \cup [1, \infty]) = \mathbf{C} \setminus (]-\infty, 0] \cup [1, +\infty[).$$

This is a simply-connected set (it is star shaped from $1/2$), so indeed all three bases of germs $\mathcal{B}_0$, $\mathcal{B}_1$ and $\mathcal{B}_\infty$ extend to bases of the solution space $\mathcal{F}(\Omega)$. We use the principal determinations of all z^μ and $(1 - z)^\nu$.

The *connection formulas* are the linear formulas relating the various bases of this space. We write them for $\mathcal{C}_0 = (F_0, G_0)$, $\mathcal{C}_1 = (F_1, G_1)$ and $\mathcal{C}_\infty = (F_\infty, G_\infty)$ in the following form:

$$F_0 = a_1 F_1 + b_1 G_1 = a_\infty F_\infty + b_\infty G_\infty,$$

$$G_0 = c_1 F_1 + d_1 G_1 = c_\infty F_\infty + d_\infty G_\infty.$$

In matricial terms:

$$\mathcal{C}_0 = \mathcal{C}_1 P_1 = \mathcal{C}_\infty P_\infty, \text{ where } P_1 = \begin{pmatrix} a_1 & c_1 \\ b_1 & d_1 \end{pmatrix} \text{ and } P_\infty = \begin{pmatrix} a_\infty & c_\infty \\ b_\infty & d_\infty \end{pmatrix}.$$

The local monodromies were previously found relative to the bases $\mathcal{B}_0$, $\mathcal{B}_1$ and $\mathcal{B}_\infty$:

$$\mathcal{B}_0^{\lambda_0} = \mathcal{B}_0 M_0, \text{ with } M_0 = \begin{pmatrix} 1 & 0 \\ 0 & e^{-2i\pi\gamma} \end{pmatrix},$$

$$\mathcal{B}_1^{\lambda_1} = \mathcal{B}_1 M_1, \text{ with } M_1 = \begin{pmatrix} 1 & 0 \\ 0 & e^{2i\pi(\gamma-\alpha-\beta)} \end{pmatrix},$$

$$\mathcal{B}_\infty^{\lambda_\infty} = \mathcal{B}_\infty M_\infty, \text{ with } M_\infty = \begin{pmatrix} e^{2i\pi\alpha} & 0 \\ 0 & e^{2i\pi\beta} \end{pmatrix}.$$

All these bases are bases of eigenvectors of the corresponding monodromy matrices. In essence, this comes from the fact that matrix D_0 relating $\mathcal{B}_0$ to $\mathcal{C}_0$ commutes with M_0, and similarly at 1 and ∞. Therefore, we also have the local monodromies relative to the bases $\mathcal{C}_0$, $\mathcal{C}_1$ and $\mathcal{C}_\infty$:

$$\mathcal{C}_0^{\lambda_0} = \mathcal{C}_0 M_0, \quad \mathcal{C}_1^{\lambda_1} = \mathcal{C}_1 M_1 \quad \text{and} \quad \mathcal{C}_\infty^{\lambda_\infty} = \mathcal{C}_\infty M_\infty.$$

Exercise 11.16. Write down the intermediate steps of the computation leading from $\mathcal{C}_0^{\lambda_0} = (\mathcal{B}_0 D_0)^{\lambda_0}$ to $\mathcal{C}_0 M_0$, and similarly at 1 and ∞.

If we can determine P_1 and P_∞, then we will be able to describe the monodromy group relative to $\mathcal{C}_0$. Indeed, we already know that $\mathcal{C}_0^{\lambda_0} = \mathcal{C}_0 M_0$, and also:

$$\mathcal{C}_0^{\lambda_1} = (\mathcal{C}_1 P_1)^{\lambda_1} = \mathcal{C}_1^{\lambda_1} P_1 = \mathcal{C}_1 M_1 P_1 = \mathcal{C}_0(P_1^{-1} M_1 P_1),$$

$$\mathcal{C}_0^{\lambda_\infty} = (\mathcal{C}_\infty P_\infty)^{\lambda_\infty} = \mathcal{C}_\infty^{\lambda_\infty} P_\infty = \mathcal{C}_\infty M_\infty P_\infty = \mathcal{C}_0(P_\infty^{-1} M_\infty P_\infty).$$

Therefore, the monodromy group relative to $\mathcal{C}_0$ will be generated by M_0, $P_1^{-1} M_1 P_1$ and $P_\infty^{-1} M_\infty P_\infty$. Moreover, from the relation $\lambda_\infty = (\lambda_0.\lambda_1)^{-1}$, we obtain:

$$(P_\infty^{-1} M_\infty P_\infty)(P_1^{-1} M_1 P_1) M_0 = I_2.$$

In the next section, we shall find an explicit connection matrix relating the explicit bases $\mathcal{B}_0$ and $\mathcal{B}_\infty$. Here, we only find connection matrices for the nonexplicit bases $\mathcal{C}_0$, $\mathcal{C}_1$ and $\mathcal{C}_\infty$. The consequence is that one cannot compute the monodromy for a given solution, because we cannot express it in those bases!

Calculation of the connection formulas. We perform analytic continuation along the loop $\lambda_1 = \lambda_0^{-1}\lambda_\infty^{-1}$, applied to the connection formula. For instance, in the first connection formula:

$$F_0 = a_1 F_1 + b_1 G_1 = a_\infty F_\infty + b_\infty G_\infty;$$

the middle expression $a_1 F_1 + b_1 G_1$ is transformed into $a_1 F_1 + b_1 e^{2i\pi(\gamma-\alpha-\beta)}G_1$ along λ_1. On the other hand, $F_0 = a_\infty F_\infty + b_\infty G_\infty$ is left invariant along λ_0^{-1}, then transformed into $a_\infty e^{-2i\pi\alpha} F_\infty + b_\infty e^{-2i\pi\beta}G_\infty$ along λ_∞^{-1}. We eventually get the "new formula":

$$a_1 F_1 + b_1 e^{2i\pi(\gamma-\alpha-\beta)}G_1 = a_\infty e^{-2i\pi\alpha}F_\infty + b_\infty e^{-2i\pi\beta}G_\infty.$$

In the same way, starting from the second connection formula

$$G_0 = c_1 F_1 + d_1 G_1 = c_\infty F_\infty + d_\infty G_\infty,$$

and noticing that G_0 is multiplied by $e^{2i\pi\gamma}$ along λ_0^{-1}, we get the "new formula":

$$c_1 F_1 + d_1 e^{2i\pi(\gamma-\alpha-\beta)}G_1 = e^{2i\pi\gamma}\left(c_\infty e^{-2i\pi\alpha}F_\infty + d_\infty e^{-2i\pi\beta}G_\infty\right).$$

Now we take some arbitrary $\sigma \in \mathbf{C}$. It will be specialized later to particular values. For each of the two pairs of formulas above (original connection formula and deduced "new formula"), we compute $e^{\sigma i\pi}$ times the original formula minus $e^{-\sigma i\pi}$ times the new formula. Taking into account the general equality

$$e^{\sigma i\pi} - e^{-\sigma i\pi} \times e^{2\tau i\pi} = 2i\sin(\sigma-\tau)\pi\, e^{\tau i\pi},$$

we end up with the following two relations:

$$a_1 \sin\sigma\pi F_1 + b_1 \sin(\sigma-(\gamma-\alpha-\beta))\pi e^{(\gamma-\alpha-\beta)i\pi}G_1$$
$$= a_\infty \sin(\sigma+\alpha)\pi\, e^{-\alpha i\pi}F_\infty + b_\infty \sin(\sigma+\beta)\pi\, e^{-\beta i\pi}G_\infty$$

and

$$c_1 \sin\sigma\pi F_1 + d_1 \sin(\sigma-(\gamma-\alpha-\beta))\pi e^{(\gamma-\alpha-\beta)i\pi}G_1$$
$$= c_\infty \sin(\sigma+\alpha-\gamma)\pi\, e^{-(\alpha-\gamma)i\pi}F_\infty + d_\infty \sin(\sigma+\beta-\gamma)\pi\, e^{-(\beta-\gamma)i\pi}G_\infty.$$

If we take $\sigma := \gamma - \alpha - \beta$ in each of these two equalities, we find

$$a_1 \sin(\gamma-\alpha-\beta)\pi\, F_1 = a_\infty \sin(\gamma-\beta)\pi\, e^{-\alpha i\pi}F_\infty + b_\infty \sin(\gamma-\alpha)\pi\, e^{-\beta i\pi}F_\infty$$

and

$$c_1 \sin(\gamma-\alpha-\beta)\pi\, F_1 = c_\infty \sin(-\beta)\pi\, e^{-(\alpha-\gamma)i\pi}F_\infty + d_\infty \sin(-\alpha)\pi\, e^{-(\beta-\gamma)i\pi}F_\infty.$$

Likewise, if we take $\sigma := 0$ in the same two equalities, we find

$$-b_1 \sin(\gamma-\alpha-\beta)\pi\, e^{(\gamma-\alpha-\beta)i\pi}\, G_1 = a_\infty \sin\alpha\pi\, e^{-\alpha i\pi}F_\infty + b_\infty \sin\beta\pi\, e^{-\beta i\pi}F_\infty$$

and

$$- d_1 \sin(\gamma - \alpha - \beta)\pi \ e^{(\gamma-\alpha-\beta)i\pi} \ G_1$$
$$= c_\infty \sin(\alpha - \gamma)\pi \ e^{-(\alpha-\gamma)i\pi} F_\infty + d_\infty \sin(\beta - \gamma)\pi \ e^{-(\beta-\gamma)i\pi} F_\infty.$$

Now we make one more special assumption:

> **All the connection coefficients** $a_1, b_1, c_1, d_1, a_\infty, b_\infty, c_\infty, d_\infty$ **are supposed to be nonzero.**

This is of course true "generically"; the opposite "degenerate" case will be discussed in the last paragraph of this section.

We then have the above two expressions of F_1 in the basis $\mathcal{C}_\infty$, and the same for G_1. Identifying, we get:

$$\frac{a_1}{c_1} = \frac{a_\infty}{c_\infty} \frac{\sin(\gamma - \beta)\pi \ e^{-\alpha i\pi}}{\sin(-\beta)\pi \ e^{-(\alpha-\gamma)i\pi}} = \frac{b_\infty}{d_\infty} \frac{\sin(\gamma - \alpha)\pi \ e^{-\beta i\pi}}{\sin(-\alpha)\pi \ e^{-(\beta-\gamma)i\pi}},$$

$$\frac{b_1}{d_1} = \frac{a_\infty}{c_\infty} \frac{\sin \alpha\pi \ e^{-\alpha i\pi}}{\sin(\alpha - \gamma)\pi \ e^{-(\alpha-\gamma)i\pi}} = \frac{b_\infty}{d_\infty} \frac{\sin \beta\pi \ e^{-\beta i\pi}}{\sin(\beta - \gamma)\pi \ e^{-(\beta-\gamma)i\pi}}.$$

Now we remember that all the basis elements $F_0, G_0, F_1, G_1, F_\infty, G_\infty$ are defined up to an arbitrary constant factor. This means that we can arbitrarily fix a_1, b_1, c_1, d_1 and one of the four other coefficients. Among the various possibilities, this was the choice of Riemann in his paper on the hypergeometrical functions:

$$P_1 = \frac{1}{\sin(\gamma - \alpha - \beta)\pi} \begin{pmatrix} \sin(\gamma - \alpha)\pi \ e^{-\gamma i\pi} & \sin \beta\pi \\ \sin \alpha\pi \ e^{-(\alpha+\beta)i\pi} & \sin(\gamma - \beta)\pi \ e^{(\gamma-\alpha-\beta)i\pi} \end{pmatrix},$$

$$P_\infty = \frac{1}{\sin(\beta - \alpha)\pi} \begin{pmatrix} \sin(\gamma - \alpha)\pi & \sin(\beta - \gamma)\pi \\ -\sin \alpha\pi & \sin \beta\pi \end{pmatrix}.$$

Theorem 11.17. *The monodromy group of $HG_{\alpha,\beta,\gamma}$ expressed in an adequate basis is generated by the matrices M_0, $P_1^{-1}M_1P_1$ and $P_\infty^{-1}M_\infty P_\infty$. These generators obey the relation:*

$$(P_\infty^{-1}M_\infty P_\infty)(P_1^{-1}M_1P_1)M_0 = I_2.$$

$\square$

Exercise 11.18. (i) By inspection, verify that the sum of the six exponents (two at each singularity) is an integer.

(ii) Prove this *a priori* by a monodromy argument.

The meaning of the nondegeneracy condition on the connection coefficients. If one of the connection coefficients $a_1, b_1, c_1, d_1, a_\infty, b_\infty, c_\infty, d_\infty$ is 0, that means that there is an element of $\mathcal{B}_0$ which is at the same time (up to a nonzero constant factor) an element of $\mathcal{B}_1$ or $\mathcal{B}_\infty$. Such a function is an eigenvector for the monodromy along λ_0 and at the same time along λ_1 or λ_∞, and therefore an eigenvector for all the monodromy. On the side of the monodromy representation $\pi_1(\mathbf{S} \setminus \{0, 1, \infty\}) \to \mathrm{GL}(\mathcal{F}(\Omega))$, this means that there is a subspace (here the line generated by the eigenvector) which is neither $\{0\}$ nor $\mathcal{F}(\Omega)$ and which is stable under the linear action of the monodromy group. Such a representation is said to be *reducible*.[1] On the side of the equation $HG_{\alpha,\beta,\gamma}$, we have a solution f such that $f^{\lambda_0} = e^{2i\pi\mu_0} f$, where $\mu_0 \in \{0, 1 - \gamma\}$ and $f^{\lambda_1} = e^{2i\pi\mu_1} f$, where $\mu_1 \in \{0, \gamma - \alpha - \beta\}$. Then $f = z^{\mu_0}(1 - z)^{\mu_1} g$, where g is at the same time uniform and of moderate growth at all singularities, whence meromorphic on $\mathbf{S}$, whence rational. Then $v := \dfrac{Df}{f} = \dfrac{Dg}{g} + \dfrac{\mu_0}{z} + \dfrac{\mu_1}{1 - z}$ is itself rational. Dividing the hypergeometrical operator by $D - v$ (the same kind of noncommutative euclidean division that we performed in the third step of the proof of Theorem 9.30), we get an equality:

$$D^2 + \frac{\gamma - (\alpha + \beta + 1)z}{z(1 - z)} D - \frac{\alpha\beta}{z(1 - z)} = (D - u)(D - v),$$

with $u, v \in \mathbf{C}(z)$; that is, the hypergeometrical differential operator is *reducible over* $\mathbf{C}(z)$.

11.5. Global monodromy using Barnes' connection formulas

Here, we give the results with incomplete explanations and no justification at all, because they require some analysis that we are not prepared for. See [**IKSY91, WW27**] for details. The main tool is *Barnes' integral representation of the hypergeometric series*:

$$F(\alpha, \beta, \gamma; z) = \frac{1}{2i\pi} \frac{\Gamma(\gamma)}{\Gamma(\alpha)\Gamma(\beta)} \int_C \frac{\Gamma(\alpha + s)\Gamma(\beta + s)\Gamma(-s)}{\Gamma(\gamma + s)} (-z)^s \, ds.$$

The line of integration C is the vertical imaginary line, followed from $-i\infty$ to $+i\infty$, with the following deviations: there must be a detour at the left to avoid $-1 + \mathbf{N}$, and there must be two detours at the right to avoid $-\alpha - \mathbf{N}$ and $-\beta - \mathbf{N}$. Using this integral representation, Barnes proved the following connection formulas, from which the monodromy is immediately

---

[1] A complete discussion of this case can be found in the books quoted at the beginning of the chapter, in particular [**AB94, IKSY91**].

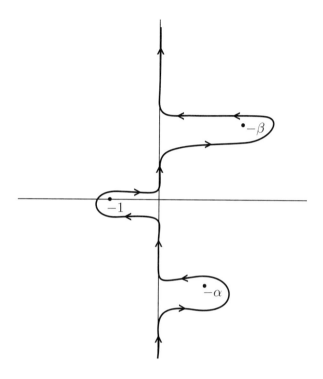

Figure 11.2. Barnes' contour of integration

deduced:

$$F(\alpha, \beta, \gamma; z) = \frac{\Gamma(\gamma)\Gamma(\beta - \alpha)}{\Gamma(\beta)\Gamma(\gamma - \alpha)}(-z)^{-\alpha}F(\alpha, \alpha - \gamma + 1, \alpha - \beta + 1; 1/z)$$

$$+ \frac{\Gamma(\gamma)\Gamma(\alpha - \beta)}{\Gamma(\alpha)\Gamma(\gamma - \beta)}(-z)^{-\beta}F(\beta, \beta - \gamma + 1, \beta - \alpha + 1; 1/z).$$

Theorem 11.19. (i) *One has* $\mathcal{B}_0 = \mathcal{B}_\infty P$, *where:*

$$P = \begin{pmatrix} e^{-i\pi\alpha}\dfrac{\Gamma(\gamma)\Gamma(\beta - \alpha)}{\Gamma(\beta)\Gamma(\gamma - \alpha)} & e^{-i\pi\alpha'}\dfrac{\Gamma(\gamma')\Gamma(\beta' - \alpha')}{\Gamma(\beta')\Gamma(\gamma' - \alpha')} \\ e^{-i\pi\beta}\dfrac{\Gamma(\gamma)\Gamma(\alpha - \beta)}{\Gamma(\alpha)\Gamma(\gamma - \beta)} & e^{-i\pi\beta'}\dfrac{\Gamma(\gamma')\Gamma(\alpha' - \beta')}{\Gamma(\alpha')\Gamma(\gamma' - \beta')} \end{pmatrix}.$$

Here, we set $\alpha' := \alpha - \gamma + 1$, $\beta' := \beta - \gamma + 1$ *and* $\gamma' := 2 - \gamma$.

(ii) *The monodromy group relative to basis* $\mathcal{B}_0$ *is generated by* M_0 *and* $P^{-1}M_\infty P$. □

Exercises

(1) Show that Fuchsian equations of the second order have the form $f'' + pf' + qf = 0$, where $p = \sum_{k=1}^{m} \dfrac{p_k}{z - a_k}$ with $p_k \in \mathbf{C}$, and where $q = \sum_{k=1}^{m} \left(\dfrac{q_k}{(z - a_k)^2} + \dfrac{r_k}{z - a_k} \right)$ with $q_k, r_k \in \mathbf{C}$ and $\sum_{k=1}^{m} r_k = 0$.

(2) Show that if a scalar differential equation has exactly three singularities on $\mathbf{S}$, then using changes of variable $1/z$ and $z + c$, one may always assume that the singularities are at $0, 1, \infty$.

(3) Show that
$$\log \frac{1 + z}{1 - z} = 2z F(1/2, 1, 3/2; z^2)$$
and that
$$\arcsin z = z F(1/2, 1/2, 3/2; z^2).$$

(4) Compute Barnes' integral with the help of Cauchy's residue formula.

Chapter 12

The global Riemann-Hilbert correspondence

12.1. The correspondence

Fix $a_1, \ldots, a_m \in \mathbf{C}$ and set $\Sigma := \{a_1, \ldots, a_m, \infty\}$; also choose $a \in \Omega := \mathbf{S} \backslash \Sigma$. Then, a well-defined monodromy representation $\pi_1(\Omega; a) \to \mathrm{GL}(\mathcal{F}_{A,a})$ is attached to each system $X' = AX$ holomorphic on Ω. Up to a choice of a fundamental matricial solution of S_A at a, we also get a matricial representation $\pi_1(\Omega; a) \to \mathrm{GL}_n(\mathbf{C})$, but this one is only defined up to conjugacy. As we already saw, one can get rid of that ambiguity by requiring that the fundamental matricial solution $\mathcal{X}$ satisfy the initial condition $\mathcal{X}(a) = I_n$, but as we also saw (Example 7.42), this might not be a good idea.

We shall restrict to systems which are regular singular at each point of Σ. To abbreviate, we shall call them RS systems (not specifying Σ, which is fixed for the whole chapter). In particular, for an RS system, A must be meromorphic on the whole of $\mathbf{S}$, thus rational: $A \in \mathrm{Mat}_n(\mathbf{C}(z))$. If $\mathcal{X}$ is a fundamental matricial solution of S_A at a, then it defines a multivalued invertible matrix which has moderate growth in sectors in the neighborhood of every point of Σ. As a consequence:

Lemma 12.1. *If $F : A \to B$, $F \in \mathrm{GL}_n(\mathcal{M}(\Omega))$, is a meromorphic equivalence on Ω between two RS systems, then it is a rational equivalence:* $F \in \mathrm{GL}_n(\mathbf{C}(z))$.

Proof. Let $\mathcal{X}, \mathcal{Y}$ be fundamental matricial solutions for A, B. Then $F\mathcal{X} = \mathcal{Y}P$ with $P \in \mathrm{GL}_n(\mathbf{C})$ and $F = \mathcal{Y}P\mathcal{X}^{-1}$ is uniform and has moderate growth near points of Σ, so that it is meromorphic at those points, thus on the whole of $\mathbf{S}$ and therefore rational. $\square$

We are going to consider rational equivalence of RS systems. We proved in Section 7.6 (under much more general assumptions) that two equivalent systems have conjugate monodromy representations. Therefore, we have a well-defined mapping:

$$\left\{ \begin{matrix} \text{rational equivalence classes} \\ \text{of RS systems} \end{matrix} \right\} \longrightarrow \left\{ \begin{matrix} \text{conjugacy classes of linear} \\ \text{representations of } \pi_1(\Omega; a) \end{matrix} \right\}.$$

This is the Riemann-Hilbert correspondence[1] in its *global* form.

Proposition 12.2. *The above mapping is injective.*

Proof. Suppose the RS systems with matrices A and B give rise to conjugate monodromy representations. We must show that they are rationally equivalent. We choose fundamental systems $\mathcal{X}$ and $\mathcal{Y}$ and write $\mathcal{M}_\lambda, N_\lambda$ as the monodromy matrices, so that $\mathcal{X}^\lambda = \mathcal{X}M_\lambda$ and $\mathcal{Y}^\lambda = \mathcal{Y}N_\lambda$ for each loop λ in Ω based at a. The assumption is that there exists $P \in \mathrm{GL}_n(\mathbf{C})$ such that $PM_\lambda = N_\lambda P$ for all λ. We put $F := \mathcal{Y}P\mathcal{X}^{-1}$. Then F is uniform:

$$F^\lambda = \mathcal{Y}^\lambda P(\mathcal{X}^\lambda)^{-1} = \mathcal{Y}N_\lambda P M_\lambda^{-1} \mathcal{X}^{-1} = \mathcal{Y}P\mathcal{X}^{-1} = F.$$

It is meromorphic on Ω and has moderate growth near points of Σ, therefore it is meromorphic on $\mathbf{S}$, thus rational. Last, from $F\mathcal{X} = \mathcal{Y}P$ one concludes as usual that $F[A] = B$. $\square$

12.2. The twenty-first problem of Hilbert

At the International Congress of Mathematicians held in 1900 in Paris, Hilbert stated 23 problems meant to inspire mathematicians for the new century. (And so they did; see the two-volume book "Mathematical developments arising from Hilbert problems" edited by Felix Browder and published by the AMS.) In the twenty-first problem, he asked "to show that there always exists a linear differential equation of the fuchsian class with given singular points and monodromy group". The problem admits various interpretations (systems or equations? of the first kind, *i.e.*, fuchsian, or regular singular? with apparent singularities or only true singularities?); see [**AB94, Del70, IKSY91, Sab93, Yos87**] as well as the above-quoted book on Hilbert problems. Some variants have a positive answer, some have a negative or conditional answer.

---

[1]Much more general versions are possible, using vector bundles on Riemann surfaces; see [**AB94, Del70, Sab93**], and some hints in Section 17.1 of Chapter 17.

Remark 12.3. The fact that the various formulations give quite different answers was long overlooked. A complete clarification and almost complete solution of the problem was first obtained by Bolibrukh and then completed in the following years by himself and others; see [**AB94**] for a clear exposition.

Avoiding these subtleties, we are going to sketch a proof of the following:

Theorem 12.4. *Any representation $\pi_1(\Omega; a) \to \mathrm{GL}_n(\mathbf{C})$ can be realized (up to conjugacy) as the monodromy representation of a system S_A which is of the first kind at $a_1, \ldots, a_m$ and which is regular singular at ∞ (see Chapter 9 for the terminology).*

Corollary 12.5. *The mapping defined in the previous section (Riemann-Hilbert correspondence) is bijective.*

Corollary 12.6. *Any RS system is equivalent to one of the forms described in the theorem.*

Remark 12.7. As a consequence, RS differential systems can be classified by purely algebraic-topological objects, the representations of the fundamental group (of which the algebraic description is perfectly known). Note however that in the simplest nontrivial case, that is, $m = n = 2$, we already get a complicated problem, that of classifying the linear two-dimensional representations of a free group on two generators. Generally speaking, the case $m = 2$, n arbitrary, is not well understood. See for instance the paper by Deligne "Le groupe fondamental de la droite projective moins trois points".

The proof of the theorem will proceed in two main steps: on $\mathbf{C}$, then at infinity.

First step: Resolution on C. Up to a rotation, we may assume that all $\Re(a_k)$ are distinct; and up to reindexing, that $\Re(a_1) < \cdots < \Re(a_m)$. Up to conjugacy of the monodromy representation, we can also change the base point and assume that $\Re(a) < \Re(a_1)$. We then define closed rectangles $R_1, \ldots, R_m$ with vertical and horizontal sides, all having the same vertical coordinates and such that: a belongs to none of the R_k; each a_k belongs to the interior of R_k and belongs to no R_l with $l \neq k$; and any two consecutive rectangles overlap, *i.e.*, they have common interior points.

We consider the prescribed monodromy representation as defined through monodromy matrices $M_1, \ldots, M_m$, each M_k corresponding to a small positive loop in $\mathbf{C}$ around a_k (and around no other a_l). We first solve the problem separately in a neighborhood of each rectangle by using

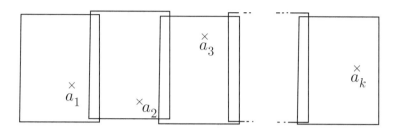

Figure 12.1. First step in global Riemann-Hilbert correspondence

the local theory of Chapter 9: this defines for each k a matrix A_k of the first kind and a fundamental matricial solution $\mathcal{X}_k$ having prescribed monodromy matrix M_k.

Suppose the problem has been solved in a neighborhood of the rectangle $R' := R_1 \cup \cdots \cup R_k$, where $k < m$. Call A' and $\mathcal{X}'$ the corresponding system (of the first kind) and fundamental matrix solution (with monodromy matrices $M_1, \ldots, M_k$). Then $\mathcal{X}'$ and $\mathcal{X}_{k+1}$ are analytic and uniform in some simply-connected neighborhood of the rectangle $R' \cap R_{k+1}$. According to Cartan's lemma stated below, there exist analytic invertible matrices H' in a neighborhood of R' and H_{k+1} in a neighborhood of R_{k+1} such that $H'\mathcal{X}' = H_{k+1}\mathcal{X}_{k+1}$ in a neighborhood of $R' \cap R_{k+1}$. This implies that $H'[A'] = H_{k+1}[A_{k+1}]$ in the same neighborhood, and so that they can be glued into a matrix of the first kind A'' in a neighborhood of $R'' := R' \cup R_{k+1}$, having as fundamental solution the glueing $\mathcal{X}''$ of $H'\mathcal{X}'$ and of $H_{k+1}\mathcal{X}_{k+1}$, which has monodromy matrices $M_1, \ldots, M_{k+1}$. Iterating, we solve the problem in a neighborhood of the rectangle $R_1 \cup \cdots \cup R_m$.

Theorem 12.8 (Cartan's lemma). *Let $K' := [a_1, a_3] + \mathrm{i}\,[b_1, b_2]$ and $K'' := [a_2, a_4] + \mathrm{i}\,[b_1, b_2]$, where $a_1 < a_2 < a_3 < a_4$ and $b_1 < b_2$, so that $K := K' \cap K'' = [a_2, a_3] + \mathrm{i}\,[b_1, b_2]$. Let F be an invertible analytic matrix in a neighborhood of K. Then there exist an invertible analytic matrix F' in a neighborhood of K' and an invertible analytic matrix F'' in a neighborhood of K'' such that $F = F'F''$ in a neighborhood of K.* □

For a proof, see [**GR09**].

Second step: Taking into account ∞. The problem has now been solved in a neighborhood U_0 of a rectangle R containing $a_1, \ldots, a_m$. Up to a translation, we can assume that this rectangle contains 0. There is an analytic contour $C \subset U_0$ containing R; this means a simple closed curve $t \mapsto C(t)$, $[0, 1] \to \mathbf{C}^*$, defined by the restriction to $[0, 1]$ of an analytic function; for a proof of the existence of such a contour, see the chapters of [**Ahl78, Car63, Rud87**] devoted to the Riemann mapping theorem. We

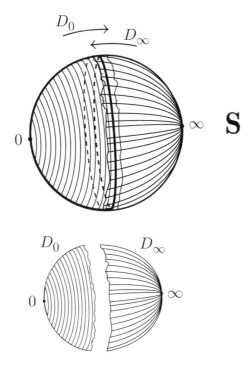

Figure 12.2. Second step in global Riemann-Hilbert correspondence

call A_0 the matrix of the first kind that solves the problem on U_0 and $\mathcal{X}_0$ a corresponding fundamental matricial solution, having monodromy matrices $M_1, \ldots, M_m$.

We also solve the problem locally at infinity in a neighborhood U_∞ of ∞, with a matrix A_∞ of the first kind and a fundamental matricial solution $\mathcal{X}_\infty$ having as monodromy matrix $M_\infty := (M_1 \cdots M_m)^{-1}$.

Now, by Jordan's theorem, $\mathbf{S} \setminus C$ has two connected components each homeomorphic to a disk (see [**Ful95**] for a proof). We call D_0 the connected component containing 0 and D_∞ the connected component containing ∞. Thus U_0 is a neighborhood of $\overline{D_0}$. There is a neighborhood U_∞ of $\overline{D_\infty}$ such that $\mathcal{X}_0$ and $\mathcal{X}_\infty$ are analytic and invertible in the neighborhood $U_0 \cap U_\infty$ of C. Now, $\mathcal{X}_0$ and $\mathcal{X}_\infty$ are multivalued, but they have the same monodromy along C, so that $M := \mathcal{X}_\infty \mathcal{X}_0^{-1}$ is uniform in a neighborhood of C. We conclude from Birkhoff's preliminary theorem, stated below, that there exist a neighborhood V_0 of $\overline{D_0}$, a neighborhood V_∞ of $\overline{D_\infty}$, an analytic invertible matrix M_0 on V_0 and an analytic matrix M_∞ on V_∞ which is invertible on $V_\infty \setminus \{\infty\}$ (but maybe not at ∞) such that $M_0 \mathcal{X}_0 = M_\infty \mathcal{X}_\infty$ on $V_0 \cap V_\infty$.

This implies that $M_0[A_0]$ and $M_\infty[A_\infty]$ coincide on $V_0 \cap V_\infty$. Glueing $M_0[A_0]$ and $M_\infty[A_\infty]$, we get the desired system with matrix A. Note that, since we are not sure that M_∞ is invertible at ∞, we cannot guarantee that A is of the first kind at ∞.

Theorem 12.9 (Preliminary theorem of Birkhoff). *With C, D_0, D_∞ as above, suppose we have an invertible analytic matrix M in a neighborhood of C. Then there exist a neighborhood V_0 of $\overline{D_0}$, a neighborhood V_∞ of $\overline{D_\infty}$, an analytic invertible matrix M_0 on V_0 and an analytic matrix M_∞ on V_∞ which is invertible on $V_\infty \backslash \{\infty\}$ (but maybe not at ∞) such that $M_0 = M_\infty M$ on $V_0 \cap V_\infty$.*

For a direct proof (without the theory of vector bundles, but using some functional analysis) see [**Bir13**]. For a proof (originally due to Röhrl) based on the theory of vector bundles on a Riemann surface, see [**For81**].

Exercises

(1) Generalize Lemma 12.1 to the case of meromorphic morphisms $F \in \mathrm{Mat}_{p,n}(\mathcal{M}(\Omega))$.

(2) We consider the following equivalence relation on pairs of matrices of $\mathrm{GL}_n(\mathbf{C})$:

$$(M, N) \sim (M', N') \Longleftrightarrow \exists P \in \mathrm{GL}_n(\mathbf{C}) : \ M' = PMP^{-1} \text{ and } N' = PNP^{-1}.$$

  (i) What is the relevance to Remark 12.7?
  (ii) Try to find a classification similar to that for matrices (such as Jordan form, or invariant factors).

(3) Redo more explicitly the first step of the proof of the classification theorem when $n = 1$.

Part 4

Differential Galois Theory

Chapter 13

Local differential Galois theory

The most characteristic feature of the monodromy action is that it preserves algebraic and differential relations (Section 5.2). We are going to axiomatize this fact and study the group of all transformations having that property. Indeed, an old saying in the field says that "differential Galois theory is what the algebra can say about the dynamics". This by the way can be extended to much more general dynamical systems than linear differential equations.

In this chapter and the following, we perform the *local* study. Therefore, equations and systems will have coefficients in the field $K := \mathbf{C}(\{z\})$ of germs of meromorphic functions at 0. This field,[1] equipped with the derivation $D := d/dz$, is a *differential field*. Note that we can (and many times will) use equivalently the derivation $\delta := z\,d/dz$.

Remark 13.1. The standard approach in this domain has for a long time been to consider extensions of K which are themselves differential *fields*. However, a different point of view begins to spread, relying on the use of differential *algebras* instead. First of all, in spite of the fact that classical Galois theory has accustomed us to the theory of fields, they are less natural in the context of differential equations. For instance, $1/\log$ is not a solution of a linear differential equation with coefficients in K (see the exercises at the end of the chapter). Another reason is that in the allied domains of difference and q-difference equations, there appear differential algebras which are not

---

[1] It was previously denoted $\mathcal{M}_0$, but in this algebraic setting we prefer the more neutral letter K.

integral rings and therefore cannot be embedded in a difference field. Since Birkhoff (see for instance [**Bir13**]), mathematicians have been trying to give as unified a treatment as possible. So here, I choose to work with algebras.

13.1. The differential algebra generated by the solutions

In the case of an equation $E_{\underline{a}}$ with coefficients $a_1, \ldots, a_n \in K$, we consider a fundamental system of solutions at some point $z_0 \neq 0$, say $\mathcal{B} = (f_1, \ldots, f_n)$, which is a basis for $\mathcal{F}_{\underline{a}, z_0}$. In order to be able to express the preservation of algebraic and differential relations, we have to "close" the space of solutions under multiplication and derivation. This is most easily done in the case of systems. (We come back to equations afterwards.)

Let $A \in \mathrm{Mat}_n(K)$ and let $\mathcal{X}$ be a fundamental matricial solution of S_A at some point $z_0 \neq 0$. We shall write $\mathcal{A}(A, z_0)$ as the K-algebra $K[\mathcal{X}]$ generated by the coefficients $x_{i,j}$ of $\mathcal{X}$. All the elements of $\mathcal{A}(A, z_0)$ are polynomial expressions in all the $x_{i,j}$ with coefficients in K. Thus, if we consider the morphism of K-algebras from $K[T_{1,1}, \ldots, T_{n,n}]$ (polynomials in n^2 indeterminates with coefficients in K) to $\mathcal{O}_{z_0}$ defined by $T_{i,j} \mapsto x_{i,j}$, the image of this morphism is $\mathcal{A}(A, z_0)$. The absence of $\mathcal{X}$ in the notation is justified by the fact that, if $\mathcal{Y}$ is another fundamental matricial solution at z_0, then $\mathcal{Y} = \mathcal{X}P$ for some $P \in \mathrm{GL}_n(\mathbf{C})$; therefore, each $y_{i,j} = \sum p_{i,k} x_{k,j}$ belongs to the algebra generated by the $x_{i,j}$, and conversely since P is invertible.

Here are the basic facts about $\mathcal{A}(A, z_0)$:

(1) As already noted, it does not depend on the choice of a fundamental matricial solution at z_0.

(2) If $A \sim B$ (meromorphic equivalence at 0), then $\mathcal{A}(A, z_0) = \mathcal{A}(B, z_0)$. Indeed, let $F \in \mathrm{GL}_n(K)$ such that $F[A] = B$ and let $\mathcal{X}$ be a fundamental system of solutions of S_A at z_0. Then $\mathcal{Y} := F\mathcal{X}$ is a fundamental system of solutions of S_B at z_0. Now, the relations $y_{i,j} = \sum f_{i,k} x_{k,j}$ and the converse relations (using F^{-1}) show that the $x_{i,j}$ and the $y_{i,j}$ generate the same K-algebra.

(3) $\mathcal{A}(A, z_0)$ is a sub-differential algebra of the differential K-algebra $\mathcal{O}_{z_0}$. Indeed, it is by definition a sub-algebra. Using Leibniz' rule, one shows that $D^{-1}(\mathcal{A}) = \{f \in \mathcal{O}_{z_0} \mid D(f) \in \mathcal{A}\}$ is a sub-algebra of $\mathcal{O}_{z_0}$, so that it is enough to check that it contains a family of generators of $\mathcal{A}(A, z_0)$, i.e., that D sends the generators of $\mathcal{A}(A, z_0)$ into itself; but this follows from the differential system since $D(x_{i,j}) = \sum a_{i,k} x_{k,j}$.

Exercise 13.2. Show rigorously that $D^{-1}(\mathcal{A})$ is a sub-algebra of $\mathcal{O}_{z_0}$.

(4) If z_1 is another point at which A is defined and γ is a path from z_0 to z_1, then analytic continuation along γ yields an isomorphism of differential K-algebras from $\mathcal{A}(A, z_0)$ to $\mathcal{A}(A, z_1)$. This isomorphism depends only on the homotopy class of γ in an open set avoiding the singularities of A.

(5) Let $E_{\underline{a}}$ be the equation $f^{(n)} + a_1 f^{(n-1)} + \cdots + a_n f = 0$ with coefficients $a_1, \ldots, a_n \in K$. Let $\mathcal{B} := (f_1, \ldots, f_n)$ be a fundamental system of solutions at some point $z_0 \neq 0$. Then the algebra $\mathcal{A}(\underline{a}, z_0) := \mathcal{A}(A_{\underline{a}}, z_0)$ is generated by $f_1, \ldots, f_n$ and their derivatives; it is enough to go up to the $(n-1)^{th}$ derivatives.

The proofs of the nontrivial properties of the following examples are given as exercises at the end of the chapter.

Examples 13.3. (1) Let $\alpha \in \mathbf{C}$. Then $zf' = \alpha f$ with $z_0 := 1$ gives $\mathcal{A}(A, z_0) = K[z^\alpha]$. The structure of this K-algebra depends on α in the following way:

- If $\alpha \in \mathbf{Z}$, then of course $\mathcal{A}(A, z_0) = K$.

- If $\alpha \in \mathbf{Q} \setminus \mathbf{Z}$, then write $\alpha = p/q$ with $p \in \mathbf{Z}$, $q \in \mathbf{N}^*$ and p, q coprime. Then $\mathcal{A}(A, z_0) = K[z^{1/q}]$, which is a field, an algebraic extension of degree q of K. (It is actually a cyclic Galois extension, with Galois group μ_q, the group of qth roots of unity in $\mathbf{C}$.)

- If $\alpha \in \mathbf{C} \setminus \mathbf{Q}$, then $\mathcal{A}(A, z_0) = K[z^\alpha]$ and z^α is transcendental over K, that is, the morphism of K-algebras from $K[T]$ to $\mathcal{A}(A, z_0)$ sending T to z^α is an isomorphism. Equivalently, the $(z^\alpha)^k$, $k \in \mathbf{N}$, form a basis of $K[z^\alpha]$.

(2) Consider the equation $zf'' + f' = 0$, and set $z_0 := 1$. Using the fact that $zf'' + f' = (zf')'$, we find readily that $\mathcal{B} := (1, \log)$ is a fundamental system of solutions. Since $1 \in K$ and $\log' \in K$, we have $\mathcal{A}(\underline{a}, z_0) = K[\log]$ and $\log$ is transcendental over K.

(3) Consider $\alpha \in \mathbf{C}$ and the equation $(\delta - \alpha)^2 f = 0$. A fundamental system of solutions at $z_0 := 1$ is $\mathcal{B} := (z^\alpha, z^\alpha \log)$. Since $\delta(z^\alpha) = \alpha z^\alpha$ and $\delta(z^\alpha \log z) = \alpha z^\alpha \log z + z^\alpha$, the algebra $\mathcal{A}(\underline{a}, z_0) = K[z^\alpha, z^\alpha \log]$ generated by $\mathcal{B}$ is stable under δ, whence, since z is invertible in K, it is stable under D.

(4) Consider the equation $f' + \frac{1}{z^2} f = 0$. A fundamental system of solutions is $(e^{1/z})$, so that $\mathcal{A}(\underline{a}, z_0) = K[e^{1/z}]$. Moreover, $e^{1/z}$ is transcendental over K.

13.2. The differential Galois group

Some of the more basic arguments we use work in any algebra equipped with a derivation, so we begin with some generalities.

A little bit of differential algebra. Let $\mathcal{A}$ be an arbitrary differential K-algebra, *i.e.*, a K-algebra equipped with a derivation $D : \mathcal{A} \to \mathcal{A}$ extending that of K. An automorphism for that structure is, by definition, an automorphism σ of a K-algebra (that is, an automorphism of the ring which is at the same time K-linear) such that $D \circ \sigma = \sigma \circ D$. If we write more intuitively f' for $D(f)$, this means that $\sigma(f') = (\sigma(f))'$. Note that an automorphism of a K-algebra automatically satisfies (by definition) $\sigma_{|K} = \mathrm{Id}_K$, that is, $\sigma(f) = f$ for all $f \in K$.

The following lemma shall make it easier to check that a particular σ is a differential automorphism.

Lemma 13.4. *Let $\mathcal{A}$ be a differential K-algebra and let σ be a K-algebra automorphism of $\mathcal{A}$. Let $f_1, \ldots, f_n$ be generators of $\mathcal{A}$ as a K-algebra. If $\sigma(f_k') = (\sigma(f_k))'$ for $k = 1, \ldots, n$, then σ is an automorphism of a differential K-algebra.*

Proof. Suppose that $\sigma(f') = (\sigma(f))'$ and $\sigma(g') = (\sigma(g))'$. Then, for $h := \lambda f + \mu g$, $\lambda, \mu \in K$, one has (by easy calculation) $\sigma(h') = (\sigma(h))'$; and the same is true for $h := fg$. Therefore, the set of those $f \in \mathcal{A}$ such that $\sigma(f') = (\sigma(f))'$ is a K-algebra containing $f_1, \ldots, f_n$, therefore it is equal to $\mathcal{A}$. $\qquad\square$

Examples 13.5. (1) Let $\Omega \in \mathbf{C}$ be a domain and $a \in \Omega$. Let $\tilde{\mathcal{O}}_a$ be the differential algebra of analytic germs at a that admit an analytic continuation along every path in Ω starting at a. Then, for every loop λ in Ω based at a, analytic continuation along λ yields a differential automorphism of $\tilde{\mathcal{O}}_a$.

(2) Let $A \in \mathrm{Mat}_n(K)$ and let $\mathcal{A} := \mathcal{A}(A, z_0)$. Analytic continuation along a loop based at z_0 (and contained in a punctured disk centered at 0 on which A is analytic) transforms a fundamental matricial solution $\mathcal{X}$ into $\mathcal{X}M$ for some $M \in \mathrm{GL}_n(\mathbf{C})$, therefore it transforms any of the generators $x_{i,j}$ of $\mathcal{A}$ into an element of $\mathcal{A}$ and therefore (by the same argument as in the "basic facts" of the previous section, *i.e.*, the preimage of $\mathcal{A}$ is a subalgebra) it sends $\mathcal{A}$ into itself. By considering the inverse loop and matrix, one sees that this is a bijection. By preservation of algebraic and differential relations, it is an automorphism of differential K-algebras.

Application to the differential Galois group. We return to the derivations D and δ of K and of differential K-algebras $\mathcal{A}(A, z_0)$, $\mathcal{A}(\underline{a}, z_0)$. A useful remark is given by the following exercise.

Exercise 13.6. With as usual $D := d/dz$ and $\delta := zd/dz$, show that for any K-algebra automorphism σ of $\mathcal{A}$:
$$\sigma D = D\sigma \Longleftrightarrow \sigma\delta = \delta\sigma.$$

In all the following examples, we use the abbreviation $\mathcal{A}$ for $\mathcal{A}(A, z_0)$ or $\mathcal{A}(\underline{a}, z_0)$.

Examples 13.7. (1) Let $zf' = \alpha f$, $\alpha \in \mathbf{C}$ and set $z_0 := 1$. We distinguish three cases:

- If $\alpha \in \mathbf{Z}$, then $\mathcal{A} = K$ and the only differential automorphism is the identity of K.

- If $\alpha \in \mathbf{Q} \setminus \mathbf{Z}$, $\alpha = p/q$, then we saw that $\mathcal{A} = K[z^{1/q}]$. By standard algebra (for instance, field theory in the book of Lang), the automorphisms of the K-algebra of $\mathcal{A}$ are defined by $z^{1/q} \mapsto jz^{1/q}$, where $j \in \mu_q$. Now it follows from the lemma above that all of them are differential automorphisms.

- If $\alpha \in \mathbf{C} \setminus \mathbf{Q}$, then we saw that $\mathcal{A} = K[z^\alpha]$ and that z^α is transcendental over K. For every differential automorphism σ of $\mathcal{A}$, $\sigma(z^\alpha)$ must be a nontrivial solution of $zf' = \alpha f$, thus $\sigma(z^\alpha) = \lambda z^\alpha$ for some $\lambda \in \mathbf{C}^*$. Conversely, this formula defines a unique automorphism σ of the K-algebra $\mathcal{A}$; and since this σ satisfies the condition $\sigma(f') = (\sigma(f))'$ for the generator z^α, by the lemma, it is a differential automorphism.

(2) Let $zf'' + f' = 0$ and $z_0 := 1$, so that $\mathcal{A} = K[\log]$ (and $\log$ is transcendental over K). From $\log' = 1/z$ one deduces that $(\sigma(\log))' = \sigma(\log') = \sigma(1/z) = 1/z$ for every differential automorphism σ. Thus, $\sigma(\log) = \log + \mu$ for some $\mu \in \mathbf{C}$. Conversely, since $\log$ is transcendental over K, this defines a unique automorphism of the K-algebra $\mathcal{A}$; and, since it satisfies the condition $\sigma(f') = (\sigma(f))'$ for the generator $\log$, by the lemma, it is a differential automorphism.

(3) Let $(\delta - \alpha)^2 f = 0$, $\alpha \notin \mathbf{Q}$ and $z_0 := 1$. Then $\mathcal{A} = K[z^\alpha, z^\alpha \log]$. As in the first example, one must have $\sigma(z^\alpha) = \lambda z^\alpha$ for some $\lambda \in \mathbf{C}^*$. Then, from $\delta(z^\alpha \log) = \alpha z^\alpha \log + z^\alpha$, we see that $f := \sigma(z^\alpha \log)$ must satisfy $\delta(f) = \alpha f + \lambda z^\alpha$, so that $g := f - \lambda z^\alpha \log$ satisfies $\delta(g) = \alpha g$, so that $g = \mu z^\alpha$ for some $\mu \in \mathbf{C}$. Therefore, we find that $\sigma(z^\alpha \log) = \lambda z^\alpha \log + \mu z^\alpha$. For the converse, assume that $\alpha \in \mathbf{C} \setminus \mathbf{Q}$. Then since z^α and $z^\alpha \log$ are algebraically independent (see the exercises at the end of the chapter), for any $(\lambda, \mu) \in \mathbf{C}^* \times \mathbf{C}$, there is a unique automorphism σ of the K-algebra $\mathcal{A}$

such that $\sigma(z^\alpha) = \lambda z^\alpha$ and $\sigma(z^\alpha \log) = \lambda z^\alpha \log + \mu z^\alpha$. Since σ satisfies the condition $\sigma(f') = (\sigma(f))'$ for the generators z^α and $z^\alpha \log$, by the lemma, it is a differential automorphism. Note that it all works as if we had computed $\sigma(\log) = \log + \nu$ and set $\mu = \lambda\nu$ but we could not do so because $\log \notin \mathcal{A}$ and $\sigma(\log)$ is therefore not defined.

Definition 13.8. The *differential Galois group* of S_A at z_0, written as $\mathrm{Gal}(A, z_0)$, is the group of all differential automorphisms of $\mathcal{A}(A, z_0)$. The *differential Galois group* of $E_{\underline{a}}$ at z_0 is $\mathrm{Gal}(A_{\underline{a}}, z_0)$.

In the examples below, we call the differential Galois group G for short.

Examples 13.9. (1) Let $zf' = \alpha f$, $\alpha \in \mathbf{C}$, and set $z_0 := 1$. If $\alpha \in \mathbf{Z}$, then $G = \{\mathrm{Id}\}$. If $\alpha \in \mathbf{Q} \setminus \mathbf{Z}$, $\alpha = p/q$, then the map $j \mapsto (f(z) \mapsto f(jz))$ is an isomorphism of μ_q with G. If $\alpha \in \mathbf{C} \setminus \mathbf{Q}$, then the map $\lambda \mapsto (f(z) \mapsto f(\lambda z))$ is an isomorphism of $\mathbf{C}^*$ with G. We write somewhat improperly $G = \mu_q$ in the first case and $G = \mathbf{C}^*$ in the second case; but it is important not only to describe the group G but also to keep track of how it operates.

(2) Let $zf'' + f' = 0$ and $z_0 := 1$; then $G = \mathbf{C}$.

(3) Let $(\delta - \alpha)^2 f = 0$ and $z_0 := 1$. Then G can be identified with $\mathbf{C}^* \times \mathbf{C}$. The element $(\lambda, \mu) \in G$ corresponds to the automorphism σ of the K-algebra $\mathcal{A} = K[z^\alpha, z^\alpha \log]$ defined by $\sigma(z^\alpha) = \lambda z^\alpha$ and $\sigma(z^\alpha \log) = \lambda z^\alpha \log + \mu z^\alpha$. In order to see $\mathbf{C}^* \times \mathbf{C}$ as a group, we must understand how to compose elements. So write $\sigma_{\lambda,\mu}$ as the automorphism just defined. Then:

$$\sigma_{\lambda',\mu'} \circ \sigma_{\lambda,\mu}(z^\alpha) = \sigma_{\lambda',\mu'}(\lambda z^\alpha)$$
$$= \lambda'\lambda z^\alpha,$$
$$\sigma_{\lambda',\mu'} \circ \sigma_{\lambda,\mu}(z^\alpha \log) = \sigma_{\lambda',\mu'}(\lambda z^\alpha \log + \mu z^\alpha)$$
$$= \lambda'\lambda z^\alpha \log + (\lambda\mu' + \lambda'\mu)z^\alpha,$$

so that the group law on $\mathbf{C}^* \times \mathbf{C}$ is:

$$(\lambda', \mu') * (\lambda, \mu) := (\lambda'\lambda, \lambda\mu' + \lambda'\mu).$$

There is yet another way to understand this group. We notice that σ is totally determined by its $\mathbf{C}$-linear action on the space of solutions. The latter has as basis $\mathcal{B} = (z^\alpha, z^\alpha \log)$ and we find $\sigma(\mathcal{B}) = \mathcal{B}M$, where $M = \begin{pmatrix} \lambda & \mu \\ 0 & \lambda \end{pmatrix}$. Therefore, G can be identified with the subgroup:

$$\left\{ \begin{pmatrix} \lambda & \mu \\ 0 & \lambda \end{pmatrix} \mid (\lambda, \mu) \in \mathbf{C}^* \times \mathbf{C} \right\} \subset \mathrm{GL}_2(\mathbf{C}).$$

The reader can check that the multiplication law coincides with the one found above.

Exercise 13.10. In the first example, how does $G = \mathbf{C}$ operate?

The basic *transcendental* facts about the differential Galois group are:

(1) The monodromy group is a subgroup of the differential Galois group. To explain more precisely what this means, we consider only the case of a system A, leaving it to the reader to deal with the case of a scalar equation. We know that both $\text{Mon}(A, z_0)$ and $\text{Gal}(A, z_0)$ are embedded into $\text{GL}(\mathcal{F}_{A,z_0})$. Identifying the two former with subgroups of the latter, the inclusion $\text{Mon}(A, z_0) \subset \text{Gal}(A, z_0)$ makes sense.

(2) If γ is a path from z_0 to z_1 and if we write ψ as the corresponding differential isomorphism from $\mathcal{A}(A, z_0)$ to $\mathcal{A}(B, z_0)$, then $\sigma \mapsto \psi \circ \sigma \circ \psi^{-1}$ is an isomorphism from the group $\text{Gal}(A, z_0)$ to the group $\text{Gal}(B, z_0)$.

Exercise 13.11. Show that two such isomorphisms from $\text{Gal}(A, z_0)$ to $\text{Gal}(B, z_0)$ differ by a conjugation by an element of $\text{Mon}(A, z_0)$.

Since the monodromy group is a subgroup of the Galois group, we should find it inside each of the examples we computed. This is indeed so:

Examples 13.12. (1) In the first example, the monodromy group is generated, as a subgroup of the Galois group identified with a subgroup of $\mathbf{C}^*$, by the factor $\lambda_0 := e^{2i\pi\alpha}$. If $\alpha \in \mathbf{Q}$, it is equal to the whole Galois group.

(2) In the second example, the monodromy group is generated, as a subgroup of the Galois group identified with $\mathbf{C}$, by the constant $\mu_0 := 2i\pi$.

(3) In the third case, the monodromy group is generated, as a subgroup of the Galois group identified with a subgroup of $\text{GL}_2(\mathbf{C})$, by the pair $(\lambda_0, \mu_0) := (e^{2i\pi\alpha}, 2i\pi e^{2i\pi\alpha})$; or, in the matricial realization, by the matrix:

$$\begin{pmatrix} e^{2i\pi\alpha} & 2i\pi e^{2i\pi\alpha} \\ 0 & e^{2i\pi\alpha} \end{pmatrix} = \begin{pmatrix} e^{2i\pi\alpha} & 0 \\ 0 & e^{2i\pi\alpha} \end{pmatrix} \begin{pmatrix} 1 & 2i\pi \\ 0 & 1 \end{pmatrix}.$$

13.3. The Galois group as a linear algebraic group

Let S_A be a system with coefficients in K, let $\mathcal{A} := \mathcal{A}(A, z_0)$ be its algebra of solutions at some point z_0 and let $\text{Gal} := \text{Gal}(A, z_0)$ be its differential Galois group. For every fundamental matricial solution $\mathcal{X}$ at z_0 and every $\sigma \in G$, the matrix $\sigma(\mathcal{X})$ has to be invertible because $\det \sigma(\mathcal{X}) = \sigma(\det \mathcal{X}) \neq 0$; and it is another fundamental matricial solution because of the following calculation:

$$(\sigma(\mathcal{X}))' = \sigma(\mathcal{X}') = \sigma(A\mathcal{X}) = A\sigma(\mathcal{X}).$$

This implies that $\sigma(\mathcal{X}) = \mathcal{X}M$ for some $M \in \text{GL}_n(\mathbf{C})$. We write M_σ as this matrix M, and so we have a map $\sigma \mapsto M_\sigma$ from Gal to $\text{GL}_n(\mathbf{C})$. Of course, we hope it is a representation. It is indeed a morphism (not an anti-morphism) of groups

$$\mathcal{X} M_{\sigma\tau} = (\sigma\tau)(\mathcal{X}) = \sigma(\tau(\mathcal{X})) = \sigma(\mathcal{X} M_\tau) = \sigma(\mathcal{X}) M_\tau = \mathcal{X} M_\sigma M_\tau$$
$$\implies M_{\sigma\tau} = M_\sigma M_\tau.$$

Moreover, it is injective: for if $M_\sigma = I_n$, then $\sigma(\mathcal{X}) = \mathcal{X}$ so that the morphism of algebras σ leaves fixed all the generators of the K-algebra $\mathcal{A}$, so that it is actually the identity of $\mathcal{A}$.

Proposition 13.13. *The map $\sigma \mapsto M_\sigma$ realizes an isomorphism of $\mathrm{Gal}(A, z_0)$ with a subgroup of $\mathrm{GL}_n(\mathbf{C})$ which contains the matricial monodromy group relative to $\mathcal{X}$.* ☐

We shall call the *matricial differential Galois group (relative to the fundamental system $\mathcal{X}$)*, or sometimes the *matricial realization of the differential Galois group (relative to the fundamental system $\mathcal{X}$)* the image of $\mathrm{Gal}(A, z_0)$ by the map $\sigma \mapsto M_\sigma$, thus a subgroup of $\mathrm{GL}_n(\mathbf{C})$.

Note that if F is a gauge equivalence of A with B and if we use $\mathcal{X}$ as above to realize the Galois group of A, then using $\mathcal{Y} := F\mathcal{X}$ for B we obtain exactly the same matricial Galois group, *i.e.*, the same subgroup of $GL_n(\mathbf{C})$. On the other hand, if we use a different fundamental matricial solution $\mathcal{X}'$ for A, $\mathcal{X}' = \mathcal{X} C$, $C \in \mathrm{GL}_n(\mathbf{C})$, then the corresponding matricial realization of the Galois group of A are conjugated by C.

Exercise 13.14. Formulate precisely the last statement and prove it.

Now, we are going to compare the matricial monodromy and Galois groups for our three favorite examples.

Examples 13.15. (1) In the first example, $n = 1$ and $\mathrm{GL}_1(\mathbf{C}) = \mathbf{C}^*$. The monodromy group is $\mathrm{Mon} = \langle e^{2i\pi\alpha} \rangle$. If $\alpha = p/q$, p, q coprime, then $\mathrm{Gal} = \mathrm{Mon} = \mu_q$. If $\alpha \notin \mathbf{Q}$, then $\mathrm{Gal} = \mathbf{C}^*$ and the inclusion $\mathrm{Mon} \subset \mathrm{Gal}$ is strict.

(2) In the second example, $n = 2$ and the matricial monodromy and Galois group are:

$$\mathrm{Mon} = \left\{ \begin{pmatrix} 1 & 2i\pi k \\ 0 & 1 \end{pmatrix} \mid k \in \mathbf{Z} \right\} \subset \mathrm{Gal} = \left\{ \begin{pmatrix} 1 & \mu \\ 0 & 1 \end{pmatrix} \mid \mu \in \mathbf{C} \right\} \subset \mathrm{GL}_2(\mathbf{C}).$$

(3) In the third example, $n = 2$; assuming again $\alpha \notin \mathbf{Q}$, the matricial monodromy and Galois group are:

$$\text{Mon} = \left\{ \begin{pmatrix} e^{2\mathrm{i}\pi\alpha k} & 2\mathrm{i}\pi k e^{2\mathrm{i}\pi\alpha k} \\ 0 & e^{2\mathrm{i}\pi\alpha k} \end{pmatrix} \mid k \in \mathbf{Z} \right\} \subset \text{Gal}$$

$$= \left\{ \begin{pmatrix} \lambda & \mu \\ 0 & \lambda \end{pmatrix} \mid \lambda \in \mathbf{C}^*, \mu \in \mathbf{C} \right\} \subset \text{GL}_2(\mathbf{C}).$$

The big difference between Mon and Gal is that the Galois group can in all cases be defined within $\text{GL}_2(\mathbf{C})$ by a set of algebraic equations:

(1) In the first example, if $\alpha = p/q$, p, q coprime, then:

$$\forall a \in \text{GL}_1(\mathbf{C}), \ a \in \text{Gal} \iff a^q = 1.$$

If $\alpha \notin \mathbf{Q}$, the set of equations is empty.

(2) In the second example:

$$\forall \begin{pmatrix} a & b \\ c & d \end{pmatrix} \in \text{GL}_2(\mathbf{C}), \ \begin{pmatrix} a & b \\ c & d \end{pmatrix} \in \text{Gal} \iff a = d = 1 \text{ and } c = 0.$$

(3) In the third example:

$$\forall \begin{pmatrix} a & b \\ c & d \end{pmatrix} \in \text{GL}_2(\mathbf{C}), \ \begin{pmatrix} a & b \\ c & d \end{pmatrix} \in \text{Gal} \iff a = d \text{ and } c = 0.$$

Except in the first very special case of a finite monodromy group, there is no corresponding description for Mon.

Theorem 13.16. *The matricial realization of the differential Galois group of A is a* linear algebraic group, *that is, a subgroup of $\text{GL}_n(\mathbf{C})$ defined by polynomial equations in the coefficients.*

Proof. We shall write $\mathcal{A} := K[x_{1,1}, \ldots, x_{n,n}]$, where the $x_{i,j}$ are the coefficients of a fundamental matricial solution $\mathcal{X}$. The matricial realization of the differential Galois group of A is the set of matrices $P \in \text{GL}_n(\mathbf{C})$ such that there is a morphism of K-algebras from $\mathcal{A}$ to itself sending $\mathcal{X}$ to $\mathcal{X}P$, that is, each generator $x_{i,j}$ to $\sum x_{i,k} p_{k,j}$. Indeed, we then get automatically from the relation $\mathcal{X}' = A\mathcal{X}$:

$$\sigma(\mathcal{X}') = \sigma(A\mathcal{X}) = A\sigma(\mathcal{X}) = A\mathcal{X}P = \mathcal{X}'P = (\sigma(\mathcal{X}))',$$

so that the relation $\sigma(x'_{i,j}) = (\sigma(x_{i,j}))'$ is true for all generators $x_{i,j}$ of $\mathcal{A}$, and so after the lemma of Section 13.2, such a σ is indeed a *differential automorphism*.

To say that $x_{i,j} \mapsto y_{i,j} := \sum x_{i,k} p_{k,j}$ comes from a morphism of K-algebras from $\mathcal{A}$ to itself is equivalent to saying that, for each polynomial relation $F(x_{1,1}, \ldots, x_{n,n}) = 0$ with coefficients in K, the corresponding relation for the $y_{i,j}$ holds: $F(y_{1,1}, \ldots, y_{n,n}) = 0$. We shall express this in a slightly different way. We call $(M_\alpha)_{\alpha \in I}$ the family of all monomials $M_\alpha(x_{1,1}, \ldots, x_{n,n}) = \prod x_{i,j}^{\alpha_{i,j}}$ in the $x_{i,j}$. Therefore, each index α is a matrix $(\alpha_{i,j})$ of exponents, and the set I is the set of such α. For each $\alpha \in I$, we then call N_α the corresponding monomial with the $x_{i,j}$ replaced by the $y_{i,j}$. It is a polynomial expression in the $x_{i,j}$, and therefore a linear combination of the M_α:

$$N_\alpha(x_{1,1}, \ldots, x_{n,n}) = M_\alpha(y_{1,1}, \ldots, y_{n,n}) = \prod_{i,j} y_{i,j}^{\alpha_{i,j}}$$

$$= \sum_\beta \Phi_{\alpha,\beta}(p_{1,1}, \ldots, p_{n,n}) M_\beta(x_{1,1}, \ldots, x_{n,n}),$$

where the $\Phi_{\alpha,\beta}(p_{1,1}, \ldots, p_{n,n})$ are themselves polynomial expressions in the $p_{i,j}$ with constant coefficients (actually, these coefficients are in $\mathbf{N}$). To be quite explicit, $\Phi_{\alpha,\beta}(p_{1,1}, \ldots, p_{n,n})$ is the coefficient of x^β in $\prod_{i,j}(\sum_k x_{i,k}p_{k,j})^{\alpha_{i,j}}$.

What we require for P is that, each time there is a linear relation $\sum \lambda_\alpha M_\alpha = 0$ with coefficients $\lambda_\alpha \in K$, the corresponding relation $\sum \lambda_\alpha N_\alpha = 0$ should hold. Now, the latter relation can be written $\sum \mu_\beta M_\beta = 0$, where $\mu_\beta := \sum_\alpha \lambda_\alpha \Phi_{\alpha,\beta}(p_{1,1}, \ldots, p_{n,n})$.

Lemma 13.17. *Let $E \subset K^{(I)}$ be a subspace of the space of all finitely supported families (λ_α). Then, for a family $(\phi_{\alpha,\beta})$ to have the property:*

$$\forall (\lambda_\alpha) \in K^{(I)}, \ setting \ \mu_\beta := \sum_\alpha \lambda_\alpha \phi_{\alpha,\beta}, \ one \ has \ (\lambda_\alpha) \in E \implies (\mu_\beta) \in E,$$

is equivalent to a family of K-linear conditions of the form $\sum c_{\alpha,\beta}^{(\gamma)} \phi_{\alpha,\beta} = 0$.

Proof. Left to the reader as a nice exercise in linear algebra. $\square$

We apply the lemma with E the set of all families with finite support $(\lambda_i)_{i \in I} \in K^{(I)}$ of coefficients of linear relations $\sum \lambda_\alpha M_\alpha = 0$ in $\mathcal{A}$. So there is a family $\Gamma^{(\gamma)}(p_{1,1}, \ldots, p_{n,n})$ of polynomials in the $p_{i,j}$ with coefficients in K such that $P \in \mathrm{Gal}$ is equivalent to $\Gamma^{(\gamma)}(p_{1,1}, \ldots, p_{n,n}) = 0$ for all γ. But we want polynomial equations with coefficients in $\mathbf{C}$. Therefore, we expand each $\Gamma^{(\gamma)} = \sum_k \Gamma_k^{(\gamma)} z^k$. If $\Gamma^{(\gamma)}(p_{1,1}, \ldots, p_{n,n}) = \sum_{\alpha,\beta}^{(\gamma)} c_{\alpha,\beta}^{(\gamma)} \Phi_{\alpha,\beta}(p_{1,1}, \ldots, p_{n,n})$ and $c_{\alpha,\beta}^{(\gamma)} = \sum_k c_{\alpha,\beta,k}^{(\gamma)} z^k$, then $\Gamma_k^{(\gamma)} = \sum_{\alpha,\beta} c_{\alpha,\beta,k}^{(\gamma)} \Phi_{\alpha,\beta}$. So in the end we obtain the characterization:

$$P \in \mathrm{Gal} \iff \forall \gamma, k \ , \ \Gamma_k^{(\gamma)}(p_{1,1}, \ldots, p_{n,n}) = 0.$$

Here, of course, the $\Gamma_k^{(\gamma)} \in \mathbf{C}[T_{1,1}, \ldots, T_{n,n}]$ so we do have polynomial equations with coefficients in $\mathbf{C}$. $\square$

Remark 13.18. What we found is an infinite family of polynomial equations. But a polynomial ring $\mathbf{C}[T_{1,1}, \ldots, T_{n,n}]$ is "noetherian", so that our family can be reduced to a finite set of polynomial equations. This is *Hilbert's basis theorem*; see the chapter on noetherian rings in [**Lan02**].

Exercises

(1) Prove that $1/\log$ is not a solution of a linear differential equation with coefficients in K.

(2) Prove that if $\alpha \in \mathbf{C} \setminus \mathbf{Q}$, then $z^\alpha, z^\alpha \log$ are algebraically independent over K, that is, the morphism of K-algebras from $K[T_1, T_2]$ to $\mathcal{A}(A, z_0)$ sending T_1 to z^α and T_2 to $z^\alpha \log$ is an isomorphism. Equivalently, the elements $(z^\alpha)^k (z^\alpha \log)^l$, $k, l \in \mathbf{N}$, form a basis of $K[z^\alpha, z^\alpha \log]$.

(3) Prove that $\log$ is transcendental over K using its monodromy.

(4) Prove that $e^{1/z}$ is transcendental over K using the growth rate when $z \to 0$ in $\mathbf{R}_+$.

(5) In the study of the Galois group of the equation $(\delta - \alpha)^2 f = 0$, we assumed that $\alpha \notin \mathbf{Q}$. What if $\alpha \in \mathbf{Q}$?

(6) For the equation $f' + z^{-2} f = 0$, prove that the Galois group is $\mathbf{C}^*$ while the monodromy group is trivial.

Chapter 14

The local Schlesinger density theorem

We are going to describe more precisely the Galois group for regular singular systems and prove that its matricial realization is the smallest algebraic subgroup of $\mathrm{GL}_n(\mathbf{C})$ containing the matricial realization of the monodromy group. This is the *Schlesinger density theorem* (here in its local form). We shall not discuss the possibility of extending this result to the global setting, nor to irregular equations; some information is given in Chapter 17, Sections 17.1 and 17.2.

In this chapter, we consider $A \in \mathrm{Mat}_n(K)$, $K = \mathbf{C}(\{z\})$, and suppose that S_A is regular singular. Then, we know that $A = F[z^{-1}A_0]$ for some $A_0 \in \mathrm{Mat}_n(\mathbf{C})$, so that $\mathcal{X} := Fz^{A_0}$ is a fundamental matricial system for S_A. From Chapter 13, especially Section 13.1, the matricial monodromy and Galois groups of S_A computed relative to $\mathcal{X}$ are *equal* to the matricial and monodromy groups of the system $X' = z^{-1}A_0X$ computed relative to its fundamental matricial solution z^{A_0}.

Therefore, we will from the start study a differential system $X' = z^{-1}AX$, where $A \in \mathrm{GL}_n(\mathbf{C})$. For the same reason, we can and will assume that A is in Jordan form (this is because a conjugation of A is also a gauge transformation of $z^{-1}A$). We shall write Mon and Gal as the matricial realizations of the monodromy and Galois group relative to z^A, which we shall simply call monodromy and Galois group (omitting "matricial realization").

14.1. Calculation of the differential Galois group in the semi-simple case

If A is semi-simple, we may (and will) assume that $A = \mathrm{Diag}(\alpha_1, \ldots, \alpha_n)$ and $\mathcal{X} = \mathrm{Diag}(z^{\alpha_1}, \ldots, z^{\alpha_n})$, so that $\mathcal{A} = K[z^{\alpha_1}, \ldots, z^{\alpha_n}]$.

We shall use the differential operator $\delta := zD$ instead of $D := d/dz$. We already saw that for any K-algebra automorphism σ of $\mathcal{A}$, the relation $\sigma D = D\sigma$ is equivalent to $\sigma\delta = \delta\sigma$. Thus, for any differential automorphism σ of $\mathcal{A}$, we deduce from the differential relation $\delta(z^{\alpha_i}) = \alpha_i z^{\alpha_i}$ that $f_i := \sigma(z^{\alpha_i})$ satisfies the same relation:

$$\delta(f_i) = \delta(\sigma(z^{\alpha_i})) = \sigma(\delta(z^{\alpha_i})) = \sigma(\alpha_i z^{\alpha_i}) = \alpha_i \sigma(z^{\alpha_i}) = \alpha_i f_i.$$

Therefore, f_i must be a constant multiple of z^{α_i}, with nonzero coefficient (since σ is an automorphism):

$$\sigma(z^{\alpha_i}) = \lambda_i z^{\alpha_i} \text{ for some } \lambda_i \in \mathbf{C}^*.$$

Thus the elements of the Galois group are diagonal matrices $\mathrm{Diag}(\lambda_1, \ldots, \lambda_n)$ $\in \mathrm{GL}_n(\mathbf{C})$. But which among these matrices are "galoisian automorphims", *i.e.*, realizations of differential automorphisms of $\mathcal{A}$? The monodromy of $\mathcal{X} = \mathrm{Diag}(z^{\alpha_1}, \ldots, z^{\alpha_n})$ along the basic loop (one positive turn around 0) is one of those galoisian automorphims. We write it (in this section) σ_0. Writing $a_j := e^{2\mathrm{i}\pi\alpha_j}$ for $j = 1, \ldots, n$, σ_0 is given by the matrix $\mathrm{Diag}(a_1, \ldots, a_n)$. So, along with its powers, it fits; but what else?

The condition was explained in the course of the proof of Theorem 13.16, Section 13.3: to say that $z^{\alpha_i} \mapsto \lambda_i z^{\alpha_i}$ comes from a morphism of K-algebras from $\mathcal{A}$ to itself is equivalent to saying that, for each polynomial relation $P(z^{\alpha_1}, \ldots, z^{\alpha_n}) = 0$ with coefficients in K, the corresponding relation for the $\lambda_i z^{\alpha_i}$ holds: $P(\lambda_1 z^{\alpha_1}, \ldots, \lambda_n z^{\alpha_n}) = 0$. So we shall have a closer look at the set of such equations:

$$I := \{P \in K[T_1, \ldots, T_n] \mid P(z^{\alpha_1}, \ldots, z^{\alpha_n}) = 0\}.$$

This is an *ideal* of $K[T_1, \ldots, T_n]$, *i.e.*, a subgroup such that if $P \in I$ and $Q \in K[T_1, \ldots, T_n]$, then $PQ \in I$. To describe it, we shall use the monodromy action of the fundamental loop: $\sigma_0(z^{\alpha_j}) = a_j z^{\alpha_j}$, where $a_j := e^{2\mathrm{i}\pi\alpha_j}$. Since the monodromy group is contained in the Galois group, we know that this σ_0 is an automorphism of $\mathcal{A}$ (this is just the principle of conservation of algebraic identities, introduced in Section 5.2 and recalled at the very beginning of Chapter 13). Therefore:[1]

$$P(z^{\alpha_1}, \ldots, z^{\alpha_n}) = 0 \implies P(a_1 z^{\alpha_1}, \ldots, a_n z^{\alpha_n}) = 0.$$

Calling ϕ the automorphism $P(T_1, \ldots, T_n) \mapsto P(a_1 T_1, \ldots, a_n T_n)$ of the K-algebra $K[T_1, \ldots, T_n]$, we can say that the subspace I of the K-linear space $K[T_1, \ldots, T_n]$ is stable under ϕ.

Lemma 14.1. *As a K-linear space, I is generated by the polynomials $T_1^{k_1} \cdots T_n^{k_n} - z^m T_1^{l_1} \cdots T_n^{l_n}$ such that $k_1, \ldots, k_n, l_1, \ldots, l_n \in \mathbf{N}$, $m \in \mathbf{Z}$ and $k_1 \alpha_1 + \cdots + k_n \alpha_n = l_1 \alpha_1 + \cdots + l_n \alpha_n + m$.*

Proof. If we replace each T_j by z^{α_j} in such a polynomial, we get:

$$(z^{\alpha_1})^{k_1} \cdots (z^{\alpha_n})^{k_n} - z^m (z^{\alpha_1})^{l_1} \cdots (z^{\alpha_n})^{l_n}$$

$$= z^{k_1 \alpha_1 + \cdots + k_n \alpha_n} - z^{l_1 \alpha_1 + \cdots + l_n \alpha_n + m} = 0.$$

Therefore these polynomials do belong to I.

To prove the converse, we use linear algebra. This could be done in the infinite-dimensional linear space $K[T_1, \ldots, T_n]$, where the usual theory works quite well (with some adaptations; see the subsection on infinite-dimensional spaces in Section C.1 of Appendix C), but for peace of mind we shall do it in finite-dimensional subspaces. So we take $d \in \mathbf{N}$ and denote by E the subspace $K[T_1, \ldots, T_n]_d$ of $K[T_1, \ldots, T_n]$ made up of polynomials of total degree $\deg P \leq d$. A basis of E is the family of monomials $T_1^{k_1} \cdots T_n^{k_n}$ with $k_1 + \cdots + k_n \leq d$. Since $\phi(T_1^{k_1} \cdots T_n^{k_n}) = a_1^{k_1} \cdots a_n^{k_n} T_1^{k_1} \cdots T_n^{k_n}$, we see that ϕ is a diagonalizable endomorphism of E. This means that E is the direct sum of its eigenspaces E_λ. Clearly, the monomials $T_1^{k_1} \cdots T_n^{k_n}$ with $k_1 + \cdots + k_n \leq d$ and such that $a_1^{k_1} \cdots a_n^{k_n} = \lambda$ form a basis of E_λ.

Now we set $F := I \cap E$. This is a subspace of E, which is stable under ϕ, so the restriction of ϕ to F is diagonalizable too. Therefore F is the direct sum of its eigenspaces F_λ. We are going to prove that each $F_\lambda = F \cap E_\lambda$ is generated by polynomials of the form stated in the theorem, and from this the conclusion will follow: for then it will be true of their direct sum F, and by letting d tend to $+\infty$ it will be true of I too.

So let $P := \sum f_{\underline{k}} T^{\underline{k}} \in F_\lambda$, where we write $\underline{k} := (k_1, \ldots, k_n)$ and $T^{\underline{k}} := T_1^{k_1} \cdots T_n^{k_n}$ for short and where the coefficients $f_{\underline{k}}$ belong to K. The sum is restricted to multi-indices $\underline{k}$ such that $k_1 + \cdots + k_n \leq d$ and $a_1^{k_1} \cdots a_n^{k_n} = \lambda$. This last monomial relation can be written $k_1 \alpha_1 + \cdots + k_n \alpha_n = c + m_{\underline{k}}$, where c is an arbitrary constant such that $e^{2i\pi c} = \lambda$ and where $m_{\underline{k}} \in \mathbf{Z}$. All this expresses the fact that $P \in E_\lambda$. The condition that $P \in F$ means that $P(z^{\alpha_1}, \ldots, z^{\alpha_n}) = 0$, *i.e.*, that $\sum f_{\underline{k}} z^{k_1 \alpha_1 + \cdots + k_n \alpha_n} = 0$, *i.e.*, that

---

[1]Note the mixture of analytic and algebraic arguments; for us, the functions z^α are characterized by transcendental properties, those of their analytic continuations.

$z^c \sum f_{\underline{k}} z^{m_{\underline{k}}} = 0$, whence $\sum f_{\underline{k}} z^{m_{\underline{k}}} = 0$. From this, selecting any particular $\underline{l}$ among the $\underline{k}$ involved:

$$P = \sum f_{\underline{k}} T^{\underline{k}} = \sum f_{\underline{k}} T^{\underline{k}} - \left(\sum f_{\underline{k}} z^{m_{\underline{k}}} \right) z^{-m_{\underline{l}}} T^{\underline{l}}$$
$$= \sum f_{\underline{k}} \left(T^{\underline{k}} - z^{m_{\underline{k}} - m_{\underline{l}}} T^{\underline{l}} \right),$$

and we are left to check that each $T^{\underline{k}} - z^{m_{\underline{k}} - m_{\underline{l}}} T^{\underline{l}}$ is of the expected form quoted in the statement of the lemma. This follows from the following computation:

$$\left. \begin{array}{l} k_1 \alpha_1 + \cdots + k_n \alpha_n = c + m_{\underline{k}} \\ l_1 \alpha_1 + \cdots + l_n \alpha_n = c + m_{\underline{l}} \end{array} \right\}$$
$$\implies k_1 \alpha_1 + \cdots + k_n \alpha_n = (m_{\underline{k}} - m_{\underline{l}}) + l_1 \alpha_1 + \cdots + l_n \alpha_n.$$

$\square$

Definition 14.2. A *replica* of $(a_1, \ldots, a_n) \in (\mathbf{C}^*)^n$ is $(\lambda_1, \ldots, \lambda_n) \in (\mathbf{C}^*)^n$ such that:

$$\forall (k_1, \ldots, k_n) \in \mathbf{Z}^n, \ a_1^{k_1} \cdots a_n^{k_n} = 1 \implies \lambda_1^{k_1} \cdots \lambda_n^{k_n} = 1.$$

The terminology is actually more standard in the theory of Lie algebras of algebraic groups, where it was introduced by Chevalley.

Theorem 14.3. *The matricial Galois group* Gal *consists of all diagonal matrices* $\mathrm{Diag}(\lambda_1, \ldots, \lambda_n) \in \mathrm{GL}_n(\mathbf{C})$ *such that* $(\lambda_1, \ldots, \lambda_n)$ *is a replica of* $(a_1, \ldots, a_n) = (e^{2i\pi\alpha_1}, \ldots, e^{2i\pi\alpha_n})$.

Proof. We saw that $\mathrm{Diag}(\lambda_1, \ldots, \lambda_n) \in$ Gal is equivalent to: $\forall P \in I$, $P(\lambda_1 z^{\alpha_1}, \ldots, \lambda_n z^{\alpha_n}) = 0$. This in turn is equivalent to the same condition restricted to the generators described in the lemma. (If it is true for these generators, it is true for all their linear combinations.) Now, restricted to the generators, the condition reads:

$$\forall \underline{k}, \underline{l} \in \mathbf{N}^n, \ \forall m \in \mathbf{Z}, \ k_1 \alpha_1 + \cdots + k_n \alpha_n = l_1 \alpha_1 + \cdots + l_n \alpha_n + m,$$

whence

$$\prod_{j=1}^{n} (\lambda_j z^{\alpha_j})^{k_j} = z^m \prod_{j=1}^{n} (\lambda_j z^{\alpha_j})^{l_j}.$$

After division on both sides by $z^{\sum k_j \alpha_j} = z^{\sum l_j \alpha_j + m}$, this in turn reads:

$$\forall \underline{k}, \underline{l} \in \mathbf{N}^n, \ (k_1 - l_1)\alpha_1 + \cdots + (k_n - l_n)\alpha_n \in \mathbf{Z} \implies \prod_{j=1}^{n} \lambda_j^{k_j} = \prod_{j=1}^{n} \lambda_j^{l_j}.$$

Last, this can be rewritten as the implication: $\forall \underline{k} \in \mathbf{Z}^n, \ k_1 \alpha_1 + \cdots + k_n \alpha_n \in \mathbf{Z} \implies \prod_{j=1}^{n} \lambda_j^{k_j} = 1$. But since $k_1 \alpha_1 + \cdots + k_n \alpha_n \in \mathbf{Z}$ is equivalent to $a_1^{k_1} \cdots a_n^{k_n} = 1$, this is the definition of a replica.

$\square$

The problem of computing the Galois group is now one in abelian group theory. We introduce:

$$\Gamma := \{\underline{k} \in \mathbf{Z}^n \mid k_1\alpha_1 + \cdots + k_n\alpha_n \in \mathbf{Z}\} = \{\underline{k} \in \mathbf{Z}^n \mid a_1^{k_1} \cdots a_n^{k_n} = 1\}.$$

By the general theory of finitely generated abelian groups (see [**Lan02**]), we know that the subgroup Γ of $\mathbf{Z}^n$ is freely generated by r elements $\underline{k}^{(1)}, \dots, \underline{k}^{(r)}$, where $r \leq n$ is the rank of the free abelian group Γ. So there are exactly r conditions to check that a given $(\lambda_1, \dots, \lambda_n) \in (\mathbf{C}^*)^n$ is a replica of $(a_1, \dots, a_n)$. These are the monomial conditions: $\lambda_1^{k_1^{(i)}} \cdots \lambda_n^{k_n^{(i)}} = 1$ for $i = 1, \dots, r$.

Example 14.4. Consider the equation $zf' = \alpha f$. Here, $\Gamma = \{k \in \mathbf{Z} \mid a^k = 1\} = \{k \in \mathbf{Z} \mid k\alpha \in \mathbf{Z}\}$, where $a := e^{2i\pi\alpha}$. If $\alpha \notin \mathbf{Q}$, then $\Gamma = \{0\}$ and $\mathrm{Gal} = \mathbf{C}^*$. If $\alpha = p/q$, p, q being coprime, then $\Gamma = q\mathbf{Z}$ and $\mathrm{Gal} = \mu_q$.

Exercise 14.5. Check these statements.

Example 14.6. We consider the system $\delta X = \mathrm{Diag}(\alpha, \beta)X$. Here, $\Gamma = \{(k, l) \in \mathbf{Z}^2 \mid k\alpha + l\beta \in \mathbf{Z}\}$. There are three possible cases according to whether $r = 0, 1$ or 2:

(1) The case $r = 2$ arises when $\alpha, \beta \in \mathbf{Q}$. Then Γ is generated by two nonproportional elements (k, l) and (k', l') and we have $\mathrm{Gal} = \{\mathrm{Diag}(\lambda, \mu) \in \mathrm{GL}_2(\mathbf{C}) \mid \lambda^k\mu^l = \lambda^{k'}\mu^{l'} = 1\}$. Note that these equations imply $\lambda^N = \mu^N = 1$, where $N := |kl' - k'l|$, so that Gal is finite. This is related to the fact that all solutions are algebraic, but we just notice this fact empirically here. More precisely, one can prove (it is a nice exercise in algebra) that Γ is either cyclic or isomorphic to the product of two cyclic groups.

Exercise 14.7. Do this nice exercise in algebra.

(2) The case $r = 0$ arises when $1, \alpha, \beta$ are linearly independent over $\mathbf{Q}$. Then $\Gamma = \{0\}$ and Gal contains all invertible diagonal matrices $\mathrm{Diag}(\lambda, \mu)$.

(3) The intermediate case $r = 1$ occurs when the $\mathbf{Q}$-linear space $\mathbf{Q} + \mathbf{Q}\alpha + \mathbf{Q}\beta$ has dimension 2. Then Γ is generated by a pair $(k, l) \neq (0, 0)$ and $\mathrm{Gal} = \{\mathrm{Diag}(\lambda, \mu) \in \mathrm{GL}_2(\mathbf{C}) \mid \lambda^k\mu^l = 1\}$. It is (again) a nice exercise in algebra to prove that this group is isomorphic to $\mathbf{C}^* \times \mu_q$, where q is the greatest common divisor of k, l.

Exercise 14.8. Find an example of a pair (α, β) for each case above.

14.2. Calculation of the differential Galois group in the general case

We now consider a system $\delta X = AX$, where $A \in \mathrm{Mat}_n(\mathbf{C})$ is in Jordan form. Thus, A is a block-diagonal matrix with k blocks $A_i = \alpha_i I_{m_i} + N_{m_i}$, where we write N_m as the nilpotent upper triangular matrix

$$
\begin{pmatrix}
0 & 1 & 0 & \cdots & 0 \\
0 & 0 & 1 & \cdots & 0 \\
\vdots & \vdots & \vdots & \ddots & \vdots \\
0 & 0 & 0 & \cdots & 1 \\
0 & 0 & 0 & \cdots & 0
\end{pmatrix}
\in \mathrm{Mat}_m(\mathbf{C}).
$$

We have $\exp(A \log z) = \mathrm{Diag}(\exp(A_1 \log z), \ldots, \exp(A_k \log z))$, i.e., $z^A = \mathrm{Diag}(z^{A_1}, \ldots, z^{A_k})$, and

$$
z^{A_i} = \exp(A_i \log z) = \exp(\alpha I_{m_i} \log z) \exp(N_{m_i} \log z),
$$

whence

$$
z^{A_i} = z^{\alpha_i} \left(I_{m_i} + (\log z) N_{m_i} + \cdots + \frac{\log^{m_i - 1} z}{(m_i - 1)!} N_{m_i}^{m_i - 1} \right).
$$

Therefore, the algebra $\mathcal{A}$ is generated by all the $z^{\alpha_i} (\log z)^l$ for $i = 1, \ldots, k$ and $0 \le l \le m_i - 1$. If $k = n$ and all $m_i = 1$, there is no log at all, but then we are in the semi-simple case of Section 14.1.

For any differential automorphism of $\mathcal{A}$, the same calculation as we did in Section 13.2 shows that $\sigma(z^{\alpha_i}) = \lambda_i z^{\alpha_i}$ and, if $m_i \ge 1$, $\sigma(z^{\alpha_i} \log z) = \lambda_i z^{\alpha_i} (\log z + \mu_i)$ for some $\lambda_i \in \mathbf{C}^*$ and $\mu_i \in \mathbf{C}$. (Beware that we slightly changed the meaning of the notation μ with respect to 13.2.) We also know that $(\lambda_1, \ldots, \lambda_k)$ must be a replica of $(e^{2i\pi\alpha_1}, \ldots, e^{2i\pi\alpha_k})$.

Now, we have two things to consider. First, what about higher powers of log? From the relation $(z^{\alpha_i} \log z)^l = (z^{\alpha_i})^{l-1}(z^{\alpha_i} (\log z)^l)$, we obtain at once that $\sigma(z^{\alpha_i} (\log z)^l) = \lambda_i z^{\alpha_i} (\log z + \mu_i)^l$ for $0 \le l \le m_i - 1$. Second thing: how are the different μ_i related? From the relation $z^{\alpha_j}(z^{\alpha_i} \log z) = z^{\alpha_i}(z^{\alpha_j} \log z)$, we obtain at once that all μ_i are equal to the same $\mu \in \mathbf{C}$.

Exercise 14.9. Fill in the details of these computations.

Therefore, any differential automorphism σ is completely determined by the equations

$$
\sigma(z^{\alpha_i} (\log z)^l) = \lambda_i z^{\alpha_i} (\log z + \mu)^l,
$$

where $(\lambda_1, \ldots, \lambda_k)$ is a replica of $(e^{2i\pi\alpha_1}, \ldots, e^{2i\pi\alpha_k})$ and where $\mu \in \mathbf{C}$. However, we must still see which choices of $(\lambda_1, \ldots, \lambda_k)$ and μ do give a differential automorphism. For this, it is enough to check that they define a K-algebra automorphism. Indeed, since the relation $\sigma(f') = (\sigma(f))'$ is satisfied by the above family of generators, it is satisfied by all elements of $\mathcal{A}$ after the lemma of Section 13.2.

By the study of the semi-simple case, we know that $R := K[z^{\alpha_1}, \ldots, z^{\alpha_k}]$ is stable under σ and that the restriction τ of σ to R is an automorphism. In the next lemma, we will show that log is transcendental over R, so that any choice of the image of log allows for an extension of τ to an automorphism of $\mathcal{A}' := R[\log]$. In particular, setting $\log \mapsto \log +\mu$, one extends τ to an automorphism σ' of $\mathcal{A}'$. But $\mathcal{A} \subset \mathcal{A}'$ and σ is the restriction of σ' to $\mathcal{A}$. We have proven (temporarily admitting the lemma):

Proposition 14.10. *The differential automorphisms of $\mathcal{A}$ are all the maps of the form $\sigma(z^{\alpha_i}(\log z)^l) = \lambda_i z^{\alpha_i}(\log z + \mu)^l$, where $(\lambda_1, \ldots, \lambda_k)$ is a replica of $(e^{2i\pi\alpha_1}, \ldots, e^{2i\pi\alpha_k})$ and where $\mu \in \mathbf{C}$ is arbitrary.* $\qquad\square$

Lemma 14.11. *Let $\alpha_1, \ldots, \alpha_k \in \mathbf{C}$ and set $R := K[z^{\alpha_1}, \ldots, z^{\alpha_k}]$. Then $\log$ is transcendental over the field of quotients L of R. Equivalently, there is no nontrivial algebraic relation $f_0 + \cdots + f_m \log^m$, with $f_i \in R$.*

Proof. We extend δ to a derivation over L by putting

$$\delta(f/g) := (g\delta(f) - f\delta(g))/g^2,$$

which is well defined. If $f/g = f_1/g_1$, then both possible formulas give the same result.

First step. Let $m + 1$ be the minimal degree of an algebraic equation of $\log$ over L. This means that $1, \ldots, \log^m$ are linearly independent over L, while $\log^{m+1} = f_0 + \cdots + f_m \log^m$, with $f_0, \ldots, f_m \in L$. Applying δ, since $\delta(\log^k) = k \log^{k-1}$, we get:

$$(m+1)\log^m = (\delta(f_0) + f_1) + \cdots + (\delta(f_{m-1}) + f_m)\log^{m-1} + \delta(f_m)\log^m,$$

so that, by the assumption of linear independence, $\delta(f_m) = m + 1$, whence $\delta(f_m - (m+1)\log) = 0$. This implies that $f_m - (m+1)\log \in \mathbf{C}$ and then that $\log \in L$.

Second step. Suppose that we have $\log \in L$, that is, $f \log = g$ with $f, g \in R$, $f \neq 0$. We use the fact that the action of the monodromy operator σ on R is semi-simple: $f = \sum f_\lambda$ (finite sum), where $\sigma(f_\lambda) = \lambda f_\lambda$; this is because f is a linear combination with coefficients in K (hence invariant under σ) of monomials in the z^{α_j}. Suppose we have written $f \log = g$ with f as "short" as possible, *i.e.*, with as few components f_λ as possible. Then, if λ_0 is one of the λ that do appear, we calculate:

$$\sigma(f)(\log + 2\mathrm{i}\pi) = \sigma(f \log) = \sigma(g)$$
$$\implies (\sigma(f) - \lambda_0 f) \log = \sigma(g) - 2\mathrm{i}\pi\sigma(f) - \lambda_0 g,$$

a shorter relation (since $\sigma(f) - \lambda_0 f$ has one less eigencomponent), except if it is trivial, in which case $\sigma(f) = \lambda_0 f$, which is therefore the only possibility.

Third step. Suppose that we have $f \log = g$ with $f, g \in R$, $f \neq 0$ and $\sigma(f) = \lambda f$. Then, applying σ and simplifying, we find $\sigma(g) - \lambda g = 2\mathrm{i}\pi\lambda f$. But this is impossible if $f \neq 0$, because $\sigma - \lambda$ sends g_λ to 0 and all other g_μ to elements of the corresponding eigenspaces. This ends the proof of the lemma and of the proposition.   □

Theorem 14.12. *Let A be in Jordan form* $\mathrm{Diag}(A_i, \ldots, A_k)$, *where* $A_i = \alpha_i I_{m_i} + N_{m_i}$. *The matricial Galois group* Gal *of the system* $X' = z^{-1}AX$ *relative to the fundamental matricial solution* z^A *is the set of matrices* $\mathrm{Diag}(\lambda_1 e^{\mu N_1}, \ldots, \lambda_k e^{\mu N_k})$, *where* $\mu \in \mathbf{C}$ *is arbitrary and where* $(\lambda_1, \ldots, \lambda_k)$ *is a replica of* $(e^{2\mathrm{i}\pi\alpha_1}, \ldots, e^{2\mathrm{i}\pi\alpha_k})$.

Proof. The automorphism σ described in the proposition transforms z^{α_j} into $\lambda_j z^{\alpha_j}$. It transforms $N_j \log z$ into $N_j(\log z + \mu)$ and, because the exponential of a nilpotent matrix is really a polynomial in this matrix, it transforms $e^{N_j \log z}$ into $e^{N_j(\log z + \mu)}$. Therefore, it transforms $z^A = \mathrm{Diag}(z^{\alpha_1} e^{N_1 \log z}, \ldots, z^{\alpha_k} e^{N_k \log z})$ into

$$\mathrm{Diag}(\lambda_1 z^{\alpha_1} e^{N_1(\log z + \mu)}, \ldots, \lambda_k z^{\alpha_k} e^{N_k(\log z + \mu)}) = z^A M,$$

where $M = \mathrm{Diag}(\lambda_1 e^{\mu N_1}, \ldots, \lambda_k e^{\mu N_k})$.   □

Corollary 14.13. *The matricial Galois group of the system* $X' = F[z^{-1}A]X$ *relative to the fundamental matricial solution* $F z^A$ *is the same group as described in the theorem.*

14.3. The density theorem of Schlesinger in the local setting

We suppose that a fundamental matricial solution $\mathcal{X}$ has been chosen for the differential system S_A, so that we have matricial realizations $\mathrm{Mon}(A) \subset \mathrm{Gal}(A) \subset \mathrm{GL}_n(\mathbf{C})$ relative to $\mathcal{X}$.

Theorem 14.14 (Local Schlesinger density theorem). *Let the system S_A be regular singular. Then $\mathrm{Gal}(A)$ is the smallest algebraic subgroup of $\mathrm{GL}_n(\mathbf{C})$ containing $\mathrm{Mon}(A)$.*

Proof. A meromorphic gauge transformation $B = F[A]$, $F \in \mathrm{GL}_n(K)$, gives rise to a fundamental matricial solution $\mathcal{Y}$ such that $F\mathcal{X} = \mathcal{Y}P$, $P \in \mathrm{GL}_n(\mathbf{C})$; then, using matricial realizations relative to $\mathcal{Y}$ of the monodromy and Galois groups of S_B, one has $\mathrm{Mon}(B) = P\mathrm{Mon}(A)P^{-1}$ and $\mathrm{Gal}(B) = P\mathrm{Gal}(A)P^{-1}$. From this, we deduce at the same time that the statement to be proved is independent from the choice of a particular fundamental matricial solution and also that it is invariant up to meromorphic equivalence. Therefore, we take the system in the form $\delta X = AX$, where $A \in \mathrm{Mat}_n(\mathbf{C})$ is in Jordan form. We keep the notation of Section 14.2. Therefore, Mon is generated by the monodromy matrix:

$$e^{2\mathrm{i}\pi A} = \mathrm{Diag}(a_1 e^{2\mathrm{i}\pi N_1}, \ldots, a_k e^{2\mathrm{i}\pi N_k}) = \underbrace{e^{2\mathrm{i}\pi A_s} e^{2\mathrm{i}\pi A_n}}_{\text{Jordan decomposition}},$$

where

$$\begin{cases} e^{2\mathrm{i}\pi A_s} &= \mathrm{Diag}(a_1 I_{m_1}, \ldots, a_k I_{m_k}) \\ \text{and} \\ e^{2\mathrm{i}\pi A_n} &= \mathrm{Diag}(e^{2\mathrm{i}\pi N_1}, \ldots, e^{2\mathrm{i}\pi N_k}). \end{cases}$$

Remember that the Jordan decomposition of an invertible matrix into its semi-simple and unipotent component was defined in the corresponding paragraph of Section 4.4. Likewise, Gal is the set of matrices of the form:

$$E(\underline{\lambda}, \mu) := \mathrm{Diag}(\lambda_1 e^{\mu N_1}, \ldots, \lambda_k e^{\mu N_k}) = \underbrace{E_s(\underline{\lambda})E_u(\mu)}_{\text{Jordan decomposition}},$$

where

$$\begin{cases} E_s(\underline{\lambda}) = \mathrm{Diag}(\lambda_1 I_{m_1}, \ldots, \lambda_k I_{m_k}) \\ \text{and} \\ E_u(\mu) = \mathrm{Diag}(e^{\mu N_1}, \ldots, e^{\mu N_k}), \end{cases}$$

where $(\lambda_1, \ldots, \lambda_k) \in (\mathbf{C}^*)^k$ is a replica of $(a_1, \ldots, a_k)$ and where $\mu \in \mathbf{C}$ is arbitrary. Moreover the two factors $E_s(\underline{\lambda}), E_u(\mu)$ commute.

We are therefore led to prove that, if $G \subset \mathrm{GL}_n(\mathbf{C})$ is an algebraic subgroup and if $e^{2\mathrm{i}\pi A} \in G$, then all matrices of the form $E(\underline{\lambda}, \mu)$ above belong to G. We shall use the following fact, a proof of which may be found in [**BorA91**]; also see Appendix C.

Proposition 14.15. *If $G \subset \mathrm{GL}_n(\mathbf{C})$ is an algebraic subgroup, then, for each $M \in G$, the semi-simple and unipotent components M_s and M_u belong to G.* $\square$

Therefore, if G is an algebraic subgroup of $\mathrm{GL}_n(\mathbf{C})$ containing $e^{2\mathrm{i}\pi A}$, then it contains $e^{2\mathrm{i}\pi A_s}$ and $e^{2\mathrm{i}\pi A_n}$. The conclusion of the theorem will therefore follow immediately from the following two lemmas.

Lemma 14.16. *If an algebraic subgroup G of $\mathrm{GL}_n(\mathbf{C})$ contains the matrix* $\mathrm{Diag}(a_1 I_{m_1}, \ldots, a_k I_{m_k})$, *then it contains all matrices of the form* $\mathrm{Diag}(\lambda_1 I_{m_1}, \ldots, \lambda_k I_{m_k})$, *where* $(\lambda_1, \ldots, \lambda_k) \in (\mathbf{C}^*)^k$ *is a replica of* $(a_1, \ldots, a_k)$.

Proof. Let $F(T_{1,1}, \ldots, T_{n,n}) \in \mathbf{C}[T_{1,1}, \ldots, T_{n,n}]$ be one of the defining equations of the algebraic subgroup G. We must prove that it vanishes on all matrices of the indicated form. If one replaces the indeterminates $T_{i,j}$ by the corresponding coefficients of the matrix $\mathrm{Diag}(T_1 I_{m_1}, \ldots, T_k I_{m_k})$, one obtains a polynomial $\Phi(T_1, \ldots, T_k) \in \mathbf{C}[T_1, \ldots, T_k]$ such that $\Phi(a_1^p, \ldots, a_k^p) = 0$ for all $p \in \mathbf{Z}$ (because G, being a group, contains all powers of $\mathrm{Diag}(a_1 I_{m_1}, \ldots, a_k I_{m_k})$) and one wants to prove that $\Phi(\lambda_1, \ldots, \lambda_k) = 0$ for all replicas $(\lambda_1, \ldots, \lambda_k)$.

We write Φ as a linear combination of monomials: $\Phi = \sum \lambda_i M_i$. Then $M_i(a_1^p, \ldots, a_k^p) = M_i(a_1, \ldots, a_k)^p$. We group the indices by packs $I(c)$ such that $M_i(a_1, \ldots, a_k) = c$ for all $i \in I(c)$. Then, if $\Lambda(c) := \sum_{i \in I(c)} \lambda_i$, we see that:

$$\forall p \in \mathbf{Z}, \; \Phi(a_1^p, \ldots, a_k^p) = \sum_c \Lambda(c) c^p = 0.$$

By classical properties of the Vandermonde determinant, this implies that $\Lambda(c) = 0$ for every c:

$$\forall c \in \mathbf{C}^*, \; \sum_{i \in I(c)} \lambda_i = 0.$$

For every relevant c (*i.e.*, such that $I(c)$ is not empty), choose a particular $i_0 \in I(c)$. Then $\Phi = \sum_c \Phi_c$, where

$$\Phi_c := \sum_{i \in I(c)} \lambda_i M_i = \sum_{i \in I(c)} \lambda_i (M_i - M_{i_0}).$$

For $i \in I(c)$, one has $M_i(a_1, \ldots, a_k) = M_{i_0}(a_1, \ldots, a_k)$ (both are equal to c); by definition, this monomial relation between the a_i remains true for any replica of $(a_1, \ldots, a_k)$, so that Φ_c vanishes on any replica, and so does Φ.   □

Lemma 14.17. *If an algebraic subgroup G of $\mathrm{GL}_n(\mathbf{C})$ contains the matrix* $\mathrm{Diag}(e^{2i\pi N_1}, \ldots, e^{2i\pi N_k})$, *then it contains all matrices of the form* $\mathrm{Diag}(e^{\mu N_1}, \ldots, e^{\mu N_k})$, *where* $\mu \in \mathbf{C}$ *is arbitrary.*

Proof. Let $F(T_{1,1}, \ldots, T_{n,n}) \in \mathbf{C}[T_{1,1}, \ldots, T_{n,n}]$ be one of the defining equations of the algebraic subgroup G. We must prove that it vanishes on all matrices of the indicated form. If one replaces the indeterminates $T_{i,j}$ by the corresponding coefficients of the matrix $\mathrm{Diag}(e^{TN_1}, \ldots, e^{TN_k})$, one obtains a polynomial $\Phi(T) \in \mathbf{C}[T]$; indeed, since the matrices N_i are nilpotent, the expressions e^{TN_i} involve only a finite number of powers of T. Moreover, since G is a group, it contains all powers $\mathrm{Diag}(e^{2i\pi N_1}, \ldots, e^{2i\pi N_k})^p = \mathrm{Diag}(e^{2i\pi p N_1}, \ldots, e^{2i\pi p N_k})$, $p \in \mathbf{Z}$, so that $\Phi(2i\pi p) = 0$ for all $p \in \mathbf{Z}$. The polynomial Φ has an infinity of roots, it is therefore trivial and $\Phi(\mu) = 0$ for all $\mu \in \mathbf{C}$, which means that $F(T_{1,1}, \ldots, T_{n,n})$ vanishes on all matrices of the indicated form, as wanted. $\quad\square$

This ends the proof of Schlesinger's theorem. $\quad\square$

14.4. Why is Schlesinger's theorem called a "density theorem"?

This is going to be a breezy introduction to affine algebraic geometry in the particular case of linear algebraic groups. For every $F(T_{1,1}, \ldots, T_{n,n}) \in \mathbf{C}[T_{1,1}, \ldots, T_{n,n}]$ and $A = (a_{i,j}) \in \mathrm{Mat}_n(\mathbf{C})$, we shall write $F(A) := F(a_{1,1}, \ldots, a_{n,n}) \in \mathbf{C}$ for short. (Do not confuse this with matrix polynomials used in reduction theory, such as the minimal and characteristic polynomials of a matrix; here, $F(A)$ is a scalar, not a matrix, and its computation does not involve the powers of A.)

Definition 14.18. Let $E \subset \mathbf{C}[T_{1,1}, \ldots, T_{n,n}]$ be an arbitrary set of polynomial equations on $\mathrm{Mat}_n(\mathbf{C})$. Then we write $V(E) := \{A \in \mathrm{Mat}_n(\mathbf{C}) \mid \forall F \in E, \ F(A) = 0\}$. The set $V(E)$ is called the *algebraic subset of* $\mathrm{Mat}_n(\mathbf{C})$ *defined by the set of equations* E.

Proposition 14.19. (i) *The subsets* $\emptyset$ *and* $\mathrm{Mat}_n(\mathbf{C})$ *are algebraic subsets.*

(ii) *If* V_1, V_2 *are algebraic subsets, so is* $V_1 \cup V_2$.

(iii) *If* (V_i) *is a (possibly infinite) family of algebraic subsets, so is* $\bigcap V_i$.

Proof. (i) One immediately checks that $\emptyset = V(\{1\})$, while $\mathrm{Mat}_n(\mathbf{C}) = V(\{0\})$.

(ii) With a little thought, one finds that $V(E_1) \cup V(E_2) = V(\{F_1.F_2 \mid F_1 \in E_1, F_2 \in E_2\})$.

(iii) One immediately checks that $\bigcap V(E_i) = V(\bigcup E_i)$. $\quad\square$

Corollary 14.20. *There is a topology on* $\mathrm{Mat}_n(\mathbf{C})$ *for which the closed subsets are exactly the algebraic subsets.* $\quad\square$

We use the abstract definition of a topology here, as a set of open subsets containing $\emptyset$ and $\mathrm{Mat}_n(\mathbf{C})$ and stable under finite intersections and arbitrary unions. Then the closed subsets are defined as the complementary subsets of the open subsets. The topology we just defined is called the *Zariski topology*. The algebraic subsets are said to be *Zariski closed*, and the closure $\overline{X}$ of an arbitrary subset X for this topology, called its *Zariski closure*, is the smallest algebraic subset containing X. Here is a way to "compute" it. Let $I(X) := \{F \in \mathbf{C}[T_{1,1}, \ldots, T_{n,n}] \mid \forall A \in X, \ F(A) = 0\}$, the set of all equations satisfied by X. Then $X \subset V(E) \iff \forall F \in E, \ \forall A \in X, \ F(A) = 0 \iff E \subset I(X)$. It follows immediately that $\overline{X} = V(I(X))$.

Another consequence is the following. We say that X is *Zariski dense* in Y if $X \subset Y \subset \overline{X}$. Then, for a subset X of Y to be Zariski dense, it is necessary and sufficient that the following condition be true: every $F \in \mathbf{C}[T_{1,1}, \ldots, T_{n,n}]$ which vanishes on X also vanishes on Y.

Now we consider the restriction of our topology to the open subset $\mathrm{GL}_n(\mathbf{C})$ of $\mathrm{Mat}_n(\mathbf{C})$. (It is open because it is the complementary subset of $V(\det)$ and $\det \in \mathbf{C}[T_{1,1}, \ldots, T_{n,n}]$.) Then one can prove that the closure in $\mathrm{GL}_n(\mathbf{C})$ of a subgroup G of $\mathrm{GL}_n(\mathbf{C})$ is a subgroup of $\mathrm{GL}_n(\mathbf{C})$ (see [**BorA91**] for a proof). Of course, this closure is then exactly what we called an algebraic subgroup and so it is the smallest algebraic subgroup of $\mathrm{GL}_n(\mathbf{C})$ containing G. The translation of the Schlesinger theorem in this language is therefore:

Corollary 14.21. *The monodromy group* Mon *is Zariski dense in the algebraic group* Gal.

Exercises

(1) Compute all powers of the nilpotent matrix N_m introduced in Section 14.2 and then give an explicit formula for z^A.

(2) Write $e^{\mu N_j}$ explicitly and find a set of equations in $\mathrm{GL}_n(\mathbf{C})$ defining the group Gal of Section 14.2.

(3) Among the classical subgroups of $\mathrm{GL}_n(\mathbf{C})$, which ones are Zariski closed? Which ones are Zariski dense?

Chapter 15

The universal (fuchsian local) Galois group

We use the same notation as in Chapter 14. So let $\mathcal{X}$ be a fundamental matricial solution of the system $X' = AX$, with $A \in \mathrm{Mat}_n(K)$. We defined the monodromy representation $\rho_A : \pi_1 \to \mathrm{GL}_n(\mathbf{C})$ by the formula:

$$[\lambda] \mapsto M_\lambda := \mathcal{X}^{-1} \mathcal{X}^\lambda.$$

Then we defined the monodromy group as:

$$\mathrm{Mon}(A) := \mathrm{Im}\ \rho_A.$$

In the *fuchsian case*, *i.e.*, when 0 is a regular singular point of the system $X' = AX$, we obtained a bijective correspondence:

$$\left\{ \begin{array}{c} \text{isomorphism classes of} \\ \text{regular singular systems} \end{array} \right\} \longleftrightarrow \left\{ \begin{array}{c} \text{isomorphism classes of} \\ \text{representations of } \pi_1 \end{array} \right\}.$$

In the context of Galois theory, we first introduced, in Section 13.1, the differential algebra $\mathcal{A} := K[\mathcal{X}]$, with its group of differential automorphisms $\mathrm{Aut}_{\mathbf{C}-algdiff}(\mathcal{A})$; then we defined the Galois group through its matricial realization as the image of the group morphism $\mathrm{Aut}_{\mathbf{C}-algdiff}(\mathcal{A}) \mapsto \mathrm{GL}_n(\mathbf{C})$ defined by the formula:

$$\sigma \mapsto \mathcal{X}^{-1}(\sigma\mathcal{X}).$$

The main difference with the monodromy representation is the following: the source of ρ_A, the group π_1, was independent of the particular system being studied; it was a "universal" group. On the other hand, the group $\mathrm{Aut}(\mathcal{A})$ is obviously related to A; it is in no way universal.

Our goal here is to construct a universal group $\hat{\pi}_1$ and, for each particular system $X' = AX$, a representation $\hat{\rho}_A : \hat{\pi}_1 \to \mathrm{GL}_n(\mathbf{C})$, in such a way that:

- The Galois group is the image of that representation: $\mathrm{Gal}(A) = \mathrm{Im}\ \hat{\rho}_A$.

- The "functor" $A \rightsquigarrow \hat{\rho}_A$ induces a bijective correspondence between isomorphism classes of regular singular systems and isomorphism classes of representations of $\hat{\pi}_1$ ("algebraic Riemann-Hilbert correspondence").

We shall be able to do this for local regular singular systems (although, as in the case of monodromy, more general results are known for global systems, as well as for irregular systems). However, we shall have to take into account the fact that the Galois group is always an *algebraic* subgroup of $\mathrm{GL}_n(\mathbf{C})$, and therefore restrict the class of possible representations to enforce this property. We begin with two purely algebraic sections.

15.1. Some algebra, with replicas

Recall Definition 14.2 of *replicas*. The following criterion is due to Chevalley.

Theorem 15.1. *Let* $(a_1, \ldots, a_n) \in (\mathbf{C}^*)^n$. *Then* $(b_1, \ldots, b_n) \in (\mathbf{C}^*)^n$ *is a replica of* $(a_1, \ldots, a_n)$ *if, and only if, there exists a group morphism* $\gamma : \mathbf{C}^* \to \mathbf{C}^*$ *such that* $\gamma(a_i) = b_i$ *for* $i = 1, \ldots, n$.

Proof. Clearly, if such a morphism γ exists, then, for any $(m_1, \ldots, m_n) \in \mathbf{Z}^n$,

$$a_1^{m_1} \cdots a_n^{m_n} = 1 \implies \gamma(a_1^{m_1} \cdots a_n^{m_n}) = 1 \implies b_1^{m_1} \cdots b_n^{m_n} = 1,$$

so that $(b_1, \ldots, b_n)$ is indeed a replica of $(a_1, \ldots, a_n)$.

Now assume conversely that $(b_1, \ldots, b_n)$ is a replica of $(a_1, \ldots, a_n)$. For $i = 1, \ldots, n$, let $\Gamma_i := \langle a_1, \ldots, a_i \rangle$ be the subgroup of $\mathbf{C}^*$ generated by $a_1, \ldots, a_i$. We are first going to construct, for each $i = 1, \ldots, n$, a group morphism $\gamma_i : \Gamma_i \to \mathbf{C}^*$ such that $\gamma_i(a_j) = b_j$ for $j = 1, \ldots, i$; these morphisms will be extensions of each other, *i.e.*, $\gamma_{i|\Gamma_{i-1}} = \gamma_{i-1}$ for $i = 2, \ldots, n$.

For $i = 1$, we know that $a^m = 1 \Rightarrow b^m = 1$, so it is an easy exercise in group theory to show that setting $\gamma_1(a^k) := b^k$ makes sense and defines a group morphism $\gamma_1 : \Gamma_1 \to \mathbf{C}^*$.

Suppose that $\gamma_i : \Gamma_i \to \mathbf{C}^*$ has been constructed and that $i < n$. Any element of Γ_{i+1} can be written ga_{i+1}^k for some $g \in \Gamma_i$ and $k \in \mathbf{Z}$; but, of course, this decomposition is not necessarily unique! However:

$$ga_{i+1}^k = g'a_{i+1}^{k'} \Rightarrow g^{-1}g' = a_{i+1}^{k-k'} \Rightarrow \gamma_i(g^{-1}g') = b_{i+1}^{k-k'}$$
$$\Rightarrow \gamma_i(g)b_{i+1}^k = \gamma_i(g')b_{i+1}^{k'},$$

so that it makes sense to set $\gamma_{i+1}(ga_{i+1}^k) := \gamma_i(g)b_{i+1}^k$ and it is (again) an easy exercise to check that this γ_{i+1} is a group morphism $\Gamma_{i+1} \to \mathbf{C}^*$ extending γ_i.

Therefore, in the end, we have $\gamma_n : \Gamma_n \to \mathbf{C}^*$ such that $\gamma_n(a_i) = b_i$ for $i = 1, \ldots, n$ and it suffices to apply the following lemma with $\Gamma := \mathbf{C}^*$ and $\Gamma' := \Gamma_n$. $\quad\square$

Exercise 15.2. Do the two exercises in group theory mentioned in the proof.

Lemma 15.3. *Let $\Gamma' \subset \Gamma$ be abelian groups and let $\gamma' : \Gamma' \to \mathbf{C}^*$ be a group morphism. Then γ' can be extended to Γ, i.e., there is a group morphism $\gamma : \Gamma \to \mathbf{C}^*$ such that $\gamma_{|\Gamma'} = \gamma'$.*

Proof. The first part of the proof relies on a mysterious principle from the theory of sets, called "Zorn's lemma" (see [**Lan02**]). We consider the set:

$$\mathcal{E} := \{(\Gamma'', \gamma'') \mid \Gamma' \subset \Gamma'' \subset \Gamma \text{ and } \gamma'' : \Gamma'' \to \mathbf{C}^* \text{ and } \gamma''_{|\Gamma'} = \gamma'\},$$

where of course Γ'' runs among subgroups of Γ and γ'' among group morphisms from Γ'' to $\mathbf{C}^*$. We define an order on $\mathcal{E}$ by setting:

$$(\Gamma''_1, \gamma''_1) \prec (\Gamma''_2, \gamma''_2) \Longleftrightarrow \Gamma''_1 \subset \Gamma''_2 \text{ and } (\gamma''_2)_{|\Gamma''_1} = \gamma''_1.$$

Then $(\mathcal{E}, \prec)$ is an *inductive ordered set*. This means that for any family $\{(\Gamma''_i, \gamma''_i)_{i \in I}\}$ of elements of $\mathcal{E}$ which is assumed to be *totally ordered, i.e.,*

$$\forall i, j \in I \ , \ (\Gamma''_i, \gamma''_i) \prec (\Gamma''_j, \gamma''_j) \text{ or } (\Gamma''_j, \gamma''_j) \prec (\Gamma''_i, \gamma''_i)$$

(such a family is called a *chain*), there is an *upper bound, i.e.,* an element $(\Gamma'', \gamma'') \in \mathcal{E}$ such that:

$$\forall i \in I, \ (\Gamma''_i, \gamma''_i) \prec (\Gamma'', \gamma'').$$

Indeed, we take $\Gamma'' := \bigcup_{i \in I} \Gamma''_i$ and check that this is a group (we have to use the fact that the family of subgroups Γ''_i is totally ordered, *i.e.,* for any two of them, one is included in the other). Then we define $\gamma'' : \Gamma'' \to \mathbf{C}^*$ such that its restriction to each Γ''_i is γ''_i (we have to use the fact that the family of elements (Γ''_i, γ''_i) is totally ordered, *i.e.,* for any two of them, one of the morphisms extends the other).

Now, since the ordered set $(\mathcal{E}, \prec)$ is inductive, Zorn's lemma states that it admits a *maximal element* (Γ'', γ''). This means that γ'' extends γ' but that it cannot be extended further. It is now enough to prove that $\Gamma'' = \Gamma$.

So assume by contradiction that there exists $x \in \Gamma \setminus \Gamma''$ and define $\Gamma''' := \langle \Gamma'', x \rangle$, the subgroup of Γ generated by Γ'' and x (it strictly contains Γ''). We are going to extend γ'' to a morphism $\gamma''' : \Gamma''' \to \mathbf{C}^*$, thereby contradicting the maximality of (Γ'', γ''). The argument is somewhat similar to the proof of the theorem (compare them!). There are two cases to consider:

(1) If $x^N \in \Gamma'' \Rightarrow N = 0$, then any element of Γ''' can be uniquely written gx^k with $g \in \Gamma''$ and $k \in \mathbf{Z}$. In this case, we choose $y \in \mathbf{C}^*$ arbitrary and it is an easy exercise in group theory to check that setting $\gamma'''(gx^k) := \gamma''(g)y^k$ makes sense and meets our requirements.

(2) Otherwise, there is a unique $d \in \mathbf{N}^*$ such that $x^d \in \Gamma'' \Leftrightarrow N \in d\mathbf{Z}$ (this is because such exponents N make up a subgroup $d\mathbf{Z}$ of $\mathbf{Z}$). Then we choose $y \in \mathbf{C}^*$ such that $y^d = \gamma''(x^d)$ (the latter is a well-defined element of $\mathbf{C}^*$). Now, any element of Γ''' can be *nonuniquely* written gx^k with $g \in \Gamma''$ and $k \in \mathbf{Z}$, and, again, it is an easy exercise in group theory to check that setting $\gamma'''(gx^k) := \gamma''(g)y^k$ makes sense and meets our requirements. □

Remark 15.4. In the theorem, the groups $\Gamma, \Gamma', \ldots$ on the left-hand side of the morphisms can be arbitrary abelian groups, but this is not true for the group $\mathbf{C}^*$ on the right-hand side. The reader can check that the decisive property of $\mathbf{C}^*$ that was used is the fact that it is *divisible*: for all $d \in \mathbf{N}^*$, the map $y \mapsto y^d$ is surjective.

15.2. Algebraic groups and replicas of matrices

Let $M \in \mathrm{GL}_n(\mathbf{C})$. (In a moment, we shall take $M := e^{2i\pi A}$, the fundamental monodromy matrix of the system $X' = z^{-1}AX$, where $A \in \mathrm{GL}_n(\mathbf{C})$.) Let $M = M_s M_u = M_u M_s$ be its Jordan decomposition and write $M_s = P\mathrm{Diag}(a_1, \ldots, a_n)P^{-1}$. It follows from Chapter 14 that the smallest algebraic group containing M (*i.e.*, the Zariski closure $\overline{\langle M \rangle}$) is the set of matrices $P\mathrm{Diag}(b_1, \ldots, b_n)P^{-1}M_u^\lambda$, where $(b_1, \ldots, b_n)$ is a replica of $(a_1, \ldots, a_n)$ and where $\lambda \in \mathbf{C}$.

Now, we introduce some new notation. For any two groups H, H', we write $\mathrm{Hom}_{gr}(H, H')$ as the set of all group morphisms $H \to H'$. When H' is moreover commutative, we can make it a group by defining, for any $f_1, f_2 \in \mathrm{Hom}_{gr}(H, H')$, the product $f_1.f_2$ as the morphism $x \mapsto f_1(x)f_2(x)$; thus, we use multiplication in H' and the fact that $f_1.f_2$ is indeed a morphism $H \to H'$ can only be proved because H' was supposed to be commutative. We shall be particularly interested in groups $\mathfrak{X}(H) := \mathrm{Hom}_{gr}(H, \mathbf{C}^*)$, some properties of which (including contravariant functoriality) are detailed in the exercises at the end of the chapter (see in particular exercise 1). Then, for every $\gamma \in \mathfrak{X}(\mathbf{C}^*) = \mathrm{Hom}_{gr}(\mathbf{C}^*, \mathbf{C}^*)$, we set:

$$\gamma(M_s) := P\mathrm{Diag}(\gamma(a_1), \ldots, \gamma(a_n))P^{-1}.$$

It is absolutely not tautological that this makes sense, *i.e.*, that the right-hand side of the equality depends on M_s only and not on the particular choice of the diagonalization matrix P; see exercise 2 at the end of this chapter. Then the criterion of Chevalley (Theorem 15.1) allows us to conclude:

Corollary 15.5. *The smallest algebraic group containing M is:*

$$\overline{\langle M \rangle} = \left\{ \gamma(M_s) M_u^\lambda \mid \gamma \in \mathrm{Hom}_{gr}(\mathbf{C}^*, \mathbf{C}^*), \lambda \in \mathbf{C} \right\}.$$

$\square$

In terms of representations, it is obvious that $\langle M \rangle$ is the image of $\rho :$ $\mathbf{Z} \to \mathrm{GL}_n(\mathbf{C}), k \mapsto M^k$ (this yields an abstract definition of the monodromy group); but it now follows that $\overline{\langle M \rangle}$ can also be obtained as the image of some representation:

$$\hat{\rho} : \begin{cases} (\gamma, \lambda) \mapsto \gamma(M_s) M_u^\lambda, \\ \hat{\pi}_1 \to \mathrm{GL}_n(\mathbf{C}), \end{cases} \qquad \text{where we put } \hat{\pi}_1 := \mathrm{Hom}_{gr}(\mathbf{C}^*, \mathbf{C}^*) \times \mathbf{C}.$$

(The reader should check that this is indeed a group morphism.) We get the following diagram, where we write π_1 for $\mathbf{Z}$:

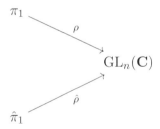

But of course, $\langle M \rangle \subset \overline{\langle M \rangle}$, *i.e.*, Im $\rho \subset$ Im $\hat{\rho}$, *i.e.*, each M^k should be expressible in the form $\gamma(M_s) M_u^\lambda$, and indeed, this is obviously true if we choose $\lambda := k$ and $\gamma : z \mapsto z^k$. Therefore, we complete the above diagram by defining a new arrow $\iota : \pi_1 \mapsto \hat{\pi}_1$ by the formula:

$$\iota(k) := \left((z \mapsto z^k), k \right) \in \mathrm{Hom}_{gr}(\mathbf{C}^*, \mathbf{C}^*) \times \mathbf{C}.$$

Note that this injective group morphism identifies $\pi_1 = \mathbf{Z}$ with a subgroup of $\hat{\pi}_1 = \mathrm{Hom}_{gr}(\mathbf{C}^*, \mathbf{C}^*) \times \mathbf{C}$. In the end, we get the following *commutative* diagram:

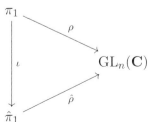

It is clear that every representation $\rho : \pi_1 \to \mathrm{GL}_n(\mathbf{C})$ gives rise to a cyclic[1] subgroup:

$$\mathrm{Im}\,\rho = \langle \rho(1) \rangle$$

of $\mathrm{GL}_n(\mathbf{C})$; and conversely, every cyclic subgroup of $\mathrm{GL}_n(\mathbf{C})$ can obviously be obtained as the image of such a representation of $\pi_1 = \mathbf{Z}$.

As for the algebraic subgroups of $\mathrm{GL}_n(\mathbf{C})$ of the form $\overline{\langle M \rangle}$, it follows from the previous discussion that all of them can be obtained as the image of a representation $\hat{\rho} : \hat{\pi}_1 \to \mathrm{GL}_n(\mathbf{C})$. (Just take the one described above.) But *the converse is false*. Not all representations are admissible and give rise to such algebraic groups. The general theory says that $\hat{\pi}_1$ is a "proalgebraic" group and that the admissible representations are the "rational" ones; this is explained in rather elementary terms in the course "Représentations des groupes algébriques et équations fonctionnelles", to be found at `http://www.math.univ-toulouse.fr/~sauloy/PAPIERS/dea08-09.pdf`. Also see Chapter 16 for some hints of the theory. We shall only illustrate this by two necessary conditions.

Note first that any representation $\hat{\rho}$ of $\hat{\pi}_1 = \mathrm{Hom}_{gr}(\mathbf{C}^*, \mathbf{C}^*) \times \mathbf{C}$ actually has two components: a representation $\hat{\rho}_s$ of $\mathrm{Hom}_{gr}(\mathbf{C}^*, \mathbf{C}^*)$, and a representation $\hat{\rho}_u$ of $\mathbf{C}$. Conversely, given the representations $\hat{\rho}_s$ and $\hat{\rho}_u$, we can recover $\hat{\rho}$ by setting $\hat{\rho}(M) := \hat{\rho}_s(M_s)\hat{\rho}_u(M_u)$; the only necessary condition is that every element of $\mathrm{Im}\,\hat{\rho}_s \subset \mathrm{GL}_n(\mathbf{C})$ commutes with every element of $\mathrm{Im}\,\hat{\rho}_u \subset \mathrm{GL}_n(\mathbf{C})$. So we shall find independent conditions on $\hat{\rho}_s$ and $\hat{\rho}_u$.

It is clear that $\hat{\rho}_u : \mathbf{C} \to \mathrm{GL}_n(\mathbf{C})$ should have the form $\lambda \mapsto U^\lambda$ for some unipotent matrix U (and conversely that any such U gives rise to a $\hat{\rho}_u$). But there are many representations that do not have this form, for instance all maps $\lambda \mapsto e^{\phi(\lambda)} I_n$, where ϕ is any nontrivial morphism from $\mathbf{C}$ to itself.

We now look for a condition on $\hat{\rho}_s$. Let Γ be any finitely generated subgroup of $\mathbf{C}^*$. Then, as shown in exercise 1 below, $\mathfrak{X}(\Gamma) := \mathrm{Hom}_{gr}(\Gamma, \mathbf{C}^*)$ can be identified with the quotient of $\mathfrak{X}(\mathbf{C}^*) = \mathrm{Hom}_{gr}(\mathbf{C}^*, \mathbf{C}^*)$ by the kernel $\mathfrak{X}(\mathbf{C}^*/\Gamma)$ of the surjective map $\mathfrak{X}(\mathbf{C}^*) \to \mathfrak{X}(\Gamma)$. Taking for Γ the subgroup generated by the eigenvalues of M, we see that every admissible representation $\hat{\rho}_s$ must be trivial on the subgroup $\mathfrak{X}(\mathbf{C}^*/\Gamma)$ of $\mathfrak{X}(\mathbf{C}^*) = \mathrm{Hom}_{gr}(\mathbf{C}^*, \mathbf{C}^*)$ for some finitely generated subgroup Γ of $\mathbf{C}^*$. (Actually, this is a sufficient

---

[1] Here, "cyclic" means "generated by one element"; in French terminology, this is called "monogène" while "cyclique" is reserved for a *finite* cyclic group, *i.e.*, one generated by an element of finite order. In this particular case, the French terminology is more logical (it implies that there are "cycles").

condition, as shown in the course quoted above.) Now, the reader will easily construct a morphism from $\mathfrak{X}(\mathbf{C}^*) = \mathrm{Hom}_{gr}(\mathbf{C}^*, \mathbf{C}^*)$ to $\mathrm{GL}_1(\mathbf{C}) = \mathbf{C}^*$ that is not admissible in this sense. (See exercise 3 below.)

Remark 15.6. The relation between the abstract group π_1 and the proalgebraic group $\hat{\pi}_1$ implies that *every representation of π_1 is the restriction of a unique "rational" representation of $\hat{\pi}_1$*; one says that $\hat{\pi}_1$ is the "proalgebraic hull" of π_1 (see Chapter 16 for details).

15.3. The universal group

The group $\hat{\pi}_1$ has been introduced in order to parameterize all algebraic groups of the form $\overline{\langle M \rangle}$. Therefore, it also parameterizes all differential Galois groups of local fuchsian systems. We now describe in more detail this parameterization.

We start from the system $X' = z^{-1}AX$, where $A \in \mathrm{GL}_n(\mathbf{C})$; we know that this restriction does not reduce the generality of our results. Let $\mathcal{A} := K[\mathcal{X}]$ be the differential algebra generated by the coefficients of the fundamental matricial solution $\mathcal{X} := z^A$. The algebra $\mathcal{A}$ is generated by multivalued functions of the form z^α, where $\alpha \in \mathrm{Sp}(A)$, and maybe functions $z^\alpha \log$ if there are corresponding nontrivial Jordan blocks. For any $(\gamma, \lambda) \in \hat{\pi}_1 = \mathrm{Hom}_{gr}(\mathbf{C}^*, \mathbf{C}^*) \times \mathbf{C}$, we know from Chapter 14 that the map $\sigma_{\gamma,\lambda}$ sending z^α to $\gamma(e^{2\mathrm{i}\pi\alpha})z^\alpha$ and $z^\alpha \log$ to $\gamma(e^{2\mathrm{i}\pi\alpha})z^\alpha(\log + 2\mathrm{i}\pi\lambda)$ is an element of $\mathrm{Aut}(\mathcal{A})$; and we also know that all elements of $\mathrm{Aut}(\mathcal{A})$ can be described in this way. Therefore, we obtain a surjective group morphism $(\gamma, \lambda) \mapsto \sigma_{\gamma,\lambda}$ from $\hat{\pi}_1$ to $\mathrm{Aut}(\mathcal{A})$. (The verification that it is indeed a group morphism is easy and left as an exercise to the reader.)

The matricial version of this morphism is obtained as follows. The fundamental monodromy matrix of $\mathcal{X} := z^A$ is $M := e^{2\mathrm{i}\pi A}$. Let $M = M_s M_u = M_u M_s$ be its Jordan decomposition. The map $\sigma \mapsto \mathcal{X}^{-1}\sigma(\mathcal{X})$ from $\mathrm{Aut}(\mathcal{A})$ to $\mathrm{GL}_n(\mathbf{C})$ is a group morphism with image the Galois group $\mathrm{Gal}(A)$. Composing it with the map $(\gamma, \lambda) \mapsto \sigma_{\gamma,\lambda}$ above, we get a representation:

$$\hat{\rho}_A : \begin{cases} (\gamma, \lambda) \mapsto \gamma(M_s)M_u^\lambda, \\ \hat{\pi}_1 \to \mathrm{GL}_n(\mathbf{C}), \end{cases} \qquad \text{where we put } \hat{\pi}_1 := \mathrm{Hom}_{gr}(\mathbf{C}^*, \mathbf{C}^*) \times \mathbf{C}.$$

Its image is the Galois group $\mathrm{Gal}(A) = \overline{\mathrm{Mon}(A)} = \overline{\langle M \rangle}$. We now state without proof (and not even complete definitions!) the algebraic version of the Riemann-Hilbert correspondence.

Theorem 15.7. *The functor $A \rightsquigarrow \hat{\rho}_A$ induces a bijective correspondence:*
$$\left\{ \begin{array}{c} \textit{isomorphism classes of} \\ \textit{regular singular systems} \end{array} \right\} \longleftrightarrow \left\{ \begin{array}{c} \textit{isomorphism classes of} \\ \textit{rational representations of } \hat{\pi}_1 \end{array} \right\}.$$
The Galois group of the system $X' = z^{-1}AX$ is Im $\hat{\rho}_A$.   □

Exercises

(1) (i) For any abelian group Γ, define $\mathfrak{X}(\Gamma) := \mathrm{Hom}_{gr}(\Gamma, \mathbf{C}^*)$ to be the set of group morphisms $\Gamma \to \mathbf{C}^*$. Show that defining a product $\star$ on $\mathfrak{X}(\Gamma)$ by the formula $(\gamma_1 \star \gamma_2)(x) := \gamma_1(x).\gamma_2(x)$ gives $\mathfrak{X}(\Gamma)$ the structure of an abelian group.

(ii) Show that any group morphism $f : \Gamma_1 \to \Gamma_2$ yields a "dual" morphism $\mathfrak{X}(f) : \gamma \mapsto \gamma \circ f$ from $\mathfrak{X}(\Gamma_2)$ to $\mathfrak{X}(\Gamma_1)$.

(iii) Check that if f is injective (resp. surjective), then $\mathfrak{X}(f)$ is surjective (resp. injective). (One of these statements is trivial, the other depends on Lemma 15.3.)

(iv) If $p : \Gamma \to \Gamma''$ is surjective with kernel Γ', show that $\mathfrak{X}(p)$ induces an isomorphism from $\mathfrak{X}(\Gamma'')$ to the kernel of the natural (restriction) map $\mathfrak{X}(\Gamma) \mapsto \mathfrak{X}(\Gamma')$ and conclude that $\mathfrak{X}(\Gamma') \simeq \mathfrak{X}(\Gamma)/\mathfrak{X}(\Gamma'')$.

(2) If $f : \mathbf{C} \to \mathbf{C}$ is an arbitrary map, if $\lambda_i, \mu_i \in \mathbf{C}$ for $i = 1, \ldots, n$ and if $P, Q \in \mathrm{GL}_n(\mathbf{C})$, then prove the following implication:
$$P\mathrm{Diag}(\lambda_1, \ldots, \lambda_n)P^{-1} = Q\mathrm{Diag}(\mu_1, \ldots, \mu_n)Q^{-1}$$
$$\Longrightarrow P\mathrm{Diag}(f(\lambda_1), \ldots, f(\lambda_n))P^{-1} = Q\mathrm{Diag}(f(\mu_1), \ldots, f(\mu_n))Q^{-1}.$$

(This remains true when $\mathbf{C}$ is replaced by an arbitrary commutative ring.)

(3) Construct a morphism from $\mathfrak{X}(\mathbf{C}^*) = \mathrm{Hom}_{gr}(\mathbf{C}^*, \mathbf{C}^*)$ to $\mathrm{GL}_1(\mathbf{C}) = \mathbf{C}^*$ that is not admissible in the sense of the text, *i.e.*, its image is not an algebraic subgroup of $\mathbf{C}^*$. (Recall that algebraic subgroups of $\mathbf{C}^*$ are itself and its finite subgroups; use exercise 1.)

Chapter 16

The universal group as proalgebraic hull of the fundamental group

16.1. Functoriality of the representation $\hat{\rho}_A$ of $\hat{\pi}_1$

To the regular singular system $S_A : X' = AX$, $A \in \mathrm{Mat}_n(\mathbf{C}(\{z\}))$, with a particular choice of a fundamental matricial solution $\mathcal{X}$ at 0, we attached a matricial monodromy representation $\rho_A : k \mapsto M^k$ of the fundamental group $\pi_1(\mathbf{C}^*, 0)$ canonically identified to $\mathbf{Z}$ (so M is the monodromy matrix of $\mathcal{X}$ along the fundamental loop) and then a representation $\hat{\rho}_A$ of the universal group $\hat{\pi}_1 = \mathrm{Hom}_{gr}(\mathbf{C}^*, \mathbf{C}^*) \times \mathbf{C}$ in $\mathrm{GL}_n(\mathbf{C})$ defined by $\hat{\rho}_A : (\gamma, \lambda) \mapsto \gamma(A_s)A_u^\lambda$, in such a way that the matricial Galois group of S_A is $\mathrm{Gal}(A) = \Im\hat{\rho}_A$. As in the Riemann-Hilbert correspondence, the important object here (the one which carries all relevant information[1]) is the representation $\hat{\rho}_A$ and not only its image $\mathrm{Gal}(A) = \Im\hat{\rho}_A$. We are going to generalize Riemann-Hilbert correspondence (in the functorial form we gave it in Chapter 10) to the functor $A \rightsquigarrow \hat{\rho}_A$.

So denote a second regular singular differential system by S_B, $B \in \mathrm{Mat}_p(\mathbf{C}(\{z\}))$, and let $F : A \to B$, $F \in \mathrm{Mat}_{p,n}(\mathbf{C}(\{z\}))$, be a morphism, i.e., $F' = BF - F\Lambda$. We fix fundamental solutions $\mathcal{X}$, $\mathcal{Y}$ of S_A, S_B and write M, N as their monodromy matrices along the fundamental loop λ_0.

---

[1]This, by the way, follows classical Galois theory where the important object is not so much the Galois group but the way it operates on roots; see for instance [**And12**, **Ram93**].

Since $F\mathcal{X}$ is a matricial solution of S_B, we can write:

$$F\mathcal{X} = \mathcal{Y}C \text{ for some } C \in \mathrm{Mat}_{p,n}(\mathbf{C}).$$

Analytic continuation of this identity along the loop λ_0 easily yields the relations:

$$(F\mathcal{X})^{\lambda_0} = (\mathcal{Y}C)^{\lambda_0} \Longrightarrow F\mathcal{X}^{\lambda_0} = \mathcal{Y}^{\lambda_0}C$$

$$\Longrightarrow F\mathcal{X}M = \mathcal{Y}NC \Longrightarrow \mathcal{Y}CM = \mathcal{Y}NC \Longrightarrow CM = NC.$$

From this, in turn, we just as easily obtain

$$CM = NC \Longrightarrow \forall k \in \mathbf{Z}, \ CM^k = N^kC$$

$$\Longrightarrow \forall g \in \pi_1(\mathbf{C}^*, 0), \ C\rho_A(g) = \rho_B(g)C,$$

which we interpret as saying that C is (the matrix of) a morphism of representations $\rho_A \to \rho_B$ (these were first defined in Section 8.4).

Lemma 16.1. *Let $M \in \mathrm{GL}_n(\mathbf{C})$, $N \in \mathrm{GL}_p(\mathbf{C})$ and $C \in \mathrm{Mat}_{p,n}(\mathbf{C})$ be such that $CM = NC$. Then, for all $f : \mathbf{C}^* \to \mathbf{C}^*$ and for all $\lambda \in \mathbf{C}$, one has:*

$$Cf(M_s) = f(N_s)C,$$

$$CM_u^\lambda = N_u^\lambda C.$$

Proof. One proof can be based on exercise F.7 in Appendix F. Another one relies on the fact that the set of pairs $(M, N) \in \mathrm{GL}_n(\mathbf{C}) \times \mathrm{GL}_p(\mathbf{C})$ satisfying $CM = NC$ is an algebraic subgroup. Actually, it is isomorphic to the group of block-diagonal matrices $\mathrm{Diag}(M, N)$ commuting with $\begin{pmatrix} I_n & 0_{n,p} \\ C & I_p \end{pmatrix}$. The conclusion then follows from Corollary 15.5, except from one point: the latter only mentions morphisms $\gamma : \mathbf{C}^* \to \mathbf{C}^*$, while the lemma to be proved involves arbitrary maps $f : \mathbf{C}^* \to \mathbf{C}^*$.

So we must check that, if $CM_s = N_sC$, then $Cf(M_s) = f(N_s)C$ for arbitrary f. We write $M_s = P\mathrm{Diag}(a_1, \ldots, a_n)P^{-1}$ and $N_s = Q\mathrm{Diag}(b_1, \ldots, b_p)Q^{-1}$, so that, setting $C' := Q^{-1}CP$, we have $C'\mathrm{Diag}(a_1, \ldots, a_n) = \mathrm{Diag}(b_1, \ldots, b_p)C'$. The argument is then the same as in exercise 2 of chapter 15: if $c'_{i,j} \neq 0$, then $a_j = b_i$, so that $f(a_j) = f(b_i)$ and one draws that $C'\mathrm{Diag}(f(a_1), \ldots, f(a_n)) = \mathrm{Diag}(f(b_1), \ldots, f(b_p))C'$, which in turns imples that $Cf(M_s) = f(N_s)C$. $\square$

Proposition 16.2. *We thus get a fully faithful functor $A \rightsquigarrow \hat{\rho}_A$ from the category of regular singular systems to the category of representations of $\hat{\pi}_1$.*

Proof. The lemma above says that this is indeed a functor, extending the one used to define Riemann-Hilbert correspondence; the proof that it is faithful mimics the proof given in that case: the map $F \mapsto C$ from $\mathrm{Mor}(S_A, S_B)$ to $\mathrm{Mor}(\hat{\rho}_A, \hat{\rho}_B)$ is a morphism of groups (actually it is **C**-linear) and its kernel is given by the condition $C = 0$, which implies $F\mathcal{X} = \mathcal{Y}C = 0$, whence

$F = 0$, *i.e.*, the kernel is trivial and the map is indeed injective, which is the definition of faithfulness.

If A, B are given, any morphism $\hat{\rho}_A \to \hat{\rho}_B$ can in particular be considered as a morphism $\rho_A \to \rho_B$ (by restriction of these representations to $\pi_1(\mathbf{C}^*, 0) \subset \hat{\pi}_1$) and this comes from a morphism $A \to B$ (fullness of the functor involved in the Riemann-Hilbert correspondence), which is easily seen to fit our needs. $\qquad\square$

Exercise 16.3. Check that the morphism $A \to B$ indeed fits our needs.

The proposition means that the functor $A \rightsquigarrow \hat{\rho}_A$ is an equivalence of the category of (local) fuchsian systems with some full subcategory of the category $\mathfrak{Rep}_{\mathbf{C}}(\hat{\pi}_1)$ of all (finite-dimensional complex) representations of $\hat{\pi}_1$. We now must characterize "admissible" representations of $\hat{\pi}_1$, *i.e.*, those which do come from a fuchsian system; remember that we found in Chapter 15 that not all representations are admissible in this sense.

16.2. Essential image of this functor

We now need to determine which representations ρ of $\hat{\pi}_1$ can be realized as $\rho = \hat{\rho}_A$ for a fuchsian system A. Since the image of $\hat{\rho}_A$ is $\mathrm{Gal}(A)$, an algebraic subgroup of $\mathrm{GL}_n(\mathbf{C})$, we immediately get a *necessary* condition on ρ (that $\Im\,\rho$ be an algebraic subgroup of $\mathrm{GL}_n(\mathbf{C})$). From exercise 3 at the end of the previous chapter, we know that this condition is not automatically satisfied.

Actually the condition can be refined: $\Im\,\rho$ has to be the Zariski closure $\overline{\langle M \rangle}$ of a cyclic subgroup of $\mathrm{GL}_n(\mathbf{C})$. But this condition is not intrinsic (it involves an unknown $M \in \mathrm{GL}_n(\mathbf{C})$); and it is maybe not sufficient. We have to take into account the very special way $\hat{\rho}_A(\gamma, \lambda)$ is computed (recalled above at the beginning of 16.1). Actually we can guess that $\hat{\pi}_1$ has some structure generalizing that of the algebraic groups $\mathrm{GL}_n(\mathbf{C})$ and $\mathrm{Gal}(A)$ and that the admissible representations should respect that structure.

Some basic observations. Let us write

$$\hat{\pi}_1 = \hat{\pi}_{1,s} \times \hat{\pi}_{1,u}, \quad \text{where} \quad \begin{cases} \hat{\pi}_{1,s} = \mathrm{Hom}_{gr}(\mathbf{C}^*, \mathbf{C}^*) \text{ ("semi-simple component")}, \\ \hat{\pi}_{1,u} = \mathbf{C} \text{ ("unipotent component")}. \end{cases}$$

Representations $\rho : \hat{\pi}_1 \to \mathrm{GL}_n(\mathbf{C})$ correspond bijectively to pairs (ρ_s, ρ_u) of representations:

$$\begin{cases} \rho_s : \hat{\pi}_{1,s} \to \mathrm{GL}_n(\mathbf{C}), \\ \rho_u : \hat{\pi}_{1,u} \to \mathrm{GL}_n(\mathbf{C}), \end{cases}$$

submitted to the unique requirement that their images $\Im \, \rho_s, \Im \, \rho_u \subset \mathrm{GL}_n(\mathbf{C})$ commute (so that $\rho := \rho_s \rho_u : g \mapsto \rho_s(g)\rho_u(g)$ would indeed define a representation).

Exercise 16.4. For a general group product $G = G_1 \times G_2$, similarly characterize representations of G in terms of those of G_1, G_2 (you should first precisely state your characterization and then prove it rigorously).

We therefore want to find criteria involving ρ_s, ρ_u separately that ensure that ρ_s has the form $\gamma \mapsto \gamma(S)$ for some semi-simple invertible matrix S; and ρ_u has the form $\lambda \mapsto U^\lambda$ for some unipotent matrix U which moreover commutes with S. Of course we have in mind that then SU is the Jordan decomposition of the monodromy matrix M from which we are able, through Riemann-Hilbert correspondence, to recover a fuchsian matrix A such that $\forall k \in \mathbf{Z}, \ \rho_A(k) = M^k$.

Exercise 16.5. Check that, if ρ_s, ρ_u are given as above and their images commute, then $SU = US$ and the fuchsian matrix A satisfies $\hat\rho_A = \rho_s \rho_u$ such that $\forall k \in \mathbf{Z} , \ \rho_A(k) = M^k$.

The answer will be the following. First, $\hat\pi_{1,u} = \mathbf{C}$ is an algebraic group and ρ_u will have to respect that structure, *i.e.*, to be a "rational" morphism. Second, $\hat\pi_{1,s} = \mathrm{Hom}_{gr}(\mathbf{C}^*, \mathbf{C}^*)$ is something a little more general, a "proalgebraic group", and ρ_s will also have to respect that structure, so that we also shall call it "rational".

Some more vocabulary from affine algebraic geometry. For some of the unproven statements in this subsection and the following, see the exercises at the end of the chapter or [**BorA91**].

The algebraic group $\mathrm{GL}_p(\mathbf{C})$ is a Zariski open subset of $\mathrm{Mat}_p(\mathbf{C})$, since it is the complement of the closed subset defined by the polynomial equation $\det \, = 0$. As such, it inherits natural coordinates $X_{i,j}, \ 1 \le i, j \le p$, and one can speak of polynomial functions, *i.e.*, restrictions to $\mathrm{GL}_p(\mathbf{C})$ of functions on $\mathrm{Mat}_p(\mathbf{C})$ defined by polynomials in $\mathbf{C}[(X_{i,j})_{1 \le i,j \le p}] = \mathbf{C}[X_{1,1}, \dots, X_{p,p}]$. Note that if we change the system of coordinates by an affine transformation the notion of polynomial function does not change, *i.e.*, it depends only on the affine space $\mathrm{Mat}_p(\mathbf{C})$ and not on the choice of a particular affine frame (*i.e.*, choice of a basis of the vector space and of an origin).

However, $\mathrm{GL}_p(\mathbf{C})$ can also be identified with a *closed* subset of the bigger affine space[2] $\mathrm{Mat}_p(\mathbf{C}) \times \mathbf{C}$. Using coordinates $X_{1,1}, \dots, X_{p,p}, T$ on this space,

---

[2]The way we defined the Zariski topology on $\mathrm{Mat}_p(\mathbf{C})$ in Section 14.4 extends to any affine space E: choosing coordinates $X_1, \dots, X_m$ on E allows one to define polynomial functions as

we have a well-defined polynomial function $1 - T \det(X_{i,j})$. The natural projection from $\mathrm{Mat}_p(\mathbf{C}) \times \mathbf{C}$ to $\mathrm{Mat}_p(\mathbf{C})$ is a bijection from the zero locus of $1 - T \det(X_{i,j})$ to $\mathrm{GL}_p(\mathbf{C})$. Using this identification, we see that a *polynomial* or *regular* function on $\mathrm{GL}_p(\mathbf{C})$ is most naturally defined as having the form $P(X_{1,1}, \ldots, X_{p,p}, 1/\det(X_{i,j}))$ for some $P \in \mathbf{C}[X_{1,1}, \ldots, X_{p,p}, T]$.

Let E be an arbitrary affine space with coordinates $X_1, \ldots, X_m$, let $F \subset E$ be a closed subset and let I be the ideal of $\mathbf{C}[X_1, \ldots, X_m]$ consisting of all $P \in \mathbf{C}[X_1, \ldots, X_m]$ that vanish on E, so that actually $E = V(I)$. Polynomial functions on F are restrictions of polynomial functions on E, that is, of elements of $\mathbf{C}[X_1, \ldots, X_m]$; and $P_1, P_2 \in \mathbf{C}[X_1, \ldots, X_m]$ define the same polynomial functions on F if, and only if, $P := P_1 - P_2$ vanishes on F, *i.e.*, if $P \in I$. Therefore the $\mathbf{C}$-algebra of polynomial functions on F is naturally identified with $\mathbf{C}[X_1, \ldots, X_m]/I$. This ring is called the *affine algebra of F* and we write it $\mathbf{A}(F)$. In particular, $\mathbf{A}(E) = \mathbf{C}[X_1, \ldots, X_m]$ and, for instance, $\mathbf{A}(\mathrm{Mat}_p(\mathbf{C})) = \mathbf{C}[X_{1,1}, \ldots, X_{p,p}]$.

In the case of $F := \mathrm{GL}_p(\mathbf{C}) \subset E := \mathrm{Mat}_p(\mathbf{C})$, one can prove that the ideal $I \subset \mathbf{C}[X_{1,1}, \ldots, X_{p,p}, T]$ is generated by $1 - T \det(X_{i,j})$, so that the affine algebra of $\mathrm{GL}_p(\mathbf{C})$ is:

$$\mathbf{A}(\mathrm{GL}_p(\mathbf{C})) = \mathbf{C}[X_{1,1}, \ldots, X_{p,p}, T]/\langle 1 - T \det(X_{i,j}) \rangle$$
$$\simeq \mathbf{C}[X_{1,1}, \ldots, X_{p,p}, 1/\Delta],$$

where $\Delta := \det(X_{i,j})$.

Examples 16.6. (1) The affine algebra of $\mathbf{C}^* = \mathrm{GL}_1(\mathbf{C})$ is $\mathbf{A}(\mathrm{GL}_1(\mathbf{C})) = \mathbf{C}[X, T]/\langle 1 - TX \rangle \simeq \mathbf{C}[X, 1/X]$.

(2) The subgroup $\mu_d \in \mathbf{C}^*$ of dth roots of unity is Zariski closed in $\mathbf{C} = \mathrm{Mat}_1(\mathbf{C})$ and the ideal of its equations is generated by $X^d - 1$, so its affine algebra is $\mathbf{A}(\mu_d) = \mathbf{C}[X]/\langle X^d - 1 \rangle$. It is reduced (it has no nonzero nilpotent elements) but it is not an integral ring.

(3) We identify the algebraic group $\mathbf{C}$ as the algebraic subgroup of upper-triangular unipotent matrices in $\mathrm{GL}_2(\mathbf{C})$; this is defined by equations $X_{2,1} = 0$ and $X_{1,1} = X_{2,2} = 1$. One can prove that the ideal of all its equations is generated by $X_{2,1}, X_{1,1} - 1, X_{2,2} - 1$, so that its affine algebra is:

$$\mathbf{A}(\mathbf{C}) = \mathbf{C}[X_{1,1}, X_{1,2}, X_{2,1}, X_{2,2}]/\langle X_{2,1}, X_{1,1} - 1, X_{2,2} - 1 \rangle \simeq \mathbf{C}[X_{1,2}].$$

---

elements of $\mathbf{C}[X_1, \ldots, X_m]$ and one readily checks that this definition does not depend on the particular choice of coordinates; then closed sets are defined to be the zero loci of arbitrary sets of polynomial equations. Everything we said in 14.4 works the same.

The isomorphism above is actually the composition of the inclusion $\mathbf{C}[X_{1,2}]$ $\to \mathbf{C}[X_{1,1}, X_{1,2}, X_{2,1}, X_{2,2}]$ with the natural projection

$$\mathbf{C}[X_{1,1}, X_{1,2}, X_{2,1}, X_{2,2}]$$
$$\to \mathbf{C}[X_{1,1}, X_{1,2}, X_{2,1}, X_{2,2}]/\langle X_{2,1}, X_{1,1} - 1, X_{2,2} - 1\rangle.$$

We now reach the most important[3] notion in elementary affine algebraic geometry. Let E, E' be two affine spaces. By choosing coordinates on E', we obtain an identification $E' \simeq \mathbf{C}^{p'}$. Therefore, any map $f : E \to E'$ can be identified with a p'-uple $(f_1, \ldots, f_{p'})$ of functions $f_i : E \to \mathbf{C}$. We say that f is a *polynomial* or *regular* map if all the f_i are polynomial (or regular) functions. This is actually independent of the choice of coordinates on E'. Now let $F \subset E$, $F' \subset E'$ be Zariski closed subsets. We say that a map $f : F \to F'$ is *polynomial* or *regular* if it is the restriction of a polynomial (or regular) map from E to E'.

Rational morphisms.

Definition 16.7. Let G, H be two algebraic groups. A *rational morphism* from G to H is a group morphism which is at the same time a regular map.

So let $\phi : G \to H$ be a group morphism, and suppose that H is given as an algebraic subgroup of $\mathrm{GL}_n(\mathbf{C}) \subset \mathrm{Mat}_n(\mathbf{C})$, so that it comes equipped with natural coordinates $X_{i,j} : H \to \mathbf{C}$. Then ϕ splits into a family $(\phi_{i,j})_{1 \leq i,j \leq n}$ of functions $\phi_{i,j} = X_{i,j} \circ \phi : G \to \mathbf{C}$ and from the definitions in the previous subsection one obtains that ϕ is rational if, and only if, all the $\phi_{i,j}$ are regular functions.

Remark 16.8. Recall that if H is commutative and if $\phi, \psi : G \to H$ are group morphisms, then $\phi\psi : g \mapsto \phi(g)\psi(g)$ is a group morphism, so in this case one can make $\mathrm{Hom}_{gr}(G, H)$ into a group. Applying this to algebraic groups, one sees easily that rational morphisms form a subgroup of $\mathrm{Hom}_{gr}(G, H)$.

Examples 16.9. (1) Rational morphisms $\mathbf{C}^* \to \mathbf{C}^*$ are Laurent polynomials $P \in \mathbf{C}[X, 1/X]$ such that $P(xy) = P(x)P(y)$ for all $x, y \in \mathbf{C}^*$ or, equivalently, $P(XY) = P(X)P(Y)$ as Laurent polynomials in two indeterminates. Identifying homogeneous parts, one finds that $P = X^k$ for some $k \in \mathbf{Z}$. Thus, the group of rational morphisms from $\mathbf{C}^*$ to itself is isomorphic to $\mathbf{Z}$; it is much smaller than the group $\mathfrak{X}(\mathbf{C}^*) = \mathrm{Hom}_{gr}(\mathbf{C}^*, \mathbf{C}^*)$ of all group morphisms from $\mathbf{C}^*$ to itself.

---

[3]It is a lesson from the twentieth-century revolution in mathematics that the significant component of a category is not so much its objects but rather its morphisms; they are the ones that carry structural information.

(2) Rational morphisms $\mathbf{C} \to \mathbf{C}$ are polynomials $P \in \mathbf{C}[X]$ such that $P(x + y) = P(x) + P(y)$ for all $x, y \in \mathbf{C}$ or, equivalently, $P(X + Y) = P(X) + P(Y)$ as polynomials in two indeterminates. One finds that only linear forms $P(X) = aX$, $a \in \mathbf{C}$, are possible.

(3) Rational morphisms $\mathbf{C} \to \mathbf{C}^*$ are polynomials $P \in \mathbf{C}[X]$ such that $P(0) = 1$ and $P(x + y) = P(x)P(y)$ for all $x, y \in \mathbf{C}$ or, equivalently, $P(X + Y) = P(X)P(Y)$ as polynomials in two indeterminates. One finds that only the constant polynomial $P = 1$ does it.

(4) Rational morphisms $\mathbf{C}^* \to \mathbf{C}$ are Laurent polynomials $P \in \mathbf{C}[X, 1/X]$ such that $P(1) = 0$ and $P(xy) = P(x) + P(y)$ for all $x, y \in \mathbf{C}^*$ or, equivalently, $P(XY) = P(X) + P(Y)$ as Laurent polynomials in two indeterminates. Only the constant polynomial $P = 0$ fits.

(5) Let $U = I_n + N$ be a unipotent matrix. Expanding U^λ with the help of the generalized Newton binomial formula yields terms $\binom{\lambda}{k} N^k$ that vanish for $k \geq n$, whence a polynomial of degree $< n$ in λ. Therefore, $\lambda \mapsto U^\lambda$ is a rational representation of $\mathbf{C}$ in $\mathrm{GL}_n(\mathbf{C})$.

The last example admits a useful converse.

Theorem 16.10. *The rational representations of $\mathbf{C}$ are the maps $\lambda \mapsto U^\lambda$, where U is a unipotent matrix.*

Proof. We already saw that any such U defines a rational representation of $\mathbf{C}$. So let $\phi : \mathbf{C} \to \mathrm{GL}_n(\mathbf{C})$ be a polynomial map such that $\phi(0) = I_n$ and $\phi(x + y) = \phi(x)\phi(y)$ for all $x, y \in \mathbf{C}$. The matrices $\phi(x)$ are pairwise commuting since $\phi(y)\phi(x) = \phi(y+x) = \phi(x+y) = \phi(x)\phi(y)$; therefore they are cotrigonalizable:

$$\phi(x) = P \begin{pmatrix} c_1(x) & \cdots & \star \\ \vdots & \ddots & \vdots \\ 0 & \cdots & c_n(x) \end{pmatrix} P^{-1}.$$

Since the c_i are rational morphisms $\mathbf{C} \to \mathbf{C}^*$, they are constant equal to 1: all $\phi(x)$ are unipotent matrices. So let $U := \phi(1)$. Since ϕ is a group morphism, $\phi(k) = U^k$ for all $k \in \mathbf{Z}$. For any $i < j$, the (i, j)-coefficients of the rational representations ϕ and $\lambda \mapsto U^\lambda$ are polynomials and they coincide on all integral values, so they are equal. $\square$

16.3. The structure of the semi-simple component of $\hat{\pi}_1$

The functor $\Gamma \rightsquigarrow \mathfrak{X}(\Gamma)$. Remember that $\hat{\pi}_{1,s} = \mathfrak{X}(\mathbf{C}^*)$, where we defined (see exercise 1 in Chapter 15) for any group Γ the group $\mathfrak{X}(\Gamma)$ as $\mathrm{Hom}_{gr}(\Gamma, \mathbf{C}^*)$, endowed with a product given by the rule: if $\phi, \psi \in \mathfrak{X}(\Gamma)$ are group morphisms $\Gamma \to \mathbf{C}^*$, then $\phi\psi \in \mathfrak{X}(\Gamma)$ is the morphism $g \mapsto \phi(g)\psi(g)$.

(The fact that this is indeed a group morphism flows from the commutativity of $\mathbf{C}^*$; the same definition would not work for $\mathrm{Hom}_{gr}(\Gamma, \mathrm{GL}_n(\mathbf{C}))$ for instance.) In this section, we are going to study the general structure of $\mathfrak{X}(\Gamma)$ and then apply our findings to $\Gamma := \mathbf{C}^*$.

As explained in exercise 1 in Chapter 15, we can make $\Gamma \rightsquigarrow \mathfrak{X}(\Gamma)$ into a contravariant functor from groups to (commutative) groups as follows. Let $f : \Gamma \to \Gamma'$ be any morphism of groups. Then, for any $\phi \in \mathfrak{X}(\Gamma')$, meaning that $\phi : \Gamma' \to \mathbf{C}^*$ is a group morphism, the composition $\phi \circ f : \Gamma \to \mathbf{C}^*$ is a group morphism. We thus define a map $\mathfrak{X}(f) : \mathfrak{X}(\Gamma') \to \mathfrak{X}(\Gamma)$, $\phi \mapsto \phi \circ f$. If $\psi \in \mathfrak{X}(\Gamma')$, i.e., $\psi : \Gamma' \to \mathbf{C}^*$ is another group morphism, then

$$\mathfrak{X}(f)(\phi\psi) = (\phi\psi) \circ f = (\phi \circ f)(\psi \circ f) = \mathfrak{X}(f)(\phi)\mathfrak{X}(f)(\psi),$$

so that $\mathfrak{X}(f) : \mathfrak{X}(\Gamma') \to \mathfrak{X}(\Gamma)$ is a morphism of groups. It is immediate that we do get a contravariant functor. Actually, because of exercise 8 at the end of this chapter, we can without loss of generality restrict the functor $\mathfrak{X}$ to the category of commutative groups. So, from now on, we shall take Γ to be commutative.

Some basic observations on admissible representations. To better understand the definitions that are going to follow, let us examine more closely the semi-simple part $\rho_s : \mathrm{Hom}_{gr}(\mathbf{C}^*, \mathbf{C}^*) = \mathfrak{X}(\mathbf{C}^*) \to \mathrm{GL}_n(\mathbf{C})$ of the admissible representation $\hat{\rho}_A = \rho_s\rho_u$. For any $\gamma \in \mathfrak{X}(\mathbf{C}^*)$, we have

$$\rho_s(\gamma) = \gamma(A_s).$$

This is entirely determined by the knowledge of the images $\gamma(c_i)$ of the eigenvalues of A_s, i.e., those of A. So call Γ_Σ the subgroup of $\mathbf{C}^*$ generated by the spectrum $\Sigma := \mathrm{Sp}(A) = \mathrm{Sp}(A_s)$ and call γ_Σ the restriction of γ to Γ_Σ. Then $\gamma(A_s) = \gamma_\Sigma(A_s)$. On the other hand, defining $\rho_{s,\Sigma} : \mathfrak{X}(\Gamma_\Sigma) \to \mathrm{GL}_n(\mathbf{C})$, we get a representation of $\mathfrak{X}(\Gamma_\Sigma)$, and ρ_s is just the composition of $\rho_{s,\Sigma}$ with the natural map $\mathfrak{X}(\mathbf{C}^*) \to \mathfrak{X}(\Gamma_\Sigma)$ coming (by contravariant functoriality) from the inclusion $\Gamma_\Sigma \subset \mathbf{C}^*$.

To summarize, we have the following situation:

- The group $\mathbf{C}^*$ is the union of all the finitely generated subgroups Γ_Σ, when Σ runs among all the finite subsets of $\mathbf{C}^*$ (and any such finite subset is the spectrum of some A).

- As a consequence, there are group morphisms from $\mathfrak{X}(\mathbf{C}^*)$ to all the groups $\mathfrak{X}(\Gamma_\Sigma)$ (the restriction maps).

- An admissible representation of $\hat{\pi}_{1,s} = \mathfrak{X}(\mathbf{C}^*)$ is one that factors through a representation of some $\mathfrak{X}(\Gamma_\Sigma)$, i.e., that comes as the composition of that representation with the natural map $\mathfrak{X}(\mathbf{C}^*) \to \mathfrak{X}(\Gamma_\Sigma)$.

This is the situation that we are next going to axiomatize.

Inverse limits. Let Γ be an arbitrary abelian group. We write I as the set of finite subsets of Γ and, for any $i \in I$ (thus $i \subset \Gamma$ is finite), we write Γ_i as the subgroup of Γ generated by i. Therefore the Γ_i are exactly all the finitely generated subgroups of Γ, and

$$\Gamma = \bigcup_{i \in I} \Gamma_i,$$

since clearly any element of Γ belongs to some i and $i \subset \Gamma_i$. Note moreover that this reunion is "directed", meaning that for any $i, j \in I$ there exists $k \in I$ such that $\Gamma_i, \Gamma_j \subset \Gamma_k$; indeed, $k := i \cup j$ will do.

We write $\prec$ as the inclusion relation on I, so that $i \prec j$ means that $i \subset j$; thus $(I, \prec)$ is a directed ordered set: for any $i, j \in I$ there exists $k \in I$ such that $i, j \prec k$.

For every $i \prec j$ in I, the inclusion $\Gamma_i \subset \Gamma_j$ induces (by contravariant functoriality; see exercise 1 in Chapter 15) a group morphism $f_i^j : \mathfrak{X}(\Gamma_j) \to \mathfrak{X}(\Gamma_i)$. Obviously, f_i^i is the identity of $\mathfrak{X}(\Gamma_i)$ and $f_i^k = f_i^j \circ f_j^k$ whenever $i \prec j \prec k$; we say that the family $(\mathfrak{X}(\Gamma_i), f_i^j)$ is *an inverse system of commutative groups*.

The meaning of the next result is explained in the course of its proof.

Proposition 16.11. *The family of maps $\mathfrak{X}(\Gamma) \to \mathfrak{X}(\Gamma_i)$ induced (via contravariant functoriality) by the inclusions $\Gamma_i \subset \Gamma$ yields an isomorphism*

$$\mathfrak{X}(\Gamma) \simeq \varprojlim \mathfrak{X}(\Gamma_i)$$

of the group $\mathfrak{X}(\Gamma)$ with the inverse limit *of the inverse system $(\mathfrak{X}(\Gamma_i), f_i^j)$.*

Proof. Let $\phi \in \mathfrak{X}(\Gamma)$. The inclusion $\Gamma_i \subset \Gamma$ induces (by contravariant functoriality) a group morphism $f_i : \mathfrak{X}(\Gamma) \to \mathfrak{X}(\Gamma_i)$. Actually, $\phi_i := f_i(\phi)$ is just the restriction $\phi_i := \phi_{|\Gamma_i} : \Gamma_i \to \mathbf{C}^*$ of $\phi : \Gamma \to \mathbf{C}^*$. We have thereby produced a family $(\phi_i)_{i \in I} \in \prod_{i \in I} \mathfrak{X}(\Gamma_i)$. This family has a particular property:

(16.1) $$\forall i, j \in I, \ i \prec j \Longrightarrow f_i^j(\phi_j) = \phi_i.$$

This just means that the restriction of ϕ_j to Γ_i is ϕ_i, which comes from the fact that both are equal to $\phi_{|\Gamma_i}$.

Conversely, every family $(\phi_i)_{i \in I} \in \prod_{i \in I} \mathfrak{X}(\Gamma_i)$ such that (16.1) is satisfied comes from a unique $\phi \in \mathfrak{X}(\Gamma)$; indeed, this simply means that morphisms $\phi_i : \Gamma_i \to \mathbf{C}^*$ which extend each other (when this makes sense) can be patched together to give a morphism ϕ from $\bigcup \Gamma_i = \Gamma$ to $\mathbf{C}^*$.

The set of all families $(\phi_i)_{i \in I} \in \prod_{i \in I} \mathfrak{X}(\Gamma_i)$ such that (16.1) is satisfied is a subgroup of the product group $\prod_{i \in I} \mathfrak{X}(\Gamma_i)$. This subgroup is called the *inverse limit*[4] of the inverse system of groups $(\mathfrak{X}(\Gamma_i), f_i^j)$ and it is written:

$$\varprojlim \mathfrak{X}(\Gamma_i) := \left\{ (\phi_i)_{i \in I} \in \prod_{i \in I} \mathfrak{X}(\Gamma_i) \mid \forall i, j \in I, \ i \prec j \implies f_i^j(\phi_j) = \phi_i \right\}.$$

The natural map $(f_i) : \phi \mapsto (f_i(\phi))_{i \in I}$ from $\mathfrak{X}(\Gamma)$ to $\prod_{i \in I} \mathfrak{X}(\Gamma_i)$ is obviously a group morphism, and it was just seen to induce a bijection from $\mathfrak{X}(\Gamma)$ to the subgroup $\varprojlim \mathfrak{X}(\Gamma_i) \subset \prod_{i \in I} \mathfrak{X}(\Gamma_i)$. That bijection is therefore an isomorphism of groups. $\qquad\qquad\qquad\qquad\qquad\qquad\qquad\qquad\qquad\qquad\qquad\qquad\qquad\quad \square$

Now, all Γ_i are finitely generated abelian groups, and we deal with this case in the next subsection.

The case of finitely generated abelian groups. We recall (see [**Lan02**]) that every finitely generated abelian group Γ is isomorphic to a direct product of cyclic groups:

$$\Gamma \simeq \mathbf{Z}^r \times (\mathbf{Z}/d_1\mathbf{Z}) \times \cdots \times (\mathbf{Z}/d_k\mathbf{Z}),$$

where the *rank* $r \in \mathbf{N}$ is uniquely determined and where $d_1, \ldots, d_k \in \mathbf{N}^*$ are uniquely determined by the condition $1 < d_1 \mid d_2 \mid \cdots \mid d_k$ (the notation means that each d_i divides d_{i+1}).

We shall reduce to the case of cyclic groups thanks to the following. Let $\Gamma := \Gamma_1 \times \cdots \times \Gamma_m$ and let p_i be the m projections from Γ to its factors Γ_i, $i = 1, \ldots, m$. Then we saw in exercise 1 in Chapter 15 that the $\mathfrak{X}(p_i)$ are injections $\mathfrak{X}(\Gamma_i) \to \mathfrak{X}(\Gamma)$, so that we can identify each $\mathfrak{X}(\Gamma_i)$ with a subgroup of $\mathfrak{X}(\Gamma)$. Dually, each natural injection $q_i : \Gamma_i \to \Gamma$ sending $g \in \Gamma_i$ to $(g_1, \ldots, g_m)$, where $g_i = g$ and all other g_j are trivial (neutral elements), gives rise to a surjective morphism $\mathfrak{X}(q_i)$ from $\mathfrak{X}(\Gamma)$ to $\mathfrak{X}(\Gamma_i)$. Last, since $p_i \circ q_i$ is obviously the identity of Γ_i, we deduce by functoriality that $\mathfrak{X}(q_i) \circ \mathfrak{X}(p_i)$ is the identity of $\mathfrak{X}(\Gamma_i)$. Actually, through the previous identifications:

---

[4]Inverse limits can (and should) be defined in a more conceptual way through their "universal property"; see [**Lan02**] or exercise 10 at the end of the chapter.

Proposition 16.12.

$$\mathfrak{X}(\Gamma) = \mathfrak{X}(\Gamma_1) \times \cdots \times \mathfrak{X}(\Gamma_m).$$

Proof. Indeed, let $\phi \in \mathfrak{X}(\Gamma)$ and let $g = (g_1, \ldots, g_m) \in \Gamma$. Then $g = q_1(g_1) \cdots q_m(g_m)$ (a commutative decomposition) and

$$\phi(g) = \prod \phi(q_i(g_i)) = \prod \phi_i(g_i),$$

where $\phi_i := \phi \circ q_i = \mathfrak{X}(q_i)(\phi) \in \mathfrak{X}(\Gamma_i)$. Conversely every m-uple $(\phi_1, \ldots, \phi_m)$ defines a $\phi \in \mathfrak{X}(\Gamma)$ through the previous calculation. $\qquad\square$

For any commutative group G, the group $\mathrm{Hom}_{gr}(\mathbf{Z}, G)$ is naturally identified with G (because a morphism $\phi : \mathbf{Z} \to G$ is totally determined by $\phi(1) \in G$). Therefore $\mathfrak{X}(\mathbf{Z}) = \mathbf{C}^*$ and, after the proposition, $\mathfrak{X}(\mathbf{Z}^r) = (\mathbf{C}^*)^r$. We are entitled to use the "=" sign and not merely the "$\simeq$" sign because there is a canonical isomorphism coming from the various identifications above:

$$(\mathbf{C}^*)^r \to \mathfrak{X}(\mathbf{Z}^r), \ (c_1, \ldots, c_r) \mapsto ((m_1, \ldots, m_r) \mapsto c_1^{m_1} \cdots c_r^{m_r}).$$

On the other hand, the canonical surjection $\mathbf{Z} \to \mathbf{Z}/d\mathbf{Z}$ induces (by contravariant functoriality) an injection $\mathfrak{X}(\mathbf{Z}/d\mathbf{Z}) \to \mathfrak{X}(\mathbf{Z})$. Here, $\mathfrak{X}(\mathbf{Z}/d\mathbf{Z})$ is identified with that subgroup of $\mathfrak{X}(\mathbf{Z})$ consisting of morphisms $\phi : m \mapsto c^m$ ($c \in \mathbf{C}^*$) which can be factorized through $\mathbf{Z} \to \mathbf{Z}/d\mathbf{Z}$; equivalently, which are trivial on $d\mathbf{Z}$; equivalently, such that $d \mapsto 1$; equivalently, such that $c^d = 1$. Therefore, through the identification of $\mathfrak{X}(\mathbf{Z})$ with $\mathbf{C}^*$ (in which ϕ corresponds to c in the calculation above), we obtain an identification of $\mathfrak{X}(\mathbf{Z}/d\mathbf{Z})$ with μ_d, the group of dth roots of unity in $\mathbf{C}^*$. To summarize:

$$X\left(\mathbf{Z}^r \times (\mathbf{Z}/d_1\mathbf{Z}) \times \cdots \times (\mathbf{Z}/d_k\mathbf{Z})\right) = (\mathbf{C}^*)^r \times \mu_{d_1} \times \cdots \times \mu_{d_k}.$$

Note that this can be identified with a subgroup of the group of diagonal matrices in $\mathrm{GL}_N(\mathbf{C})$, $N = r+k$. Therefore, for any finitely generated abelian group Γ, the group $\mathfrak{X}(\Gamma)$ is, in a natural way, an algebraic group.

Lemma 16.13. *Let f be a group morphism from $\Gamma := \mathbf{Z}^r \times (\mathbf{Z}/d_1\mathbf{Z}) \times \cdots \times (\mathbf{Z}/d_k\mathbf{Z})$ to $\Gamma' := \mathbf{Z}^s \times (\mathbf{Z}/e_1\mathbf{Z}) \times \cdots \times (\mathbf{Z}/e_l\mathbf{Z})$. Then $\mathfrak{X}(f) : \mathfrak{X}(\Gamma') \to \mathfrak{X}(\Gamma)$ is a rational morphism of algebraic groups from $(\mathbf{C}^*)^s \times \mu_{e_1} \times \cdots \times \mu_{e_l}$ to $(\mathbf{C}^*)^r \times \mu_{d_1} \times \cdots \times \mu_{d_k}$.*

Proof. It is enough to check that for individual "factors" $\mathfrak{X}(\mathbf{Z}/e\mathbf{Z}) \to \mathfrak{X}(\mathbf{Z}/d\mathbf{Z})$ coming from a morphism $\mathbf{Z}/d\mathbf{Z} \to \mathbf{Z}/e\mathbf{Z}$, where d is 0 or one of the d_i and e is 0 or one of the e_j. Note that any morphism $\mathbf{Z}/d\mathbf{Z} \to \mathbf{Z}/e\mathbf{Z}$ has the form $\bar{a} \mapsto \overline{ka}$ for some $k \in \mathbf{Z}$ such that $\overline{kd} = 0$ in $\mathbf{Z}/e\mathbf{Z}$, i.e., such that $e|kd$. (Equivalently, k is a multiple of $e/pgcd(e, d)$.) But then the dual morphism $\mu_e \to \mu_d$ (with the convention that $\mu_0 = \mathbf{C}^*$) is $z \mapsto z^k$. $\qquad\square$

A first consequence of this lemma is that, if Γ is a finitely generated abelian group, all isomorphisms from Γ to some group $\mathbf{Z}^r \times (\mathbf{Z}/d_1\mathbf{Z}) \times \cdots \times (\mathbf{Z}/d_k\mathbf{Z})$ induce on $\mathfrak{X}(\Gamma)$ the same algebraic group structure. A second consequence is that $\Gamma \rightsquigarrow \mathfrak{X}(\Gamma)$ is a contravariant functor from the category of finitely generated abelian groups (and group morphisms) to the category of algebraic groups (and rational morphisms).

Remark 16.14. For any algebraic group G define its "dual" $\mathbf{X}(G)$ as the group of rational morphisms $G \to \mathbf{C}^*$. It can be shown that if $G = \mathfrak{X}(\Gamma)$, Γ a finitely generated abelian group, then $\mathbf{X}(G)$ is Γ. Conversely, one can characterize intrinsically the class of algebraic groups G that are isomorphic to $\mathfrak{X}(\mathbf{X}(G))$ (see [**BorA91**, Chap. III] or Appendix E); in the case of the base field $\mathbf{C}$, these are just the so-called "diagonalizable" groups. Therefore, we have defined an anti-equivalence of the category of diagonalizable algebraic groups (and rational morphisms) with the category of finitely generated abelian groups (and group morphisms).

$\hat{\pi}_{1,s}$ is a proalgebraic group. A *proalgebraic group* is an inverse limit of algebraic groups. More precisely, we must have an inverse system (G_i, f_i^j) of groups as above (see exercise 10 in this chapter for the general definition) such that the G_i are algebraic groups and the f_i^j are rational morphisms, and a family of group morphisms $f_i : G \to G_i$ that induce an isomorphism of groups from G to the inverse limit $\varprojlim G_i$.

Examples 16.15. (1) Any algebraic group is a proalgebraic group in an obvious way.

(2) If Γ is any commutative group and (Γ_i) is the family of all its finitely generated subgroups, the maps $\mathfrak{X}(\Gamma) \to \mathfrak{X}(\Gamma_i)$ yield an isomorphism of $\mathfrak{X}(\Gamma)$ with the inverse limit of all the $\mathfrak{X}(\Gamma_i)$ and this makes $\mathfrak{X}(\Gamma)$ a proalgebraic group.

It is important to understand that the inverse system (G_i, f_i^j) and the family (f_i) are data belonging to the structure. However, as we shall see in a moment, different such data can define the same structure. But first, we define (rational) morphisms of proalgebraic groups; we do this in two steps.

(1) Let G' be a proalgebraic group, with structural data $(G'_k, {f'}^l_k)$ and (f'_k). Let G be an algebraic group. Then, to any group morphism $\phi : G \to G'$ is associated the family of group morphisms $\phi_k := f'_k \circ \phi : G \to G'_k$. Note that each ϕ_k links two algebraic groups G and G'_k; so we say that ϕ is a rational morphism if, and only if, all ϕ_k are. (Also note that, by the universal property, any family (ϕ_k) uniquely defines a ϕ.)

(2) Now we take G to be a proalgebraic group, with structural data (G_i, f_i^j) and (f_i). Then a group morphism $\phi : G \to G'$ is said to be rational if it factors as $\phi = \psi \circ f_i$ for some i and some rational $\psi : G_i \to G'$ (the latter makes sense according to the first step since G_i is algebraic).

Note that this allows us to say when two structural data (G_i, f_i^j) and (f_i) (on the one hand) and (G'_k, f'^l_k) and (f'_k) (on the other hand), given for the same group G, define the same structure of proalgebraic group on G; we just require that the identity map of G be an isomorphism.

Exercise 16.16. Translate this into more concrete compatibility conditions between the two sets of structural data. (Caution: to obtain symmetric conditions, you must take into account the morphism requirement from one structure to the other *and back!*)

16.4. Rational representations of $\hat{\pi}_1$

In this section and the next one we shall describe all rational representations of $\hat{\pi}_1$ and achieve the extension of Riemann-Hilbert correspondence to differential Galois theory. We begin with the description of the rational representations of the semi-simple component $\hat{\pi}_{1,s} = \mathfrak{X}(\mathbf{C}^*)$.

Lemma 16.17. *The rational representations of μ_d are the maps $j^k \mapsto S^k$ where j is a generator of μ_d (i.e., a dth primitive root of unity) and where $S \in \mathrm{GL}_n(\mathbf{C})$ is such that $S^d = I_n$.*

Proof. Note that any map $\mu_d \to \mathbf{C}$ is a regular map so we just have to look for morphisms $\mu_d \to G$, with $G := \mathrm{GL}_n(\mathbf{C})$ here; but, μ_d being cyclic of order d, the said morphisms are exactly the maps described. $\square$

Exercise 16.18. Let $S \in \mathrm{GL}_n(\mathbf{C})$. Prove that $S^d = I_n$ is equivalent to: S is semi-simple and $\mathrm{Sp}\, S \subset \mu_d$.

Lemma 16.19. *The rational representations of $\mathbf{C}^*$ are the maps $x \mapsto P\mathrm{Diag}(x^{k_1}, \ldots, x^{k_n})P^{-1}$, where $P \in \mathrm{GL}_n(\mathbf{C})$ and $k_1, \ldots, k_n \in \mathbf{Z}$.*

Proof. The maps described are clearly rational representations. Conversely, let $\rho : \mathbf{C}^* \to \mathrm{GL}_n(\mathbf{C})$ be a rational representation. We can write:

$$\rho(x) = x^k(M_0 + \cdots + M_p x^p),$$

where $k \in \mathbf{Z}$, $p \in \mathbf{N}$ and $M_0, \ldots, M_p \in \mathrm{Mat}_n(\mathbf{C})$, $M_0, M_p \neq 0$.

Since $(xy)^{-k}\rho(xy) = x^{-k}y^{-k}\rho(x)\rho(y) = (x^{-k}\rho(x))(y^{-k}\rho(y))$, we have:

$$\forall x, y \in \mathbf{C}^*, \ M_0 + \cdots + M_p(xy)^p = (M_0 + \cdots + M_p x^p)(M_0 + \cdots + M_p y^p).$$

It follows that $M_i^2 = M_i$ and $M_i M_j = 0$ for $i, j = 0, \ldots, p$, $i \neq j$. Moreover, $\phi(1) = M_0 + \cdots + M_p = I_n$. We thus have a complete family of orthogonal projectors, and general linear algebra says there exists $P \in \mathrm{GL}_n(\mathbf{C})$ such that, writing r_i as the rank of M_i for $i = 0, \ldots, p$:

$$M_i = P\mathrm{Diag}(0, \ldots, I_{r_i}, \ldots, 0)P^{-1}.$$

The conclusion follows. $\square$

Exercise 16.20. Prove the lemma by using the fact that the $\rho(x)$ form a commuting family of invertible matrices, so they are cotrigonalizable.

Theorem 16.21. *Let Γ be a finitely generated subgroup of $\mathbf{C}^*$. The rational representations of $\mathfrak{X}(\Gamma)$ are the maps $\gamma \mapsto \gamma(S)$, where $S \in \mathrm{GL}_n(\mathbf{C})$ is semi-simple and such that $\mathrm{Sp}\ S \subset \Gamma$.*

Proof. By the structure theorem of finitely generated abelian groups, we can write:

$$\Gamma = \langle x_1, \ldots, x_r, y_1, \ldots, y_k \rangle,$$

where $\langle x_1, \ldots, x_r \rangle$ is free abelian of rank r, where each y_i has order $d_i \in \mathbf{N}^*$ and where the sum $\langle x_1, \ldots, x_r, y_1, \ldots, y_k \rangle$ of the $\langle x_i \rangle$ and the $\langle y_j \rangle$ is direct. Thus:

$$\mathfrak{X}(\Gamma) = \prod_{i=1}^r \mathfrak{X}(\langle x_i \rangle) \times \prod_{j=1}^k \mathfrak{X}(\langle y_j \rangle).$$

The restriction of $\rho : \mathfrak{X}(\Gamma) \to \mathrm{GL}_n(\mathbf{C})$ to each of the $\mathfrak{X}(\langle x_i \rangle) \simeq \mathbf{C}^*$, $\mathfrak{X}(\langle y_j \rangle) \simeq \mu_{d_j}$ has the form indicated in the previous lemmas, whence pairwise commuting semi-simple matrices $S_1, \ldots, S_r, S_1', \ldots, S_k' \in \mathrm{GL}_n(\mathbf{C})$. Then $S := S_1 \cdots S_r S_1' \cdots S_k' \in \mathrm{GL}_n(\mathbf{C})$ is semi-simple and solves the problem. $\square$

Exercise 16.22. Rigorously show the last statement.

16.5. Galois correspondence and the proalgebraic hull of $\hat{\pi}_1$

Rational representations of $\hat{\pi}_1 = \mathrm{Hom}_{gr}(\mathbf{C}^*, \mathbf{C}^*) \times \mathbf{C}$ will be defined as pairs (ρ_s, ρ_u) of a rational representation ρ_u of the algebraic group $\mathbf{C}$, say in $\mathrm{GL}_n(\mathbf{C})$, and a rational representation ρ_s (that is, a rational morphism as defined above) of the proalgebraic group $\hat{\pi}_{1,s} = \mathrm{Hom}_{gr}(\mathbf{C}^*, \mathbf{C}^*) = \mathfrak{X}(\mathbf{C}^*)$ in the same $\mathrm{GL}_n(\mathbf{C})$, with the additional requirement that $\Im\ \rho_u$ and $\Im\ \rho_s$ commute (see Section 16.2). According to Sections 16.2 and 16.4, we obtain:

Theorem 16.23. *The rational representations of $\hat{\pi}_1 = \mathrm{Hom}_{gr}(\mathbf{C}^*, \mathbf{C}^*) \times \mathbf{C}$ are exactly the maps of the form*

$$(\gamma, \lambda) \mapsto \gamma(S)U^\lambda,$$

where $S, U \in \mathrm{GL}_n(\mathbf{C})$, S is semi-simple, U is unipotent and S, U commute.
□

We have now solved the problem of the essential image that was raised in Section 16.2.

Corollary 16.24. *The functor $A \rightsquigarrow \hat{\rho}_A$ is an equivalence of the category of fuchsian systems (local at 0) with the category of rational representations of $\hat{\pi}_1$.*

In this statement, rational representations are viewed as a full subcategory of the category of all representations; said otherwise, morphisms of rational representations are simply defined as morphisms of plain representations, without reference to the proalgebraic structure.

Consider now the following setting. We have a group G and a proalgebraic group G', together with a group morphism $f : G \to G'$. Then $\rho \mapsto \rho \circ f$ sends rational representations ρ of G' to representations $\rho \circ f$ of G; actually, it induces a functor between the corresponding categories. In the case that this functor is an equivalence of categories, we say that G' is the *proalgebraic hull* of G (it is implicitly understood that the morphism f belongs to the structure of G' as a proalgebraic hull). The universal property expressed by the fact that we have an equivalence of categories implies that G' is unique up to rational isomorphism. We write it G^{alg}.

Corollary 16.25. *The inclusion $\pi_1 \subset \hat{\pi}_1$ makes the latter the proalgebraic hull of the former.*

Proof. Indeed, representations of π_1 are equivalent to fuchsian systems (local at 0) by Riemann-Hilbert correspondence, thus to rational representations of $\hat{\pi}_1$ by the previous corollary.   □

Corollary 16.26. *The proalgebraic hull of $\mathbf{Z}$ is*

$$\mathbf{Z}^{alg} = \mathrm{Hom}_{gr}(\mathbf{C}^*, \mathbf{C}^*) \times \mathbf{C}$$

together with the map $\iota : \mathbf{Z} \to \mathbf{Z}^{alg}$ given by

$$\iota(k) := \left((z \mapsto z^k), k \right) \in \mathrm{Hom}_{gr}(\mathbf{C}^*, \mathbf{C}^*) \times \mathbf{C}.$$

Exercises

(1) Show how to deduce lemma 16.1 from exercise F.7 in Appendix F.

(2) Let E be a finite-dimensional affine space E over $\mathbf{C}$.
   (i) Define polynomial functions over E. (You may use coordinates, but then you will have to prove that your definition is independent of a particular choice.) Check that they form a $\mathbf{C}$-algebra $\mathbf{A}(E)$.
   (ii) Define (with proofs) the Zariski topology on E. You will denote $V(I)$ to be the closed subset defined by an ideal I.
   (iii) Let I_0 be an ideal of $\mathbf{A}(E)$ and $F := V(I_0)$. Let $I \subset \mathbf{A}(E)$ be the ideal of all polynomial functions vanishing on F. Check that $I_0 \subset I$ and that $V(I) = F$.
   (iv) Show that the algebra $\mathbf{A}(F)$ of polynomial functions on F is $\mathbf{A}(E)/I$.

(3) With the above notation, check that[5] $\sqrt{I_0} \subset I$, where the *radical* $\sqrt{I_0}$ of I_0 is the set of $f \in \mathbf{A}(E)$ such that $f^m \in I_0$ for some $m \in \mathbf{N}$.

(4) With the above notation, check that if $E = \mathbf{C}$, with the identification $\mathbf{A}(E) = \mathbf{C}[X]$, and $I_0 = \langle 1 - X^d \rangle$, then $F = \mu_d$ and $I = I_0$.

(5) (i) Let f be a nontrivial polynomial function on the affine space E. Write $V(f) := f^{-1}(0) \subset E$ (thus a Zariski closed subset) and $D(f) := E \setminus D(f)$. Show that there is a natural bijection of the closed subset F of $E \times \mathbf{C}$ given by equation $1 - Tf$ with $D(f)$.
   (ii) Show that $\mathbf{A}(F) = \mathbf{A}(E)[T]/\langle 1 - Tf \rangle = \mathbf{A}(E)[1/f]$.

(6) Let G, H be two algebraic groups, H being commutative. Prove that the product of two rational morphisms $G \to H$ is rational.

(7) (i) Find all Laurent polynomials $P \in \mathbf{C}[X, 1/X]$ such that $P(xy) = P(x)P(y)$ for all $x, y \in \mathbf{C}^*$.
   (ii) Find all polynomials $P \in \mathbf{C}[X]$ such that $P(x + y) = P(x) + P(y)$ for all $x, y \in \mathbf{C}$.
   (iii) Find all polynomials $P \in \mathbf{C}[X]$ such that $P(0) = 1$ and $P(x+y) = P(x)P(y)$ for all $x, y \in \mathbf{C}$.

(8) (i) In any group Γ define the *commutator* or *derived* subgroup $[\Gamma, \Gamma]$ as the group generated by all the commutators $[x, y] := xyx^{-1}y^{-1}$ for $x, y \in \Gamma$. Prove that $[\Gamma, \Gamma]$ is an invariant subgroup.
   (ii) One defines the *abelianization* of Γ as follows:

$$\Gamma^{ab} := \Gamma/[\Gamma, \Gamma],$$

---

[5]The converse inclusion is true but more difficult; this is the celebrated *nullstellensatz* of Hilbert (see [**BorA91**, **Lan02**]).

where the subgroup $[\Gamma, \Gamma]$ is the group generated by all the commutators $[x, y] := xyx^{-1}y^{-1}$ for $x, y \in \Gamma$. Prove that $[\Gamma, \Gamma]$ is an invariant subgroup.

$$\Gamma^{ab} := \Gamma/[\Gamma, \Gamma].$$

Check that Γ^{ab} is abelian and that any group morphism from Γ to an abelian group factors uniquely through the canonical surjection $p : \Gamma \to \Gamma^{ab}$.

(iii) In particular, the group morphism $\mathfrak{X}(p) : \mathfrak{X}(\Gamma^{ab}) \to \mathfrak{X}(\Gamma)$ is a bijection.

(9) (i) Describe the general form of a group morphism f from $\mathbf{Z}^r \times (\mathbf{Z}/d_1\mathbf{Z}) \times \cdots \times (\mathbf{Z}/d_k\mathbf{Z})$ to $\mathbf{Z}^s \times (\mathbf{Z}/e_1\mathbf{Z}) \times \cdots \times (\mathbf{Z}/e_l\mathbf{Z})$.

(ii) Explicitly describe $\mathfrak{X}(f)$.

(10) Let $(I, \prec)$ be a directed ordered set (this means that for any $i, j \in I$ there exists $k \in I$ such that $i, j \prec k$). Let $(G_i)_{i \in I}$ be a family of groups and, for $i \prec j$ in I, group morphisms $f_i^j : G_j \to G_i$ such that $f_i^i = \mathrm{Id}_{G_i}$ for all $i \in I$ and $f_i^k = f_i^j \circ f_j^k$ whenever $i \prec j \prec k$. We then say that (G_i, f_i^j) is an inverse system of groups.[6]

(i) Prove that the *inverse limit* of the inverse system (G_i, f_i^j),

$$G := \varprojlim G_i := \left\{ (g_i)_{i \in I} \in \prod_{i \in I} G_i \mid \forall i, j \in I, \ i \prec j \implies f_i^j(g_j) = g_i \right\},$$

is a subgroup of $\prod_{i \in I} G_i$ and that the maps $f_i : G \to G_i$ obtained by restricting the ith projection $\prod_{i \in I} G_i \to G_i$ to G are group morphisms such that $f_i^j \circ f_j = f_i$ whenever $i \prec j$.

(ii) Let G' be a group and $(f_i')_{i \in I}$ a family of group morphisms $f_i' : G' \to G_i$ such that $f_i'^j \circ f_j' = f_i'$ whenever $i \prec j$. Show that there is a unique group morphism $f : G' \to G$ such that $f_i' = f_i \circ f$ for all $i \in I$. This is the *universal property of the inverse limit*.

(11) Interpret an inverse system as a contravariant functor. (You will have to consider I as a category, where the objects are the elements of I and there is a morphism $i \to j$ whenever $i \prec j$.) Then define the inverse limit of any contravariant functor by stating its required universal property, and give examples.

(12) Let G be an algebraic group and I an arbitrary set. Show how to make G^I a proalgebraic group.

---

[6]There is a similar notion in any category; and "direct systems" can be defined the same way, by just "reversing arrows".

Chapter 17

Beyond local fuchsian differential Galois theory

The goal of this chapter is to mention what comes next in differential Galois theory. We very superficially present what seems to be the most natural topics beyond those covered in this course, along with appropriate bibliographical references.

The most complete published reference for differential Galois theory is undoubtedly the book [**vS03**] by two experts, Marius van der Put and Michael Singer. However, it is somewhat "algebraized to death" (to mimic Chevalley's words about his own version of the theory of algebraic curves) and it sometimes tends to conceal fundamental ideas of a more transcendental nature. So I should rather warmly recommend [**CR09**] by Jose Cano and Jean-Pierre Ramis as an antidote to an overdose of algebra. Also, an excellent concise introduction, with a point of view related to ours but at a more advanced level is presented by Claude Mitschi and David Sauzin in the first four chapters of [**MS16**].

Before considering more specific topics, I would like to mention texts emphasizing more one of the two directions (algebraic or transcendental) between which I have been trying to keep an equilibrium.

Transcendental trend. A very nice survey of the history and theory of meromorphic differential equations is offered by Varadarajan [**Var96**]. A quite complete book along these lines (also covering some related topics) was written by two masters, Ilyashenko and Yakovenko [**IY08**]. More specifically related to Riemann-Hilbert correspondence are [**IKSY91**] and [**AB94**], the latter giving all the refinements of the "true" twenty-first Hilbert problem (which we commented on at the beginning of Chapter 9).

I have carefully avoided going into geometric considerations involving Riemann surfaces and vector bundles. Yet, even at the elementary level chosen for the present course, they allow for a deeper understanding of what goes on. The foundations can be learned in the justly famous book by Deligne [**Del70**]. A more didactic presentation, yet going quite far into some modern applications, is the book by Sabbah [**Sab02**].

Algebraic trend. Algebraic differential Galois theory has been expounded in some nice books. We only mention [**Mor13**], which is aimed at an important result in integrability theory, the theorem of Morales-Ramis; and [**CH11**], which is a very complete course. More references are to be found in these two books, as well as in all the books and papers by Mitschi and by Singer mentioned in this chapter.

17.1. The global Schlesinger density theorem

Most of what we did was of a local nature: germs at 0, functions in $\mathbf{C}(\{z\})$, etc. The natural extension to a global setting could be to consider some open subset Ω of the Riemann sphere and some discrete subset Σ of Ω; then, considering differential equations or systems analytic on Ω but with singularities in Σ, their global study (on Ω) would involve as a first step the local studies at all points of Σ.

The original formulation of the problem was that of Riemann in his study of the hypergeometric equation (Chapter 11); there, the coefficients are rational, that is, meromorphic over the Riemann sphere. Already in that case we had a problem of coordinate with the point at infinity. This (and the general experience of real or complex differential geometry) suggests that the global theory should rather involve differential equations over a general Riemann surface X equipped with some prescribed discrete subset of singularities Σ, and that the formulation should be coordinate-free.

The reader will find in [**AB94, Del70, Sab93**] a formulation of the Riemann-Hilbert correspondence in that framework. Regular singularities can be defined and characterized and one obtains an equivalence of the category of differential equations or systems with regular singularities on Σ with the category of representations of $\pi_1(X \setminus \Sigma)$. The image of the representation attached to a system A is the monodromy group $\text{Mon}(A)$.

The framework adopted in most books is even more general. In our definition of a system, solutions are column vectors and initial conditions are points in $\mathbf{C}^n$, which amounts to fixing the space of initial conditions along with a particular basis. Letting space and basis vary, one can get a coordinate-free formulation in which the "trivial bundle" $X \times \mathbf{C}^n$ is replaced with an arbitrary "holomorphic vector bundle" over X. Then to define a differential system, instead of using $A(z)dz$ (for some coordinate z on X) one uses an object originating in differential geometry and called a "connection" (not directly related to connection matrices however).

To do differential Galois theory, one takes as a base field $K := \mathcal{M}(X)$ (the differential field of meromorphic functions on X) and as an extension algebra or field the one generated by the solutions. The corresponding group of differential automorphisms is the Galois group $\text{Gal}(A)$, an algebraic subgroup of the full linear group containing $\text{Mon}(A)$, and one can then prove:

Theorem 17.1 (Global Schlesinger density theorem). *For a regular singular system, $\text{Mon}(A)$ is Zariski-dense in $\text{Gal}(A)$.* □

Note that, once the formalism has been properly set up, the deduction of this theorem from the local Schlesinger theorem is rather easy.

17.2. Irregular equations and the Stokes phenomenon

Irregular equations. The study of irregular equations is essentially of a local nature, so we shall consider equations or systems with coefficients in $\text{Mat}_n(\mathbf{C}(\{z\}))$ and assume that 0 is not a regular singular point (which, as we already saw, is a lot easier to check with scalar equations than with systems). Then it is not generally true that the monodromy group is Zariski-dense in the Galois group. Actually, not even the basic form of Riemann-Hilbert correspondence is valid; along with the monodromy representation, some additional transcendental information is required to be able to recover the equation or system. The missing transcendental invariants come under two forms:

(1) Components of the Galois group called "exponential tori", actually connected diagonalizable algebraic groups, thus products of $\mathbf{C}^*$-factors (see Appendix E). These $\mathbf{C}^*$-factors reveal the presence of "wildly growing exponentials" (such as $e^{1/z}$) inside the solutions. These exponentials are not necessarily seen by the monodromy. An example will be shown in the next subsection.

(2) Unipotent matrices called "Stokes operators" which also do not belong to the monodromy group. They reveal the presence of divergent series. An example will be shown in the last subsection of this section.

At any rate, a satisfying theory has been developed. One of its culminating points is due to Ramis; first a Riemann-Hilbert correspondence exists, where the fundamental group is replaced by the "wild fundamental group" (actually a rather strange object which mixes the fundamental group, the exponential tori and a free Lie algebra generated by the Stokes operators). Then Ramis' density theorem states that the images of the wild fundamental group are Zariski-dense in the corresponding Galois groups [**CR09**, **vS03**].

Exponential tori: an example. The scalar equation of order 1 (or equivalently the system of rank 1) $z^2 f' + f = 0$ can be written, using Euler's differential operator $\delta := z \, d/dz$, as $\delta f + z^{-1} f = 0$, which shows that it is irregular at 0. A fundamental solution is $e^{1/z}$, which is uniform, so that the monodromy group is trivial.

To compute the Galois group, we take $K := \mathbf{C}(\{z\})$ and $\mathcal{A} := K[e^{1/z}]$. From the equation $\delta f + z^{-1} f = 0$ one obtains that for any differential automorphism σ the function $g := \sigma(f)$ satisfies $\delta g + z^{-1} g = 0$, from which one deduces that $g = \lambda f$, $\lambda \in \mathbf{C}^*$. Conversely, since $e^{1/z}$ is transcendental over K, this indeed defines a differential automorphism and one concludes that the Galois group is isomorphic to $\mathbf{C}^*$; it is actually realized as $\mathbf{C}^* = \mathrm{GL}_1(\mathbf{C})$ and the monodromy group, being trivial, is certainly not Zariski-dense in it.

Stokes operators: an example. Although the Stokes phenomenon was first observed in relation with the asymptotic study of irregular solutions of the Bessel equations, the whole story of irregular equations is closely related with that of divergent series which starts with Euler (for more about this, the nicest introduction is undoubtedly [**Ram93**]). So we shall start with the attempt by Euler to give a meaning to the expression

$$0! - 1! + 2! - 3! + \cdots ,$$

a wildly divergent series! One way to do this is to define it as the value at $z = 1$ of the "function" defined by the *Euler series*

$$\hat{f}(z) := 0!z - 1!z^2 + 2!z^3 - 3!z^4 + \cdots = \sum_{n \geq 0} (-1)^n n! z^{n+1},$$

a wildly diverging power series with radius of convergence 0. The hat on $\hat{f}(z)$ is here because it is commonly associated with formal objects. Euler sought to give a "true" analytic meaning to $\hat{f}$. Various ways to give an analytic meaning to such formal objects have been devised since the time of Euler, the domain bears the general name of "resummation of divergent series". Beyond [**Ram93**], you can look at the books [**BorE28, Har91**].

There are various ways to attack the problem, one of them is to notice that $\hat{f}$ satisfies a first order nonhomogeneous linear equation:

$$(17.1) \qquad z^2 f' + f = z \text{ or equivalently } z\delta f + f = (z\delta + 1)f = z.$$

To apply our theory, we transform it into a homogeneous equation. Since $\delta z = z$, *i.e.*, $(\delta - 1)z = 0$, we easily get:

$$(17.2) \qquad (\delta - 1)(z\delta + 1)f = 0.$$

Let us admit that, instead of our "fake" solution $\hat{f}$ we can find a "true" solution ϕ to (17.1) (this will be done in a moment). Since the underlying first order homogeneous equation (not to be confused with (17.2)) is $(z\delta + 1)f = z^2 f' + f = 0$, which has $e^{1/z}$ as fundamental solution, we see that $(e^{1/z}, \phi)$ is a basis of solutions of (17.2).

Computing the Galois group is a purely algebraic problem. From the previous subsection we already know that any differential automorphism σ must send $e^{1/z}$ to $\lambda e^{1/z}$ for some $\lambda \in \mathbf{C}^*$. As for the image of ϕ, say $\psi := \sigma(\phi)$, applying σ to (17.1) tells us that ψ is a solution too, so that $\psi - \phi = \mu e^{1/z}$ for some $\mu \in \mathbf{C}$. Last, one proves that $e^{1/z}$ and ϕ are algebraically independent over K and deduces that any such λ, μ will do (*i.e.*, will indeed define a differential automorphism σ) so that the Galois group, realized in the basis $(e^{1/z}, \phi)$, is the group of all matrices $\begin{pmatrix} \lambda & \mu \\ 0 & 1 \end{pmatrix}$, $\lambda \in \mathbf{C}^*$, $\mu \in \mathbf{C}$. Note that this is the semi-direct product of the "torus" $\{\mathrm{Diag}(\lambda, 1) \mid \lambda \in \mathbf{C}^*\}$ by the normal unipotent subgroup of all matrices $\begin{pmatrix} 1 & \mu \\ 0 & 1 \end{pmatrix}$, $\mu \in \mathbf{C}$. In the general case, the structure is more intricate.

Now we find ϕ, thus giving an analytic meaning to the formal power series $\hat{f}$. For this, we apply Lagrange's method of variation of constants. After some changes of variables to make the solution look nice, we find:

$$\phi(z) := \int_0^\infty \frac{e^{-\xi/z}}{1 + \xi} \, d\xi.$$

This indeed defines a holomorphic function in the half-plane $H_+ : \Re(z) > 0$ which is moreover a solution of (17.1). It is related to $\hat{f}$ in at least two ways.

First, substituting the expansion

$$\frac{e^{-\xi/z}}{1+\xi} = \sum_{k=0}^{n-1}(-\xi)^k e^{-\xi/z} + \frac{(-\xi)^n e^{-\xi/z}}{1+\xi}$$

in the integral above, one finds that

$$\phi(z) := \sum_{k=0}^{n-1}(-1)^k k! z^{k+1} + R_n(z),$$

where

$$R_n(z) = \int_0^\infty \frac{(-\xi)^n e^{-\xi/z}}{1+\xi}\,d\xi = z^{n+1}\int_0^{\infty z^{-1}} \frac{(-u)^n e^{-u}}{1+zu}\,du.$$

One checks that $R_n(z) = O(z^{n+1})$ when $z \to 0$ in H_+, so that $\hat{f}$ is the asymptotic expansion of ϕ there.

The second relation goes back to the theory of "resummation of divergent series". A process called *Borel-Laplace transform* works as follows:

(1) Formally replace each z^{n+1} in $\hat{f}$ by $\xi^n/n!$. This yields:

$$\mathcal{B}\hat{f}(\xi) = \sum_{n\geq 0}(-1)^n \xi^n = \frac{1}{1+\xi}.$$

(2) Apply the Laplace transformation $\mathcal{B}\hat{f}(\xi)$ to obtain $\phi := \mathcal{L}\mathcal{B}\hat{f}$:

$$\phi(z) = \int_0^\infty e^{-\xi/z}\mathcal{B}\hat{f}(\xi)\,d\xi.$$

It is to be noted that, in the Laplace transformation, the half-line $(0, +\infty)$ $= \mathbf{R}_+^*$ used for integration can be replaced by other half-lines $(0, +\infty d) = \mathbf{R}_+^* d$ in other directions $d \in \mathbf{C}^*/\mathbf{R}_+^*$. The only prohibited direction is $\mathbf{R}_+^* d = \mathbf{R}_-^*$, because one must avoid the singularity $\xi = -1$ of $\mathcal{B}\hat{f}(\xi)$ in the line of integration. Accordingly, we shall write:

$$\phi_d := \int_0^{+\infty d} e^{-\xi/z}\mathcal{B}\hat{f}(\xi)\,d\xi.$$

This function is analytic in the half-plane H_d bisected by d, it is a solution of (17.1) and it admits the asymptotic expansion $\hat{f}$ in H_d.

To some extent, the various solutions ϕ_d can be patched. Indeed, for any two such functions, we know two things about the difference $\phi_d - \phi_{d'}$: it is solution of $z\delta f + f = 0$, so it has the form $Ce^{1/z}$, and it is asymptotic to 0 in $H_d \cap H_{d'}$. However the function $e^{1/z}$ has various asymptotic behaviors when z approaches 0. In particular, if $\Re(z) > 0$ it grows wildly, while if $\Re(z) < 0$ it decays quickly; $e^{1/z}$ is asymptotic to 0 in the left half-plane H_-

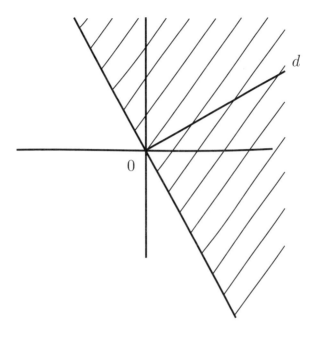

Figure 17.1. Half-plane of convergence of ϕ_d

but not in the right half-plane H_+. As a consequence, if $H_d \cap H_{d'}$ intersects H_+, then the only $Ce^{1/z}$ asymptotic to H in $H_d \cap H_{d'}$ is 0, so that $\phi_d = \phi_{d'}$ on $H_d \cap H_{d'}$ and the solutions $\phi_d, \phi_{d'}$ can be patched to give a solution on a bigger domain. We leave it to the reader to figure out the extent to which one can enlarge the domain.

On the opposite side, if $H_d \cap H_{d'}$ is contained in H_-, then $\phi_{d'} = \phi_d + Ce^{1/z}$ with a possibility that $C \neq 0$. For instance, if we take $d = d+$ just above $\mathbf{R}_-$ and $d = d-$ just under $\mathbf{R}_-$, there is a possible jump $Ce^{1/z} = \phi_+ - \phi_-$. This jump can be computed exactly using the Cauchy residue formula on the following contour: d_+ starting from 0, then an "arc of circle at infinity", then d_- going back to 0. One then finds:

$$\phi_+ - \phi_- = 2\pi e^{1/z}.$$

We interpret this as saying that while jumping over the prohibited direction of summation $\mathbf{R}_-^*$, the basis $(e^{1/z}, \phi)$ has been transformed by the Stokes operator with matrix $\begin{pmatrix} 1 & 2\pi \\ 0 & 1 \end{pmatrix}$ (with the idea that the Stokes operator does not affect $e^{1/z}$).

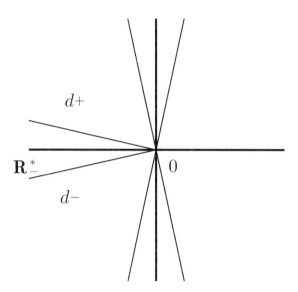

Figure 17.2. The Stokes jump over $\mathbf{R}^*_-$

Now the reader can check that the "exponential torus" made up of all the $\begin{pmatrix} \lambda & 0 \\ 0 & 1 \end{pmatrix}$, $\lambda \in \mathbf{C}^*$, and the Stokes matrix $\begin{pmatrix} 1 & 2\pi \\ 0 & 1 \end{pmatrix}$ together generate a group which is Zariski-dense in the Galois group; this is an instance of Ramis' density theorem.

17.3. The inverse problem in differential Galois theory

The inverse problem asks when an algebraic group is the differential Galois group of some differential system. In this form it is much too vague and some restrictions are required in order to make it more interesting: local or global? fuchsian or irregular?, etc. Also the approaches to solve the problem have been quite diverse: algebraic or transcendental or somewhere in between.

An easy example requires us to use a local system fuchsian at 0; or a system with constant coefficients, which amounts to the same thing according to Chapter 13. Then we found that each differential Galois group must be a subgroup of $\mathrm{GL}_n(\mathbf{C})$ "topologically generated by one element", meaning that there is $M \in \mathrm{GL}_n(\mathbf{C})$ such that G is the Zariski closure of the cyclic group $\langle M \rangle$; and this is obviously a necessary and sufficient condition.

The reader will find various formulations and solutions of the inverse problem in the survey "Direct and inverse problems in differential Galois theory" by Michael Singer, which appears as an appendix in [**Kol99**]. More recent work by Ramis and by Mitschi and Singer is described in the Bourbaki talk by van der Put [**van98**]. The content of this talk is expounded in a reworked form in [**vS03**], but the Bourbaki talk seems to me easier to read as the main lines are clearly shown there. Along the same, very algebraic, lines, the reader will profitably refer to other texts by Singer, such as [**Sin90a**, **Sin90b**, **Sin09**].

17.4. Galois theory of nonlinear differential equations

This subject is much more difficult! For a long time, only the work by J.-F. Pommaret was known, but it seems that his theory never reached a state of maturity. In recent years, there have been advances mainly by Bernard Malgrange, Hiroshi Umemura and Pierre Cartier. It seems that the theory of Malgrange is, by far, the most developed (it led to the succesful solution of some long-standing problems); it relies heavily on differential geometry. See [**Mal02**] for an exposition. The theory of Umemura has a much more algebraic foundation, but he proved himself in [**Ume08**] that it is equivalent to the theory of Malgrange. At the moment (summer 2015), it seems that the theory of Cartier has given rise only to prepublications and talks.

Another proof of the surjectivity of
$\exp : \mathrm{Mat}_n(\mathbf{C}) \to \mathrm{GL}_n(\mathbf{C})$

We give here a proof of independent interest of the surjectivity of the exponential map from $\mathrm{Mat}_n(\mathbf{C})$ to $\mathrm{GL}_n(\mathbf{C})$. So let $M \in \mathrm{GL}_n(\mathbf{C})$ and let $\log$ be a determination of the logarithm defined on some open subset of $\mathbf{C}^*$ containing $\mathrm{Sp}\, M$. This allows us in particular to define, for any $\lambda \in \mathrm{Sp}\, M$, the $\mathcal{C}^\infty$ function $x \mapsto \lambda^x := e^{x \log \lambda}$ on $\mathbf{R}$ in such a way that $\forall x, y \in \mathbf{R}$, $\lambda^{x+y} = \lambda^x \lambda^y$.

Using this, we are going to construct a $\mathcal{C}^\infty$ function $x \mapsto M^x$ from $\mathbf{R}$ to $\mathrm{Mat}_n(\mathbf{C})$ with the following properties:

- $\forall x, y \in \mathbf{R}$, $M^{x+y} = M^x M^y$;
- for $x = p \in \mathbf{N}$, we have $M^x = M^p$, the usual p^{th} power.

Note that, assuming the first property, the second one will follow if $M^0 = I_n$ and $M^1 = M$; and then we shall actually have $M^x \in \mathrm{GL}_n(\mathbf{C})$ for all $x \in \mathbf{R}$ and M^p will have the usual meaning for all $p \in \mathbf{Z}$. Such a function $x \mapsto M^x$ is called a *one-parameter subgroup* of $\mathrm{GL}_n(\mathbf{C})$.[1]

Admitting for now that the one-parameter subgroup $x \mapsto M^x$ has been obtained, let us see how it solves our problem. Derivating the relation

---

[1] Actually, one-parameter subgroups exist under more general conditions: if $G \subset \mathrm{GL}_n(\mathbf{C})$ is any connected algebraic group and if $M \in G$, one can define $x \mapsto M^x$ as above with values in G. This is actually valid for all so-called "Lie subgroups" of $\mathrm{GL}_n(\mathbf{C})$.

$M^{y+x} = M^y M^x$ with respect to y and then evaluating at $y = 0$ gives:

$$\forall x \in \mathbf{R}, \ \frac{d}{dx} M^x = A M^x, \text{ where } A := \left(\frac{dM^x}{dx} \right)_{|x=0}.$$

Since $M^0 = I_n$, we know from Section 1.5 that $M^x = e^{xA}$ for all $x \in \mathbf{R}$. Taking $x = 1$ yields the desired relation $e^A = M$.

In order to define our one-parameter subgroup, we first need some notation. Write $\chi_M(T) = (T - \lambda_1)^{r_1} \cdots (T - \lambda_m)^{r_m}$ as the characteristic polynomial (with $\lambda_1, \ldots, \lambda_m \in \mathbf{C}^*$ pairwise distinct and $r_1, \ldots, r_m \in \mathbf{N}^*$). For every $p \in \mathbf{N}$ we perform the euclidean division

$$T^p = Q_p \chi_M + R_p,$$

and write $R_p = a_0(p) + \cdots + a_{n-1}(p) T^{n-1}$. From the Cayley-Hamilton theorem, $\chi_M(M) = 0$, so that:

$$M^p = a_0(p) I_n + \cdots + a_{n-1}(p) M^{n-1}.$$

We intend to be able to replace $p \in \mathbf{N}$ by $x \in \mathbf{R}$ in this equality.

Lemma A.1. *Let E be the space of $\mathcal{C}^\infty$ functions over $\mathbf{R}$ generated by the n functions $x \mapsto x^j \lambda_i^x$, $i = 1, \ldots, m$, $j = 0, \ldots, r_i - 1$.*

(i) These functions form a basis of E.

(ii) The restriction map $E \to \mathbf{R}^{\mathbf{N}}$ which sends a function $f \in E$ to the sequence $(f(p))_{p \in \mathbf{N}}$ is injective.

(iii) For every $f \in E$ and $c \in \mathbf{C}$, the function $x \mapsto f(x+c)$ belongs to E.

Proof. These are classical facts in elementary real analysis, and the proof is left to the reader. $\qquad\square$

We can therefore identify E with its image in $\mathbf{R}^{\mathbf{N}}$ and the sentence $a_0, \ldots, a_{n-1} \in E$ has a meaning. We shall prove the corresponding fact in a moment. Assuming this has been done, we can consider each sequence $(a_j(p))_{p \in \mathbf{N}}$, $j = 0, \ldots, n-1$, as the restriction to $\mathbf{N}$ of a $\mathcal{C}^\infty$ function $a_j \in E$ on $\mathbf{R}$. Therefore, we can define:

$$\forall x \in \mathbf{R}, \ M^x := a_0(x) I_n + \cdots + a_{n-1}(x) M^{n-1}.$$

By construction, this is a $\mathcal{C}^\infty$ function from $\mathbf{R}$ to $\mathrm{Mat}_n(\mathbf{C})$ and for $x = p \in \mathbf{N}$, we have $M^x = M^p$, the usual p^{th} power. We now prove the first required property of one-parameter subgroups. Call $m_{i,j}(x)$ the (i,j)-coefficient of M^x. It is a linear combination of $a_0, \ldots, a_{n-1} \in E$, so it belongs to E. We are going to prove:

$$\forall x, y \in \mathbf{R}, \ m_{i,j}(x+y) = \sum_{k=1}^{n} m_{i,k}(x) m_{k,j}(y).$$

This will imply the required property: $\forall x, y \in \mathbf{R}$, $M^{x+y} = M^x M^y$. The proof of the above equality goes in two steps:

(1) The equality is obviously verified for $x = p, y = q$, $p, q \in \mathbf{N}$, since then it amounts to $M^{p+q} = M^p M^q$ (usual powers of M). So if we fix $y = q$, we have two functions $m_{i,j}(x+q)$ and $\sum_{k=1}^{n} m_{i,k}(x) m_{k,j}(q)$ of x that coincide on $\mathbf{N}$. But the reader will check that both functions belong to E (the first one by (iii) of the lemma, the second one because E is a vector space), so, after (ii) of the lemma, they are equal.

(2) We have proved, for any $x \in \mathbf{R}$ and any $q \in \mathbf{N}$, the equality $m_{i,j}(x+q) = \sum_{k=1}^{n} m_{i,k}(x) m_{k,j}(q)$. So we fix $x \in \mathbf{R}$ and consider the functions $m_{i,j}(x+y)$ and $\sum_{k=1}^{n} m_{i,k}(x) m_{k,j}(y)$ of y. Both belong to E and they coincide on $\mathbf{N}$, so they are equal on $\mathbf{R}$.

It remains to show our contention that $a_0, \dots, a_{n-1} \in E$. We do this in a constructive way. In the equality

$$T^p = Q_p \chi_M + R_p,$$

we know that each λ_i is a root of χ_M of multiplicity r_i, so $(Q_p \chi_M)^{(j)}(\lambda_i) = 0$ for $j = 0, \dots, r_i - 1$. Therefore, if in the above equality we take the jth derivative for $j < r_i$ and evaluate at λ_i, we find the equality

$$p(p-1) \cdots (p-j+1) \lambda_i^{p-j} = R_p^{(j)}(\lambda_i).$$

As a function of p, the left-hand side is an element of E. Call it $f_{i,j}$ (again with $i = 1, \dots, m$, $j = 0, \dots, r_i - 1$). But the system of n equations $R_p^{(j)}(\lambda_i) = f_{i,j}$ can be solved uniquely for the $(n-1)$ unknowns $a_0, \dots, a_{n-1}$. This can be seen either by writing it down explicitly (which we encourage the reader to do), and this allows for an explicit computation of the functions a_i and then of M^x and of A; or by a standard argument from linear algebra running as follows: the composite linear map

$$(a_0, \dots, a_{n-1}) \mapsto R := a_0 + \cdots + a_{n-1} T^{n-1} \mapsto (R^{(j)}(\lambda_i))_{\substack{i=1,\dots,m \\ j=0,\dots,r_i-1}}$$

is injective, because, if R has zero image, then it is a multiple of χ_M while having degree $\deg R \leq n - 1$, so it is 0. Since the source and target of this map have dimension n, it is bijective, and therefore there is an explicit expression of $a_0, \dots, a_{n-1}$ as linear combinations of the $R^{(j)}(\lambda_i)$, $i = 1, \dots, m$, $j = 0, \dots, r_i - 1$.

Whatever was the argument used, we conclude that $a_0, \ldots, a_{n-1} \in E$ which ends the proof of the surjectivity of the exponential of matrices.

A complete (if simple) example. Let $M := \begin{pmatrix} \lambda & 1 & 0 \\ 0 & \lambda & 0 \\ 0 & 0 & \mu \end{pmatrix}$, $\lambda, \mu \in \mathbf{C}^*$,

$\lambda \neq \mu$. Then $\chi_M(T) = (T - \lambda)^2(T - \mu)$ and E is generated by the three functions $x \mapsto \lambda^x$, $x \mapsto x\lambda^x$ and $x \mapsto \mu^x$ (which presupposes a choice of $\log \lambda, \log \mu$). We write, for all $p \in \mathbf{N}$:

$$T^p = Q_p(T)(T - \lambda)^2(T - \mu) + a_0(p) + a_1(p)T + a_2(p)T^2.$$

Substitutions of λ, μ (along with one derivation in the case of λ) yield the system of equations:

$$a_0 + \lambda a_1 + \lambda^2 a_2 = \lambda^p,$$

$$a_1 + 2\lambda a_2 = p\lambda^{p-1},$$

$$a_0 + \mu a_1 + \mu^2 a_2 = \mu^p.$$

Solving for a_0, a_1, a_2, we get:

$$a_2(p) = \frac{1}{\mu - \lambda} \left(\frac{\mu^p - \lambda^p}{\mu - \lambda} - p\lambda^{p-1} \right),$$

$$a_1(p) = p\lambda^{p-1} - 2\lambda a_2(p),$$

$$a_0(p) = \lambda^p - \lambda a_1(p) - \lambda^2 a_2(p).$$

Then we get

$$M^x = a_0(x)I_2 + a_1(x)M + a_2(x)M^2 = \begin{pmatrix} \lambda^x & x\lambda^{x-1} & 0 \\ 0 & \lambda^x & 0 \\ 0 & 0 & \mu^x \end{pmatrix},$$

whence

$$A := \left(\frac{dM^x}{dx} \right)_{|x=0} = \begin{pmatrix} \log \lambda & \lambda^{-1} & 0 \\ 0 & \log \lambda & 0 \\ 0 & 0 & \log \mu \end{pmatrix}.$$

Another construction of the logarithm of a matrix

For a small enough matrix $A \in \mathrm{Mat}_n(\mathbf{C})$, the logarithm of $I_n + A$ can be defined as the power series $\sum_{k \geq 1}(-1)^{k+1}A^k/k$. However, power series have a domain of definition more limited than the functions they define. As Rudin notes in [**Rud91**], the integral formulas of Cauchy theory (Section 3.3) are more adaptable. We shall sketch a way to extend the definition of any holomorphic function to some set of matrices. This procedure (which actually makes sense in any complex Banach algebra) is known, among other names, as "symbolic calculus" or "holomorphic functional calculus". See [**Bou07**, **Rud91**] for a more systematic exposition.

We start with the following case of the Cauchy formula given in Corollary 3.25: let Ω be a simply-connected domain, $f \in \mathcal{O}(\Omega)$ and γ a loop in $\Omega \setminus \{a\}$ looping once around $a \in \Omega$ (*i.e.*, $I(a,\gamma) = 1$). Then

$$f(a) = \frac{1}{2i\pi} \int_\gamma \frac{f(z)}{(z-a)}\, dz.$$

The reader is warmly invited to deduce the following easy consequence. If $A = PDP^{-1}$, $P \in \mathrm{GL}_n(\mathbf{C})$ and $D = \mathrm{Diag}(a_1,\ldots,a_n)$, where $a_1,\ldots,a_n \in \Omega$, and if γ loops once around all the a_i, then

$$f(A) = \frac{1}{2i\pi} \int_\gamma f(z)\,(zI_n - A)^{-1}\, dz,$$

where, as usual, $f(A) := P\mathrm{Diag}(f(a_1), \ldots, f(a_n))P^{-1}$ (we already saw in exercise 2 in Chapter 15 that this depends on A and f only, not on the choice of a particular diagonalization). Since semi-simple matrices are dense in $\mathrm{Mat}_n(\mathbf{C})$, we need a way to extend this equality by continuity.

So we fix a simply-connected domain Ω and a simple contour inside Ω. Giving this contour a positive orientation, we define a loop γ. Call $U \subset \Omega$ the set of those $a \in \Omega$ such that $I(a, \gamma) = 1$. This is a domain. One can prove (it is not very difficult, and you should try it) that the set M_U of those matrices A such that $\mathrm{Sp}A \subset U$ is open in $\mathrm{Mat}_n(\mathbf{C})$. We then define, for every $A \in M_U$:

$$f_\gamma(A) := \frac{1}{2\mathrm{i}\pi} \int_\gamma f(z)\,(zI_n - A)^{-1}\,dz.$$

This is a continuous function on M_U. Moreover, if γ is continuously deformed into another loop γ' in Ω which loops once around $\mathrm{Sp}A$, then $f_\gamma(A) = f_{\gamma'}(A)$, so that we can write more simply $f(A) := f_\gamma(A) = f_{\gamma'}(A)$. Obviously, f is now defined over all M_Ω.

Now it can be proven that relations among holomorphic functions on Ω are still satisfied by their extensions to M_Ω:

$$\forall f, g \in \mathcal{O}(\Omega),\ \forall \lambda \in \mathbf{C},\ \forall A \in M_\Omega : \begin{cases} \lambda f(A) = (\lambda f)(A), \\ f(A) + g(A) = (f + g)(A), \\ f(A)g(A) = (fg)(A). \end{cases}$$

This extends to functional relations; under proper conditions on the domains of definition,

$$g(f(A)) = (g \circ f)(A).$$

If f is an entire function and $f = \sum a_k z^k$ is its power series expansion at 0, then $f(A) = \sum a_k A^k$. For instance, $\exp A$ as defined by functional calculus is the same as before. The reader should think about how this statement extends to power series with finite radius of convergence. (Conditions on the matrix should involve as well their norm as the modulus of its eigenvalues; but, for a Banach algebra norm, these conditions are not independent!)

Therefore, if we take Ω to be the cut plane $\mathbf{C} \setminus \mathbf{R}_-$ and the principal determination log of the logarithm, we see that for any matrix A having no real negative eigenvalues, the matrix

$$\log A := \frac{1}{2\mathrm{i}\pi} \int_\gamma (\log z)\,(zI_n - A)^{-1}\,dz$$

is indeed a logarithm of A. (One can always draw a loop γ in $\mathbf{C} \setminus \mathbf{R}_-$ that winds once around the whole of SpA.)

Example B.1. Let $A := \begin{pmatrix} 1 & 1 \\ 0 & 1 \end{pmatrix}$ and let γ be in $\mathbf{C} \setminus \mathbf{R}_-$ winding once around 1. Then

$$\log_\gamma A = \frac{1}{2i\pi} \int_\gamma (\log z)\,(zI_n - A)^{-1}\,dz$$

$$= \frac{1}{2i\pi} \int_\gamma (\log z) \begin{pmatrix} z-1 & -1 \\ 0 & z-1 \end{pmatrix}^{-1} dz$$

$$= \frac{1}{2i\pi} \int_\gamma \begin{pmatrix} (\log z)/(z-1) & (\log z)/(z-1)^2 \\ 0 & (\log z)/(z-1) \end{pmatrix} dz$$

$$= \begin{pmatrix} 0 & 1 \\ 0 & 0 \end{pmatrix}.$$

Exercise B.2. The *resolvent* of A is defined by the formula:

$$R(z, A) := (A - zI_n)^{-1}.$$

It is defined for $z \notin \mathrm{Sp}A$ and takes values in $\mathbf{C}[A]$. Also,

$$R(z, A) - R(w, A) = (z - w)R(z, A)R(w, A) = (z - w)R(w, A)R(z, A)$$

and

$$R(z, A) - R(z, B) = R(z, A)(B - A)R(z, B).$$

Note that there is no commutation in the last equality.

Jordan decomposition in a linear algebraic group

We intend to give a proof here of the following fact, which we used at a crucial point in Chapter 14.

Theorem C.1. *If $G \subset \mathrm{GL}_n(\mathbf{C})$ is an algebraic subgroup, then, for each $M \in G$, the semi-simple component M_s (along with its replicas) and the unipotent component M_u (along with its powers) belong to G.*

This statement normally belongs to the general theory of linear algebraic groups [**BorA91**] but a direct proof is possible relying on some standard linear algebra. Due to the importance of Jordan decomposition (and of the theory of replicas) as a tool in differential Galois theory, we found it useful to present this proof here. On our way, we completely prove the Dunford-Jordan decomposition theorem.

C.1. Dunford-Jordan decomposition of matrices

This will be considered as a direct application of the following.

Dunford-Jordan decomposition of endomorphisms. Let K be an algebraically closed field, E a finite-dimensional linear space over K and $\phi \in \mathrm{End}_K(E)$ an endomorphism of E. For every $\lambda \in K$, we write:

$$E(\lambda) := \bigcup_{m \geq 0} \mathrm{Ker}(\phi - \lambda \mathrm{Id}_E)^m.$$

If $\lambda \in \mathrm{Sp}\,\phi$, this is the characteristic subspace of E associated to the eigenvalue λ; if $\lambda \notin \mathrm{Sp}\,\phi$, then $E(\lambda) = \{0\}$. The characteristic space decomposition of E is:

$$E = \bigoplus_{\lambda \in K} E(\lambda) = \bigoplus_{\lambda \in \mathrm{Sp}\,\phi} E(\lambda).$$

Any linear map on E can be uniquely defined by its restriction to each of the characteristic spaces $E(\lambda)$. In this way, we can define $\phi_s \in \mathrm{End}_K(E)$ to be, on each $E(\lambda)$, the homothety $x \mapsto \lambda x$; then we set $\phi_n := \phi - \phi_s$. The following facts are now obvious:

- ϕ_s is semi-simple;
- ϕ_n is nilpotent;
- $\phi = \phi_s + \phi_n$;
- $\phi_s \phi_n = \phi_n \phi_s$.

We shall prove in a moment that every endomorphism commuting with ϕ automatically commutes with ϕ_s and ϕ_n. From this, one obtains that the above conditions uniquely characterize ϕ_s and ϕ_n. Indeed, if $\phi = \phi_s' + \phi_n'$ is another such decomposition, we see that $\phi_s - \phi_s' = \phi_n' - \phi_n$ is at the same time semi-simple (as a difference of two commuting semi-simple endomorphisms) and nilpotent (as a difference of two commuting nilpotent endomorphisms). This implies of course that it is 0, whence the unicity.

To prove the above statement about commutation, we denote $\mathrm{Sp}\,\phi = \{\lambda_1, \ldots, \lambda_k\}$, so that the characteristic and minimal polynomials of ϕ can respectively be written:

$$\chi_\phi(T) = \prod_{i=1}^{k} (T - \lambda_i)^{r_i},$$

$$\mu_\phi(T) = \prod_{i=1}^{k} (T - \lambda_i)^{s_i},$$

with $1 \leq s_i \leq r_i$ and $\sum r_i = n := \dim_K E$. Then $E(\lambda_i) = \mathrm{Ker}(\phi - \lambda_i \mathrm{Id}_E)^{s_i}$.

Now, for each $i = 1, \ldots, k$, let $P_i \in K[T]$ be a polynomial such that $P_i(\lambda_i) = 1$ and $P_i(\lambda_j) = 0$ for $j \neq i$ (these are the polynomials involved in the Lagrange interpolation formula). Then the polynomial $P := \sum_{i=1}^{k} \lambda_i P_i$ is such that the restriction of $P(\phi)$ to $E(\lambda_i)$ is the homothety with ratio λ_i, i.e., $P(\phi) = \phi_s$; and therefore, if $Q := T - P$, then $Q(\phi) = \phi_n$. As a consequence, as said before, every endomorphism commuting with ϕ automatically commutes with ϕ_s and ϕ_n.

We now leave to the reader as an easy exercise to deduce the existence of properties of the decompositions $M = M_s + M_n$ for $M \in \mathrm{Mat}_n(K)$ and $M = M_s M_u$ for $M \in \mathrm{GL}_n(K)$ stated and used in Section 4.4.

Stable subspaces. Let $E' \subset E$ be a ϕ-stable subspace. Then it is $P(\phi)$-stable for every $P \in K[T]$, so it is ϕ_s-stable and ϕ_n-stable. The restriction of ϕ_n to E' is obviously nilpotent. Since the separated polynomial $\prod_{i=1}^{k} (T - \lambda_i)$ kills ϕ_s, it also kills its restriction to E', which is therefore semi-simple. And of course, these (respectively semi-simple and nilpotent) endomorphisms commute and have sum $\phi' := \phi_{|E'}$. The following facts are now immediate:

- the Dunford decomposition $\phi' = \phi'_s + \phi'_n$ is obtained by restriction: ϕ'_s is the restriction of ϕ_s to E' and ϕ'_n is the restriction of ϕ_n to E';
- $E'(\lambda) = E' \cap E(\lambda)$;
- $E' = \bigoplus (E' \cap E(\lambda))$.

Remark C.2. The assumption that K is algebraically closed can be replaced by the assumption that the characteristic polynomial $\chi_\phi(T)$ splits in K. Moreover, there is a more general theory that does not even assume this: there, one simply defines "semi-simple" as "diagonalizable over some extension field" and exactly the same conclusions hold (with obvious adaptations).

Infinite-dimensional spaces. Now E is an arbitrary linear space over K (not required to be finite dimensional). Let $\phi \in \mathrm{End}_K(E)$. If E', E'' are any two finite-dimensional (resp. ϕ-stable) subspaces of E, then so is $E' + E''$. Moreover, for any $x \in E$, we see that x belongs to a finite-dimensional ϕ-stable subspace if, and only if, the subspace $\sum K\phi^i(x)$ spanned by the iterated images $\phi^i(x)$ is itself finite dimensional. We conclude that the following conditions on E and ϕ are equivalent:

- E is a union of finite-dimensional ϕ-stable subspaces;
- every $x \in E$ belongs to a finite-dimensional ϕ-stable subspace.

Moreover, in this case, E is a directed union of finite-dimensional ϕ-stable subspaces, meaning that $E = \bigcup E_i$, where each E_i is a finite-dimensional ϕ-stable subspace and where for any indices i, j, there is k such that $E_i + E_j \subset E_k$. (This condition often proves to be useful.) We then say that ϕ is a *locally finite* endomorphism. Note that if ϕ is locally finite and injective, all the restrictions $\phi_{|E_i}$ are bijective (injective endomorphisms in finite dimension) so that ϕ is an automorphism.

Examples C.3. Let E have a denumerable basis $(e_n)_{n\in\mathbf{N}}$. Then, if ϕ is given by $\phi(e_i) := a_i e_i + b_i e_{i-1}$, it is locally finite. On the other hand, ψ given by $\psi(e_i) := e_{i+1}$ is not.

We shall keep the previous notation: for $\lambda \in K$, we write $E(\lambda) := \bigcup_{m\geq 0} \mathrm{Ker}(\phi - \lambda\mathrm{Id}_E)^m$. If $\lambda \in \mathrm{Sp}\ \phi$, *i.e.*, if $E(\lambda)$ is not trivial, this is the characteristic subspace of E associated to the eigenvalue λ. Using the previous subsection on stable subspaces, one easily proves the following facts. Assuming that ϕ is locally finite:

- $E = \bigoplus_{\lambda\in K} E(\lambda) = \bigoplus_{\lambda\in\mathrm{Sp}\ \phi} E(\lambda)$;

- $\phi = \phi_s + \phi_n$, where ϕ_s is diagonalizable, ϕ_n is locally nilpotent (for every $x \in E$, there exists k such that $\phi_n^k(x) = 0$) and they commute;

- the previous decomposition is unique;

- any endomorphism commuting with ϕ automatically commutes with ϕ_s and ϕ_n.

It is not however generally the case that ϕ_s and ϕ_n can be expressed as polynomials in ϕ.

Then if $E' \subset E$ is a ϕ-stable subspace, the restriction of ϕ to E' is locally finite and the above decomposition restricts nicely to E' as in the previous subsection.

Last, all this can be extended to an automorphism ϕ and a decomposition $\phi = \phi_s\phi_u$, where ϕ_u is locally unipotent. However, there is of course no simple matricial interpretation.

An application. The following application will be used in the next section. Let $E := K[X_1,\ldots,X_n,Y]$ and let $a_1,\ldots,a_n \in K^*$ and $c \in K$. The automorphism

$$\phi : F(X_1,\ldots,X_n,Y) \mapsto F(a_1X_1,\ldots,a_nX_n,Y+c)$$

is locally finite since the finite-dimensional space E_d of polynomials P of total degree deg $P \leq d$ is ϕ-stable.

We shall use multi-indices $\underline{i} \in \mathbf{N}^n$ and the shorthand for monomials $\underline{X}^{\underline{i}} := X_1^{i_1} \cdots X_n^{i_n}$. For every $a \in K^*$, write:

$$I(a) := \{(i_1, \ldots, i_n) \in \mathbf{N}^n \mid a_1^{i_1} \cdots a_n^{i_n} = a\}.$$

Then the characteristic space associated to a is:

$$E(a) = \bigoplus_{\underline{i} \in I(a)} \underline{X}^{\underline{i}} K[Y].$$

The restriction of ϕ to $E(a)$ is the map:

$$\sum_{\underline{i} \in I(a)} \underline{X}^{\underline{i}} P_{\underline{i}}(Y) \mapsto a \sum_{\underline{i} \in I(a)} \underline{X}^{\underline{i}} P_{\underline{i}}(Y + c).$$

Clearly, if $c = 0$, then ϕ is semi-simple and $\phi = \phi_s$ is entirely character-ized[1] by $X_i \mapsto a_i X_i$, $Y \mapsto Y$. More generally, the multiplicative Dunford decomposition of ϕ is described by:

$$\phi_s : \begin{cases} X_i \mapsto a_i X_i, \\ Y \mapsto Y, \end{cases}$$

$$\phi_u : \begin{cases} X_i \mapsto X_i, \\ Y \mapsto Y + c. \end{cases}$$

C.2. Jordan decomposition in an algebraic group

Here, we shall moreover assume that K has characteristic 0. Maybe the reader will have fun finding out where exactly this assumption is necessary. We begin with a useful lemma. Here again we set $E := K[X_1, \ldots, X_n, Y]$.

Lemma C.4. *Let $a_1, \ldots, a_n, c \in K^*$ (so here we require that $c \neq 0$). Then*

$$E' := \{F \in E \mid \forall k \in \mathbf{Z}, \ F(a_1^k, \ldots, a_n^k, kc) = 0\}$$

is the ideal of $E = K[X_1, \ldots, X_n, Y]$ generated by the binomials $\underline{X}^{\underline{i}} - \underline{X}^{\underline{j}}$ with multi-exponents $\underline{i}, \underline{j} \in \mathbf{N}^n$ such that $\underline{a}^{\underline{i}} = \underline{a}^{\underline{j}}$.

Of course, we use the shorthand $\underline{a}^{\underline{i}} := a_1^{i_1} \cdots a_n^{i_n}$. Let us show right away how to deduce from the lemma the Jordan decomposition in an algebraic group.

Theorem C.5. *Let $M \in \mathrm{GL}_n(K)$, let $M = M_s M_u$ be its multiplicative Dunford decomposition and let $F \in K[X_{1,1}, \ldots, X_{n,n}]$ such that:*

$$\forall k \in \mathbf{Z}, \ F(M^k) = 0.$$

Then $F(N) = 0$ for all N replicas of M_s and also for all $N = M_u^c$, $c \in K$.

---

[1] Note that ϕ is actually a K-algebra automorphism, as we will find ϕ_s and ϕ_u to be.

Proof. In the case that $M_s = P\text{Diag}(a_1, \ldots, a_n)P^{-1}$, apply the lemma to the polynomial:

$$G(X_1, \ldots, X_n, Y) := F\left(P\text{Diag}(X_1, \ldots, X_n)P^{-1}M_u^Y\right).$$

$\square$

Corollary C.6 (Jordan decomposition). *If $G \subset \text{GL}_n(K)$ is an algebraic subgroup and if $M \in G$, then $M_s, M_u \in G$.*

Proof. Indeed, for every polynomial equation F defining G, one has $F(M^k) = 0$ for all $k \in \mathbf{Z}$. $\square$

Now we prove the lemma.

Proof. It is clear that the set E' described in the statement of the lemma is an ideal and that all the indicated binomials belong to E'.

We know from the application at the end of the previous section that the K-algebra automorphism ϕ of E defined by $X_i \mapsto a_i X_i$, $Y \mapsto Y + c$ is locally finite. It is obvious from the definition that E' is stable under ϕ. We use the same notation $I(a)$, $E(a)$ as in the just mentioned application. There we have completely described $E(a)$ and also the components ϕ_s, ϕ_u of ϕ. By the results of the previous section, E' is the direct sum of the $E'(a) = E' \cap E(a)$. We shall now describe these subspaces.

Let $F = \sum\limits_{\underline{i} \in I(a)} \underline{X}^{\underline{i}} P_{\underline{i}}(Y)$. Then:

$$F \in E' \iff \forall k \in \mathbf{Z}, \; a^k \sum_{\underline{i} \in I(a)} F_{\underline{i}}(kc) = 0 \iff \sum_{\underline{i} \in I(a)} F_{\underline{i}} = 0.$$

Then, choosing an arbitrary $\underline{i}_0 \in I(a)$, such an F can be written:

$$F = \sum_{\underline{i} \in I(a)} \left(F_{\underline{i}} - F_{\underline{i}_0}\right).$$

But each $F_{\underline{i}} - F_{\underline{i}_0}$ clearly belongs to the ideal generated by all binomials $\underline{X}^{\underline{i}} - \underline{X}^{\underline{j}}$ such that $\underline{a}^{\underline{i}} = \underline{a}^{\underline{j}}$. $\square$

Remark C.7. The proof actually shows that E' is the $K[Y]$-module generated by all these binomials.

Tannaka duality
without schemes

Tannaka[1] theory (in the sense we shall give it here) originated in the desire to classify "motives", a fancy and mysterious object appearing in algebraic geometry, as representations of an algebraic or proalgebraic group. There were at that time various duality theorems that allowed one to recover a group from the category of its representations and Grothendieck was in search of an intrinsic description of such categories. Actually, the approach of Grothendieck (to be explained further below, at least as far as useful here) originates in his previous unification of classical Galois theory and of homotopy theory by seeing them both as categories of actions of a group (respectively the Galois group and the fundamental group) endowed with a multiplicative structure (the fiber product). This is beautifully expounded in [**DD05**].

Before going into tannakian duality, let us quote some precursors. In all the following examples, one associates to some object X (itself a set with some additional structure) a set X^* of maps defined over X; X^* is moreover endowed with some algebraic structure (group, linear space, algebra, etc.). Then one seeks to recover X (along with its additional structure) from the knowledge of X^* (along with its algebraic structure).

    (1) Since Descartes, we know the affine plane ("euclidean plane") through coordinates on it, that is, through the ring of functions $f(x, y)$.

---

[1] A large part of this appendix is reproduced almost word for word from [**Sau16**].

(2) From the first or second year of university, we know how to study a vector space V through linear forms on it, that is, through its dual V^*. If $\dim V < \infty$, one can recover V as $(V^*)^*$ (in the general case, one only has an embedding of V into $(V^*)^*$).

(3) In the case of an infinite-dimensional topological vector space V, we must consider its "continuous dual" V° (space of continuous linear forms) and maybe add some conditions on V to be able to recover it as $V = (V^\circ)^\circ$.

(4) Associate to a compact topological space X the $\mathbf{R}$-algebra $\mathcal{C}(X, \mathbf{R})$ of continuous functions $X \to \mathbf{R}$. Then X can be recovered as the "maximal spectrum" of $\mathcal{C}(X, \mathbf{R})$, *i.e.*, the set of its maximal ideals, the topology being defined in a somewhat similar manner as Zariski topology.

(5) An affine algebraic variety V is entirely determined by its algebra of regular functions $\mathbf{A}(V)$ (see any book on elementary affine algebraic geometry, such as [**Sha77**, **BorA91**]).

(6) A finite abelian group G is determined by its dual, the group $G^\vee :=$ $\mathrm{Hom}_{gr}(G, \mathbf{U})$ of its "characters" ($\mathbf{U}$ denotes the group of complex numbers with modulus 1); there is a canonical isomorphism from G to $(G^\vee)^\vee$.

(7) For a topological group G, the dual $G^\vee$ is the group of continuous morphisms from G to $\mathbf{U}$. If G is locally compact and abelian, then there is again a canonical isomorphism from G to $(G^\vee)^\vee$; this is called Pontriaguine duality.

Tannakian duality started with an attempt to generalize Pontriaguine duality to nonabelian groups. But it is certainly impossible in general to recover a nonabelian group from its morphisms to $\mathbf{C}^*$, even if one adds some topological conditions; indeed, many nontrivial compact groups admit no nontrivial continuous morphisms to $\mathbf{C}^*$. So the first version of the Tannaka theorem showed how to recover a compact group from the category of all its continuous representations (morphisms to $\mathbf{C}^*$ are representations of rank 1). Then Krein extended it to the Tannaka-Krein theorem, which states an intrinsic characterization of categories that are equivalent to the category of continuous representations of a compact group. A nice exposition is given in [**Che99**] of the side of this theorem nearest to the theory of algebraic groups.

In the sequel of this appendix, we shall give the name "Tannaka duality" to the rather different version invented by Grothendieck and developed by Saavedra and Deligne [**DM82**]. Its introduction in differential Galois theory came when Deligne realized how to obtain the differential Galois group (along with more structure) in this way [**Del90**, **vS03**]. We shall give a hint of it in the last section.

D.1. One weak form of Tannaka duality

We start from the question: can we recover a group G from the category of its finite-dimensional complex representations? (We shall from now on omit the words "finite-dimensional complex".) Note that in the case of an abelian group one must use the fact that (isomorphism classes of) one-dimensional representations, *i.e.*, morphisms from G to $\mathbf{C}^*$, come with a multiplicative structure: the dual $\mathrm{Hom}_{gr}(G, \mathbf{C}^*)$ is a group. In the general case, the tensor product of representations will yield this multiplicative structure. So we shall start with the following initial data:

(1) Let G be an "abstract" group, meaning that we consider no additional structure on G. Write $\mathcal{C}$ as the category of all its (finite-dimensional complex) representations. This is a $\mathbf{C}$-linear abelian category, meaning that there is a natural notion of exact sequences with usual properties and that all sets $\mathrm{Mor}_{\mathcal{C}}(\rho_1, \rho_2)$ of morphisms in $\mathcal{C}$ are complex vector spaces (and composition is bilinear).

(2) The category $\mathcal{C}$ comes equipped with a tensor product having good algebraic properties: associativity, commutativity, existence of a neutral element (all this up to natural isomorphisms). Moreover each object has a dual also having good properties with respect to exact sequences and tensor product.

(3) There is a functor ω from $\mathcal{C}$ to the category $\mathrm{Vect}_{\mathbf{C}}^{f}$ of finite-dimensional complex vector space, which associates to every representation $\rho : G \to \mathrm{GL}_{\mathbf{C}}(V)$ its underlying space V. This functor is exact, faithful and compatible with the tensor products in $\mathcal{C}$ and $\mathrm{Vect}_{\mathbf{C}}^{f}$.

Now let $g \in G$ and $\rho : G \to \mathrm{GL}_{\mathbf{C}}(V)$ be a representation, so that $\omega(\rho) = V$. Then $g_\rho := \rho(g) \in \mathrm{GL}_{\mathbf{C}}(\omega(\rho))$ and the family of all the $g_\rho \in \mathrm{GL}_{\mathbf{C}}(\omega(\rho))$ has the following properties:

(1) It is functorial; more precisely, $\rho \rightsquigarrow g_\rho$ is an isomorphism of functors $\omega \mapsto \omega$.

(2) It is tensor-compatible, *i.e.*, under the identification of $\omega(\rho_1 \otimes \rho_2)$ with $\omega(\rho_1) \otimes \omega(\rho_2)$, one has $g_{\rho_1 \otimes \rho_2} = g_{\rho_1} \otimes g_{\rho_2}$.

We write γ_g as the family of all the g_ρ and summarize these properties by writing $\gamma_g \in \mathrm{Aut}^{\otimes}(\omega)$, the group of tensor automorphisms of ω. In our particular case, $\mathcal{C}$ being the category of representations of G, this group is written $G^{alg} := \mathrm{Aut}^{\otimes}(\omega)$ and called the *proalgebraic hull* of G. Indeed, it can be seen to be the inverse limit of a family of algebraic groups, the Zariski closures $\overline{\rho(G)} \subset \mathrm{GL}_{\mathbf{C}}(V)$, and this is what we called (in Chapter 16) a proalgebraic group.[2] In the end, what we obtain in this way is not G itself but a proalgebraic group G^{alg} together with a group morphism $G \to G^{alg}$ which satisfies a particular universal property that we now explain.

Note that since G^{alg} is the inverse limit of the algebraic groups $\overline{\rho(G)} \subset \mathrm{GL}_{\mathbf{C}}(V)$, it comes with a particular family of representations $G^{alg} \to \overline{\rho(G)} \to \mathrm{GL}_{\mathbf{C}}(V)$. These can be directly characterized as the "rational" representations of G^{alg}. The word "rational" here means that they are in some sense compatible with the proalgebraic structure on G^{alg} and the algebraic structure on $\mathrm{GL}_{\mathbf{C}}(V)$. The universal property can then be stated as follows: the category of representations of G is equivalent to the category of rational representations of G^{alg}, and this equivalence is tensor compatible. The reader will profitably compare this construction with the definition of the proalgebraic hull in Chapter 16.

Example D.1. Take $G := \mathbf{Z}$. Any representation of G is equivalent to a matricial representation and a matricial representation $G \to \mathrm{GL}_n(\mathbf{C})$ is totally determined by the image $A \in \mathrm{GL}_n(\mathbf{C})$ of $1 \in \mathbf{Z}$. If moreover $B \in \mathrm{GL}_p(\mathbf{C})$ encodes another representation, a morphism $A \to B$ is a matrix $F \in \mathrm{Mat}_{p,n}(\mathbf{C})$ such that $FA = BF$. The definition of tensor product requires some care; see [**Sau03**] and the references quoted there as [26] and [31]. Also see the next section, where we shall summarize the corresponding constructions.

Note that this example has a meaning in differential Riemann-Hilbert and Galois theory: in the local study of analytic differential equations at $0 \in \mathbf{C}$, we found that the category of fuchsian differential equations (those such that 0 is a regular singular point) is equivalent to the category of representations of the local fundamental group (this is one form of the local Riemann-Hilbert correspondence). Since this local fundamental group is $\pi_1(\mathbf{C}^*) = \mathbf{Z}$, this is the category above and it is equivalent to the category of rational representations of $\mathbf{Z}^{alg}$. In this sense, $\mathbf{Z}^{alg}$ is the universal local Galois group for fuchsian differential equations; all the other Galois groups are its images under all various rational representations (see Chapter 16).

D.2. The strongest form of Tannaka duality

The complete version of Tannaka theory gives a criterion to recognize if a tensor category is equivalent to the category of rational representations of a proalgebraic group, and, if it is, a process to compute that group. We

---

[2]Actually Tannaka theory allows us to define $\mathrm{Aut}^{\otimes}(\omega)$ as an "affine group scheme" but we shall not need this [**DM82**].

shall only sketch it. To begin with, we have a category $\mathcal{C}$ which we suppose to be abelian and **C**-linear. We suppose that $\mathcal{C}$ has a tensor product with properties similar to the ones stated in the previous paragraph (commutativity, associativity, neutral object, duals, etc.). Last, there is a faithful exact functor $\omega : \mathcal{C} \to \text{Vect}_{\mathbf{C}}^{f}$ which is moreover tensor compatible. This is called a "fiber functor" (because of its origins in homotopy theory; see the historical introduction at the beginning of this appendix).

Now as before we can define the group $\text{Gal}(\mathcal{C}, \omega) := \text{Aut}^{\otimes}(\omega)$ of tensor compatible automorphisms of our fiber functor. It can then be proven [**DM82**] that this is a proalgebraic group and that the tensor category $\mathcal{C}$ is equivalent to the category of rational representations of $\text{Gal}(\mathcal{C}, \omega)$. The equivalence can be made explicit as follows. An element of $\text{Gal}(\mathcal{C}, \omega)$ is a family (γ_M) indexed by objects M of $\mathcal{C}$ and each $\gamma_M \in \text{GL}_{\mathbf{C}}(\omega(M))$ (and of course it is subject to various axioms). Then to each object M of $\mathcal{C}$ we associate the representation $\text{Gal}(\mathcal{C}, \omega) \to \text{GL}_{\mathbf{C}}(\omega(M))$ which sends an element of $\text{Gal}(\mathcal{C}, \omega)$ to the corresponding component $\gamma_M \in \text{GL}_{\mathbf{C}}(\omega(M))$. This is the effect of our equivalence functor on objects. The effect on morphisms can be easily deduced by a little playing in pure category theory!

It is proved in [**DM82**] that the choice of the fiber functor is unessential; indeed, it is analogous to the choice of a base point when computing a fundamental group. Therefore, with a slight abuse of notation, our group could be written $\text{Gal}(\mathcal{C})$.

Example D.2. We already tackled the case of the category $\mathcal{C}$ of representations of **Z**. Suppose we define a new category[3] $\mathcal{C}'$ having the same objects $A \in \text{GL}_n(\mathbf{C})$ (where n is not fixed); but if $B \in \text{GL}_p(\mathbf{C})$, morphisms $A \to B$ are matrices $F \in \text{Mat}_{p,n}(\mathbf{C}\{z\})$ such that $F(qz)A = BF(z)$. It can be (easily) proven that then F is a Laurent polynomial, thus defined everywhere on $\mathbf{C}^*$. Among the morphisms in $\mathcal{C}'$ we have the morphisms in $\mathcal{C}$, so that in some sense $\mathcal{C}$ embeds into $\mathcal{C}'$.

We fix some point $a \in \mathbf{C}^*$ and define a fiber functor ω_a sending the object $A \in \text{GL}_n(\mathbf{C})$ to $\mathbf{C}^n$, and the morphism $F : A \in \text{GL}_n(\mathbf{C})$ to $F(a) \in \text{Mat}_{p,n}(\mathbf{C})$. You should prove that this is indeed a fiber functor and that (through the embedding mentioned above) it restricts to ω on $\mathcal{C}$.

Now any tensor automorphism of ω_a is a family (γ_M) and can be considered as well as a tensor automorphism of ω; only, since $\mathcal{C}'$ has more

---

[3]The motivation of this example comes from the Galois theory of fuchsian q-difference equations; see [**Sau03**, **Sau16**].

morphisms, it is constrained by more functoriality conditions so that in the end $\mathrm{Gal}(\mathcal{C}')$ is a subgroup of $\mathrm{Gal}(\mathcal{C})$. (This is actually a special case of a general theorem in [**DM82**].)

It is proved in [**Sau03**], but you should try it as a pleasant exercise, that $\mathrm{Gal}(\mathcal{C}')$ is the subgroup of those $(\gamma, \lambda) \in \mathrm{Hom}_{gr}(\mathbf{C}^*, \mathbf{C}^*) \times \mathbf{C} = \mathbf{Z}^{alg} = \mathrm{Gal}(\mathcal{C})$ such that $\gamma(q) = 1$, so that we get a natural identification $\mathrm{Gal}(\mathcal{C}') = \mathrm{Hom}_{gr}(\mathbf{C}^*/q^{\mathbf{Z}}, \mathbf{C}^*) \times \mathbf{C}$.

D.3. The proalgebraic hull of Z

In this section, we seek to apply Tannaka duality to the category $\mathcal{C} := \mathfrak{Rep}_{\mathbf{C}}(\mathbf{Z})$ of finite-dimensional complex representations of $\mathbf{Z}$ and to compute $\mathrm{Gal}(\mathcal{C})$ according to the previous definitions. We shall only sketch the constructions and arguments, expecting the reader to fill in the missing details. Reading (and understanding) this section requires some knowledge about tensor products, so we briefly summarize the basics in the first subsection; skip it if you already know all this.

Tensor products of vector spaces and linear maps. Let K be an arbitrary commutative field. To any two vector spaces E, F over K, one associates their *tensor product* $E \otimes_K F$, together with a K-bilinear map $E \times F \to E \otimes_K F$, $(x, y) \mapsto x \otimes y$. There are two ways to characterize the tensor product:

- Abstractly, it is required to satisfy the following universal property: for every vector space G and every K-bilinear map $\phi : E \times F \to G$, there is a unique linear map $f : E \otimes_K F \to G$ such that $\phi(x, y) = f(x \otimes y)$ for all $x \in E$, $y \in F$.

- Concretely: if (u_i) is any basis of E and (v_j) any basis of F, then the doubly indexed family $(u_i \otimes v_j)$ is a basis of $E \otimes_K F$.

Note, as a consequence of the concrete characterization, that if E and F are finite dimensional, so is $E \otimes_K F$ and its dimension is $\dim_K(E \otimes_K F) = (\dim_K E)(\dim_K F)$.

Now, if $f : E \to E'$ and $g : F \to F'$ are two K-linear maps, one can easily deduce from either characterization that there is a unique K-linear map $f \otimes g : E \otimes_K F \to E' \otimes_K F'$ such that $(f \otimes g)(x \otimes y) = f(x) \otimes g(y)$ for all $x \in E$, $y \in F$. Then it is equally easy to prove for instance that, if $f' : E' \to E''$ and $g' : F' \to F''$ are two K-linear maps, then $(f' \circ f) \otimes (g' \circ g) = (f' \otimes g') \circ (f \otimes g)$.

Most standard properties of tensor products flow easily from the above facts. Actually, the most nontrivial part of the basic theory is to prove that tensor products do exist; this is done by an effective construction (see [**Lan02**]).

The tannakian category $\mathfrak{Rep}_{\mathbf{C}}(G)$**.** For a general "abstract group" G (meaning that we do not take into account any structure on G), objects of $\mathfrak{Rep}_{\mathbf{C}}(G)$ are group morphisms $\rho : G \to \mathrm{GL}_{\mathbf{C}}(E)$, E a finite-dimensional complex space. Morphisms from $\rho : G \to \mathrm{GL}_{\mathbf{C}}(E)$ to $\rho' : G \to \mathrm{GL}_{\mathbf{C}}(E')$ are linear maps $f : E \to E'$ intertwining E and E', *i.e.*, $\rho'(g) \circ f = f \circ \rho(g)$ for all $g \in G$. For any two objects, the set of morphisms is in a natural way a $\mathbf{C}$-linear space, composition is bilinear, etc.; moreover there is a natural definition and construction of "exact sequences" with standard properties: our category is "$\mathbf{C}$-linear abelian".

Then, we must define some additional structure on $\mathfrak{Rep}_{\mathbf{C}}(G)$:

(1) The "unit element", denoted $\underline{1}$, will be the trivial representation $G \to \mathrm{GL}_{\mathbf{C}}(\mathbf{C}) = \mathbf{C}^*$ mapping all $g \in G$ to $1 \in \mathbf{C}^*$.

(2) For any two representations $\rho : G \to \mathrm{GL}_{\mathbf{C}}(E)$ and $\rho' : G \to \mathrm{GL}_{\mathbf{C}}(E')$, the tensor product is the representation $\rho \otimes \rho' : G \to \mathrm{GL}_{\mathbf{C}}(E \otimes_{\mathbf{C}} E')$ sending every $g \in G$ to $\rho(g) \otimes \rho'(g) \in \mathrm{GL}_{\mathbf{C}}(E \otimes_{\mathbf{C}} E')$.

(3) For any two representations $\rho : G \to \mathrm{GL}_{\mathbf{C}}(E)$ and $\rho' : G \to \mathrm{GL}_{\mathbf{C}}(E')$, the "internal Hom" is the representation $\underline{\mathrm{Hom}}(\rho, \rho') : G \to \mathrm{GL}_{\mathbf{C}}(\mathrm{Hom}_{\mathbf{C}}(E, E'))$ sending every $g \in G$ to the map $f \mapsto \rho'(g) \circ f \circ \rho(g)^{-1}$; this map does belong to $\mathrm{GL}_{\mathbf{C}}(\mathrm{Hom}_{\mathbf{C}}(E, E'))$.

(4) The "dual" of the object ρ is the object $\rho^* := \underline{\mathrm{Hom}}(\rho, \underline{1})$.

(5) The "forgetful functor" ω from $\mathfrak{Rep}_{\mathbf{C}}(G)$ to $\mathrm{Vect}_{\mathbf{C}}^f$ sends $\rho : G \to \mathrm{GL}_{\mathbf{C}}(E)$ to E and $f : \rho \to \rho'$ to $f : E \to E'$.

In this general context, all axioms of "$\mathbf{C}$-linear neutral tannakian categories" are verified (see [**DM82**] for their list). For instance, the tensor product is, in some sense, associative; tensor and internal Hom are "adjoint", meaning that $\mathrm{Mor}(\rho \otimes \rho', \rho'') = \mathrm{Mor}(\rho, \underline{\mathrm{Hom}}(\rho', \rho''))$ (*i.e.*, there is a canonical and functorial isomorphism); and all objects are "reflexive", *i.e.*, they can be identified to their bidual. Last but not least, ω is a fiber functor: it is $\mathbf{C}$-linear, exact, faithful and it preserves tensor products.

Then the group $G^{alg} := \mathrm{Gal}(\mathcal{C}, \omega) := \mathrm{Aut}^{\otimes}(\omega)$ is well defined as a proalgebraic group and the category $\mathfrak{Rep}_{\mathbf{C}}(G)$ can be identified with the category of rational representations of G^{alg}. This identification can be realized in a very concrete way. We are going to define a morphism $\iota : G \to G^{alg}$. Then, associating to each rational representation ρ of G^{alg} the composition $\rho \circ \iota$ will yield the desired functor, indeed an equivalence of tensor categories.

To define ι, we start from some fixed $g \in G$. For every object $\rho : G \to \mathrm{GL}_{\mathbf{C}}(E)$, the map $\rho(g) \in \mathrm{GL}_{\mathbf{C}}(E)$ is a linear morphism from $E = \omega(\rho)$ to itself. The family of all $\rho(g)$ (while ρ runs among the objects of $\mathfrak{Rep}_{\mathbf{C}}(G)$) is a morphism $\iota(g)$ of the functor ω to itself. Then one checks that it is a tensor automorphism, $i.e.$, $\iota(g) \in \mathrm{Aut}^{\otimes}(\omega)$; and also, now letting g vary in G, that $g \mapsto \iota(g)$ is a group morphism $\iota : G \to G^{alg}$.

The tannakian category $\mathcal{C} = \mathfrak{Rep}_{\mathbf{C}}(\mathbf{Z})$. A representation of $\mathbf{Z}$ in a space E is completely characterized by the image ϵ of 1, which must be an (otherwise arbitrary) automorphism of E. In this way, we can identify the objects of $\mathfrak{Rep}_{\mathbf{C}}(\mathbf{Z})$ with pairs (E, ϵ), where E is a finite-dimensional complex vector space and where $\epsilon \in \mathrm{GL}_{\mathbf{C}}(E)$. We shall describe the tannakian category $\mathcal{C}$ with this model, $i.e.$, its objects are such pairs. Then morphisms from (E, ϵ) to (E', ϵ') are linear maps $f : E \to E'$ such that $\epsilon' \circ f = f \circ \epsilon$. As for the additional structures on $\mathcal{C}$:

(1) The unit element $\underline{1}$ is $(\mathbf{C}, 1)$ (with the identification of $\mathrm{GL}_1(\mathbf{C})$ with $\mathbf{C}^*$ and of $\mathrm{Id}_{\mathbf{C}}$ with 1).

(2) The tensor product of $(E, \epsilon) \otimes (E', \epsilon')$ is $(E \otimes_{\mathbf{C}} E', \epsilon \otimes \epsilon')$.

(3) The internal Hom, $\underline{\mathrm{Hom}}((E, \epsilon), (E', \epsilon'))$, is $(\mathrm{Hom}(E, E'), \phi)$ where $\phi : f \mapsto \epsilon' \circ f \circ \epsilon^{-1}$.

(4) The dual of (E, ϵ) is (E^*, ϵ^*), where E^* is the dual of E and ϵ^* is the contragredient of ϵ (inverse of the transpose map).

(5) The forgetful functor ω sends (E, ϵ) to E.

Matricial description of $\mathcal{C}$. If $E = \mathbf{C}^n$, then $\mathrm{GL}_{\mathbf{C}}(E) = \mathrm{GL}_n(\mathbf{C})$. Morphisms from $E = \mathbf{C}^n$ to $F = \mathbf{C}^p$ are in natural bijective correspondence with rectangular matrices in $\mathrm{Mat}_{p,n}(\mathbf{C})$ and the composition of morphisms corresponds to the product of matrices. Since every E is isomorphic to some $\mathbf{C}^n$, we see that a category equivalent to $\mathcal{C}$ can be entirely described in terms of matrices. Its objects will be elements of $\mathrm{GL}_n(\mathbf{C})$ (n running in $\mathbf{N}$); its morphisms from $A \in \mathrm{GL}_n(\mathbf{C})$ to $B \in \mathrm{GL}_p(\mathbf{C})$ will be elements P of $\mathrm{Mat}_{p,n}(\mathbf{C})$ such that $BF = FA$. As for the additional structures, we can (and will) realize them as follows:

(1) The unit element $\underline{1}$ is $1 \in \mathrm{GL}_1(\mathbf{C}) = \mathbf{C}^*$.

(2) The tensor product of $\mathbf{C}^n$ with $\mathbf{C}^p$ is $\mathbf{C}^{np}$, where the tensor product basis of the canonical basis $(e_1, \ldots, e_n)$ of $\mathbf{C}^n$ with the canonical

basis $(f_1, \ldots, f_p)$ of $\mathbf{C}^p$ is linearly ordered as:

$$(g_1, \ldots, g_{np}) := (e_1 \otimes f_1, \ldots, e_1 \otimes f_p, \ldots, e_n \otimes f_1, \ldots, e_n \otimes f_p).$$

Then the tensor product of two matrices is naturally defined.

(3) The dual of M is $M^* := {}^t M^{-1}$.

(4) The internal Hom, $\underline{\mathrm{Hom}}(M, M')$, is constructed as $M^* \otimes M'$.

(5) The forgetful functor ω sends $M \in \mathrm{GL}_n(\mathbf{C})$ to $\mathbf{C}^n$.

From this point on, the computation of $\mathrm{Aut}^\otimes(\omega)$ can be translated into a computational problem: how to associate to every $M \in \mathrm{GL}_n(\mathbf{C})$ a $\gamma_M \in \mathrm{GL}_n(\mathbf{C})$ in such a way that:

(1) If $M \in \mathrm{GL}_n(\mathbf{C})$ and $N \in \mathrm{GL}_p(\mathbf{C})$, for every $P \in \mathrm{Mat}_{p,n}(\mathbf{C})$ such that $NP = PM$, one has $\gamma_N P = P\gamma_M$.

(2) If $M \in \mathrm{GL}_n(\mathbf{C})$ and $N \in \mathrm{GL}_p(\mathbf{C})$, then $\gamma_{M \otimes N} = \gamma_M \otimes \gamma_N$.

This is a nice (if rather long) exercise in linear algebra. A complete solution is to be found in the appendix of `arXiv:math/0210221v1` where, in total agreement with the results of Chapter 16, one computes $\mathbf{Z}^{alg}$ as $\mathrm{Hom}_{gr}(\mathbf{C}^*, \mathbf{C}^*) \times \mathbf{C}$ and the morphism $\mathbf{Z} \to \mathbf{Z}^{alg}$ as the map:

$$\iota : \begin{cases} \mathbf{Z} \to \mathrm{Hom}_{gr}(\mathbf{C}^*, \mathbf{C}^*) \times \mathbf{C}, \\ k \mapsto ((z \mapsto z^k), k). \end{cases}$$

A representation of $\mathbf{Z}$ encoded by the matrix $A \in \mathrm{GL}_n(\mathbf{C})$ then gives rise to the following representation of $\mathbf{Z}^{alg}$:

$$(\gamma, \lambda) \mapsto \gamma(A_s)A_u^\lambda,$$

where $A = A_s A_u$ is the Dunford decomposition, $\gamma(A_s)$ is obtained by replacing each eigenvalue c by $\gamma(c)$ and A_u^λ is well defined because A_u is unipotent.

D.4. How to use tannakian duality in differential Galois theory

We give only a very superficial view here (for complete details see [**Del90**, **vS03**]); moreover we only tackle the local problem.

In the text as well as in the previous sections of this appendix, we relied on Riemann-Hilbert correspondence, which provides us with an equivalence between the category of fuchsian differential systems (local at 0) and the category of representations of $\pi_1(\mathbf{C}^*)$. Since moreover we know the algebraic structure of $\pi_1(\mathbf{C}^*)$ (it is $\mathbf{Z}$!), we could, in the previous section, apply tannakian duality to the latter category to find the proalgebraic group $\hat{\pi}_1$ as the proalgebraic hull of $\pi_1(\mathbf{C}^*)$. This involved additional structures on the category of representations of $\pi_1(\mathbf{C}^*)$ (tensor product, etc.).

However we can try instead to apply tannakian duality directly to the category of fuchsian differential systems (local at 0), *i.e.*, to describe the relevant additional structures directly. We shall presently do so, keeping however the same elementary point of view as before (the report [**Del90**] and the book [**vS03**] take the point of view of group schemes).

We shall use the following description of our category: the objects are matrices $A(z) \in \mathrm{Mat}_n(\mathbf{C}(\{z\}))$ required to have a regular singularity at 0; the morphisms from $A(z) \in \mathrm{Mat}_n(\mathbf{C}(\{z\}))$ to $B(z) \in \mathrm{Mat}_p(\mathbf{C}(\{z\}))$ are matrices $F(z) \in \mathrm{Mat}_{p,n}(\mathbf{C}(\{z\}))$ such that $F' = BF - FA$. Now we go for the additional structure.

(1) The unit element $\underline{1}$ is $0 \in \mathrm{Mat}_1(\mathbf{C}) = \mathbf{C}$.

(2) The tensor product of $A(z) \in \mathrm{Mat}_n(\mathbf{C}(\{z\}))$ with $B(z) \in \mathrm{Mat}_p(\mathbf{C}(\{z\}))$ is $A(z) \otimes I_p + I_n \otimes B(z)$ made into a matrix in $\mathrm{Mat}_{np}(\mathbf{C}(\{z\}))$ through the reordering rules explained in the subsection "Matricial description of $\mathcal{C}$" in the previous section. Indeed, these rules yield a natural identification of $\mathrm{Mat}_n(\mathbf{C}(\{z\})) \otimes \mathrm{Mat}_p(\mathbf{C}(\{z\}))$ with $\mathrm{Mat}_{np}(\mathbf{C}(\{z\}))$.

(3) The dual of A is $A^* := -^t A$.

(4) The internal Hom, $\underline{\mathrm{Hom}}(A, B)$, is constructed as $A^* \otimes B$.

(5) For each $a \in \mathbf{C}^*$, the fiber functor ω_a sends $A(z) \in \mathrm{Mat}_n(\mathbf{C}(\{z\}))$ to $\mathbf{C}^n$ and $F : A \to B$ to $F(a) : \mathbf{C}^n = \omega_a(A) \to \mathbf{C}^p = \omega_a(B)$.

We shall not try to prove that all this really works but rather give some motivation. The point a used for the fiber functor should be considered as a base point where initial conditions will be taken. The tensor product has been defined in such a way that, if $\mathcal{X}$ is a fundamental matricial solution of A and $\mathcal{Y}$ is a fundamental matricial solution of B at a, then $\mathcal{X} \otimes \mathcal{Y}$ is a fundamental matricial solution of $A \otimes I_p + I_n \otimes B$ at a. This is readily proved using the rules on tensor product given at the beginning of D.3:

$$(\mathcal{X} \otimes \mathcal{Y})' = \mathcal{X}' \otimes \mathcal{Y} + \mathcal{X} \otimes \mathcal{Y}' = A\mathcal{X} \otimes \mathcal{Y} + \mathcal{X} \otimes B\mathcal{Y}$$
$$= (A \otimes I_p)(\mathcal{X} \otimes \mathcal{Y}) + (I_n \otimes B)(\mathcal{X} \otimes \mathcal{Y}) = (A \otimes I_p + I_n \otimes B)(\mathcal{X} \otimes \mathcal{Y}).$$

We used Leibniz' rule for the differentiation of $\mathcal{X} \otimes \mathcal{Y}$, which is a consequence of the fact that all coefficients of $\mathcal{X} \otimes \mathcal{Y}$ are products $x_{i,j} y_{k,l}$. We also need to check that, if $\mathcal{X}(a) = I_n$ and $\mathcal{Y}(a) = I_p$, then $(\mathcal{X} \otimes \mathcal{Y})(a) = I_{np}$; but this comes from the identification of $I_n \otimes I_p$ with I_{np}.

For any object A endowed with the fundamental matricial solution that has the value I_n at a, any loop in $\mathbf{C}^*$ based at a defines a monodromy matrix $M_A \in \mathrm{GL}_n(\mathbf{C})$ which we can consider as a linear automorphism of

$\mathbf{C}^n = \omega_a(A)$. Then it is not too difficult to prove that $A \rightsquigarrow M_A$ is a tensor compatible automorphism of the fiber functor, thus an element of $\mathrm{Aut}^\otimes(\omega_a)$. From this, we deduce a group morphism from $\pi_1(\mathbf{C}^*, a)$ to $\mathrm{Aut}^\otimes(\omega_a)$. In the end, one finds that this is exactly our map $\pi_1 \to \hat{\pi}_1$.

Note that there is a slight problem with our constructions above: we have evaluated various matricial functions (fundamental matricial solution $\mathcal{X}$, morphism F) at a, although nothing guarantees that they are defined there. There are two possible ways out of this difficulty:

(1) We can restrict the construction to the equivalent subcategory of systems $X' = AX$ such that $A(z) \in z^{-1}\mathrm{Mat}_n(\mathbf{C})$, which is in essence what we did in the text; everything then works fine.

(2) We can consider that all this is defined for some subcategory depending on a, say $\mathcal{C}_a$. Then one shows that the proalgebraic groups $\mathrm{Gal}(\mathcal{C}_a)$ can all be canonically identified to some unique well-defined $\mathrm{Gal}(\mathcal{C})$ that does not depend on the choice of a base point a.

Actually, an even better realization of the second choice mentioned above is the following: for any discrete subset Σ of X, one defines a full subcategory $\mathcal{C}_\Sigma$ of systems having all their singularities inside Σ. For such a category, any base point $a \notin \Sigma$ will do. The whole Galois group is the inverse limit of all Galois groups obtained this way.

Remark D.3. In the theory of "q-difference equations" $X(qz) = A(z)X(z)$, one cannot rely on analytic continuation and the Riemann-Hilbert correspondence cannot be defined directly. Instead one follows an approach similar to the one above to construct a universal proalgebraic Galois group (in the local setting this is the group described in Example D.2) and then one extracts from it a Zariski-dense discrete group of which the Galois group is therefore the proalgebraic hull; that Zariski-dense discrete group is a natural candidate for playing the role of the fundamental group in Riemann-Hilbert correspondence. See [**Sau03, Sau16, RS15**].

Duality for diagonalizable algebraic groups

This appendix is an attempt to explain what was alluded to in Remark 16.14 in Chapter 16. Everything that follows is proved in [**BorA91**] in a more general setting.

Recall that for any "abstract" group Γ (meaning that we do not take into account any possible additional structure on Γ) we defined the group $\mathfrak{X}(\Gamma) := \mathrm{Hom}_{gr}(\Gamma, \mathbf{C}^*)$, thus obtaining a contravariant functor from abstract groups to proalgebraic groups (see the exercises of Chapter 15 and the whole of Chapter 16). We also noted that we could, without loss of generality, restrict to the case of abelian Γ, which we shall always do in this appendix. Lastly recall that, if Γ is finitely generated, then $\mathfrak{X}(\Gamma)$ is actually an algebraic group.

E.1. Rational functions and characters

Rational functions on an algebraic group. The affine algebra (recall that this term was introduced in Section 16.2) of the full linear group is:

$$\mathbf{A}(\mathrm{GL}_n(\mathbf{C})) = \mathbf{C}[X_{1,1}, \ldots, X_{n,n}, 1/\Delta],$$

where $\Delta := \det(X_{i,j})_{1 \leq i,j \leq n} \in \mathbf{C}[X_{1,1}, \ldots, X_{n,n}]$. Let $G \subset \mathrm{GL}_n(\mathbf{C})$ be an algebraic subgroup and let I be the ideal of all $f \in \mathbf{A}(\mathrm{GL}_n(\mathbf{C}))$ that vanish all over G, *i.e.*, the kernel of the morphism of $\mathbf{C}$-algebras $\mathbf{A}(\mathrm{GL}_n(\mathbf{C})) \to \mathbf{A}(G)$.

Since by definition a rational function on $\mathrm{GL}_n(\mathbf{C})$ or on G is the restriction of a polynomial function of $\mathrm{Mat}_n(\mathbf{C})$, the morphisms of $\mathbf{C}$-algebras $\mathbf{A}(\mathrm{Mat}_n(\mathbf{C})) \to \mathbf{A}(\mathrm{GL}_n(\mathbf{C}))$ and $\mathbf{A}(\mathrm{Mat}_n(\mathbf{C})) \to \mathbf{A}(G)$ are both surjective, so $\mathbf{A}(\mathrm{GL}_n(\mathbf{C})) \to \mathbf{A}(G)$ also is, in a natural way:

$$\mathbf{A}(G) = \mathbf{A}(\mathrm{GL}_n(\mathbf{C}))/I.$$

This can also be expressed by the "exact sequence"

$$0 \to I \overset{i}{\to} \mathbf{A}(\mathrm{GL}_n(\mathbf{C})) \overset{p}{\to} \mathbf{A}(G) \to 0,$$

meaning that the injective map i identifies I with the kernel of the surjective map p.

Characters of an algebraic group. We call *characters*[1] of an algebraic group G the rational morphisms $G \to \mathbf{C}^*$. These form a subgroup $\mathbf{X}(G)$ of the group $\mathfrak{X}(G)$ of all group homomorphisms $G \to \mathbf{C}^*$. Since a rational map $G \to \mathbf{C}^*$ is the same thing as an invertible rational function on G, we have:

$$\mathbf{X}(G) := \mathrm{Hom}_{gralg}(G, \mathbf{C}^*) = \{\chi \in \mathbf{A}(G)^* \mid \forall g, g' \in G,\ \chi(gg') = \chi(g)\chi(g')\}.$$

Obviously the definition of $\mathbf{X}(G)$ can be extended to proalgebraic groups and if $G = \varprojlim G_i$ we find that $\mathbf{X}(G)$ can be identified with $\bigcup \mathbf{X}(G_i)$; this is because of the way we defined rational maps on a proalgebraic group.

Examples E.1. (1) Characters on $G := \mathbf{C}^*$ have the form $x \mapsto x^m$, $m \in \mathbf{Z}$. Here $\mathbf{A}(G) = \mathbf{C}[X, 1/X]$ and $\mathbf{X}(G) = \{X^m \mid m \in \mathbf{Z}\}$ as a subset of $\mathbf{A}(G)$, which we can canonically identify with $\mathbf{Z}$.

(2) Characters on $G := \mu_d$ have the form $x \mapsto x^m$, $m \in \mathbf{Z}$, but since $x^d = 1$ identically on μ_d, we can rather take $m \in \mathbf{Z}/d\mathbf{Z}$. Here $\mathbf{A}(G) = \mathbf{C}[X]/\langle X^d - 1 \rangle = \mathbf{C}[x]$, where x denotes the class of X and $\mathbf{X}(G) = \{x^m \mid m \in \mathbf{Z}/d\mathbf{Z}\}$ as a subset of $\mathbf{A}(G)$, which we can canonically identify with $\mathbf{Z}/d\mathbf{Z}$.

(3) Write $\mathbf{D}_n$ as the subgroup of diagonal matrices in $\mathrm{GL}_n(\mathbf{C})$, which can naturally be identified with $(\mathbf{C}^*)^n$. The ideal I of $G := \mathbf{D}_n$ in $\mathrm{GL}_n(\mathbf{C})$ is generated by all the $X_{i,j}$, $1 \le i, j \le n$, $i \ne j$. Calling X_i the class of $X_{i,i}$ in $\mathbf{A}(\mathrm{GL}_n(\mathbf{C}))/I$, we find that here

$$\mathbf{A}(G) = \mathbf{C}[X_1, \ldots, X_n, 1/X_1 \cdots X_n] = \mathbf{C}[X_1, \ldots, X_n, 1/X_1, \ldots, 1/X_n].$$

Then $\mathbf{X}(G) = \{\underline{X}^{\underline{m}} := X_1^{m_1} \cdots X_n^{m_n} \mid (m_1, \ldots, m_n) \in \mathbf{Z}^n\}$ as a subset of $\mathbf{A}(G)$, and we can canonically identify it to $\mathbf{Z}^n$.

Note that in all these examples $\mathbf{X}(G)$ generates the $\mathbf{C}$-algebra $\mathbf{A}(G)$.

---

[1] Beware that the same word is used with various meanings in other contexts.

The case of $\mathfrak{X}(\Gamma)$. Let Γ be a finitely generated abelian group, so that we have an exact sequence

$$0 \to \Lambda \to \mathbf{Z}^n \to \Gamma \to 0.$$

As before, this exact sequence means that $\Gamma = \mathbf{Z}^n/\Lambda$. From the general properties of the contravariant functor $\Gamma \rightsquigarrow \mathfrak{X}(\Gamma)$, we get another exact sequence

$$0 \to \mathfrak{X}(\Gamma) \to \mathfrak{X}(\mathbf{Z}^n) \to \mathfrak{X}(\Lambda) \to 0,$$

which means that $\mathfrak{X}(\mathbf{Z}^n) \to \mathfrak{X}(\Lambda)$ is surjective with kernel $\mathfrak{X}(\Gamma)$. So we see that any such $\mathfrak{X}(\Gamma)$ is isomorphic as an algebraic group to a subgroup of $\mathfrak{X}(\mathbf{Z}^n) = (\mathbf{C}^*)^n$. More precisely, $\mathfrak{X}(\Gamma)$ is thereby identified with the subgroup of $(\mathbf{C}^*)^n$ defined by the equations $x_1^{m_1} \cdots x_n^{m_n} = 1$, where $(m_1, \ldots, m_n)$ runs in Λ.

Remark E.2. From the structure theorem of finitely generated abelian groups, one may take $\Lambda = \{0\}^r \times d_1\mathbf{Z} \times \cdots \times d_k\mathbf{Z}$, $n = r + k$, $1 < d_1 | \cdots | d_k$, so that:

$$\Gamma = \mathbf{Z}^n/\Lambda \simeq \mathbf{Z}^r \times \mathbf{Z}/d_1\mathbf{Z} \times \cdots \times \mathbf{Z}/d_k\mathbf{Z}.$$

Identifying $\mathfrak{X}(\mathbf{Z}^n) = (\mathbf{C}^*)^n$ and $\mathfrak{X}(\Lambda) = (\mathbf{C}^*)^k$, the map $\mathfrak{X}(\mathbf{Z}^n) \to \mathfrak{X}(\Lambda)$ is $(x_1, \ldots, x_r, y_1, \ldots, y_k) \mapsto (y_1^{d_1}, \ldots, y_k^{d_k})$ so that its kernel $\mathfrak{X}(\Gamma)$ is realized as the subgroup $(\mathbf{C}^*)^r \times \mu_{d_1} \times \cdots \times \mu_{d_k}$ of $(\mathbf{C}^*)^n$.

E.2. Diagonalizable groups and duality

Subgroups of the diagonal group. Let $G \subset \mathbf{D}_n$ be an algebraic subgroup and I the ideal of $\mathbf{A}(\mathbf{D}_n)$ consisting of those f that vanish on G; by the same argument as in the previous section,

$$\mathbf{A}(G) = \mathbf{A}(\mathbf{D}_n)/I = \mathbf{C}[X_1, \ldots, X_n, 1/X_1, \ldots, 1/X_n]/I.$$

Then using the same kind of technique as in the proof of the lemma of Section C.2, one can prove that I is generated by all $X_1^{m_1} \cdots X_k^{m_k} - 1$ that vanish on G, *i.e.*, such that $(m_1, \ldots, m_n) \in \Lambda$, where Λ is some subgroup of $\mathbf{Z}^n = \mathbf{X}(\mathbf{D}_n)$.

In the same way, we have an exact sequence

$$0 \to \Lambda \to \mathbf{X}(\mathbf{D}_n) \to \mathbf{X}(G) \to 0,$$

i.e., $\mathbf{X}(G) = \mathbf{X}(\mathbf{D}_n)/\Lambda$ under the identification $\mathbf{Z}^n = \mathbf{X}(\mathbf{D}_n)$.

Algebraic groups isomorphic to algebraic subgroups of $\mathbf{D}_n$ are called *diagonalizable*. One can prove the following more or less along the same lines.

Theorem E.3. *The commutative algebraic group G is diagonalizable if, and only if, $\mathbf{X}(G)$ generates the $\mathbf{C}$-algebra $\mathbf{A}(G)$.* $\square$

Duality. Starting from Γ, we have an obvious map $\Gamma \times \mathfrak{X}(\Gamma) \to \mathbf{C}^*$ which sends (g, f) to $f(g)$. By the usual "bidualizing" process, this induces a group morphism $\Gamma \to \mathrm{Hom}_{gr}(\mathfrak{X}(\Gamma), \mathbf{C}^*)$, and it is easy to check that its image is actually contained in $\mathbf{X}(\mathfrak{X}(\Gamma))$. The previous arguments imply:

Theorem E.4. *The natural map $\Gamma \to \mathbf{X}(\mathfrak{X}(\Gamma))$ is an isomorphism.* $\square$

Starting from a diagonalizable group G, we have an obvious map $G \times \mathbf{X}(G) \to \mathbf{C}^*$ which sends (g, f) to $f(g)$. By the usual "bidualizing" process, this induces a group morphism $G \to \mathrm{Hom}_{gr}(\mathbf{X}(G), \mathbf{C}^*) = \mathfrak{X}(\mathbf{X}(G))$, and it is easy to check that this map is rational. The previous arguments imply:

Theorem E.5. *The natural map $G \to \mathfrak{X}(\mathbf{X}(G))$ is an isomorphism.* $\square$

To avoid confusion in the following duality table, we systematically write additively the abstract abelian groups (and of course multiplicatively the algebraic subgroups of $\mathbf{D}_n$). Then we have a correspondence:

abstract finitely generated abelian group $\longleftrightarrow$ diagonalizable algebraic group

$$\Gamma = \mathbf{X}(G) \longleftrightarrow G = \mathfrak{X}(\Gamma)$$

$$\mathbf{Z} \longleftrightarrow \mathbf{C}^*$$

$$\mathbf{Z}/d\mathbf{Z} \longleftrightarrow \mu_d$$

$$\mathbf{Z}^n \longleftrightarrow \mathbf{D}_n$$

torsion free group $\longleftrightarrow$ connected group

Connected diagonalizable groups are called *tori*. They are all isomorphic to some $\mathbf{D}_n$.

Extension to proalgebraic groups. Using the fact that any Γ is the union of its finitely generated subgroups Γ_i, we found the proalgebraic structure on $\mathfrak{X}(\Gamma)$:

$$\mathfrak{X}(\Gamma) = \varprojlim \mathfrak{X}(\Gamma_i).$$

Conversely, for every commutative proalgebraic group $G = \varprojlim \Gamma_i$, we obtain from the definition of rational morphisms that the $\mathbf{X}(G_i)$ are a directed family of finitely generated subgroups and that

$$\mathbf{X}(G) = \bigcup \mathbf{X}(G_i).$$

Using this and the above results, we deduce that there is a duality between abstract abelian groups and those commutative proalgebraic groups which are inverse limits of diagonalizable groups.

Revision problems

The problems in the first two sections are slightly adapted versions of the final exams that were given at the end of this course. In each case, all documents were allowed (but this book did not exist then). The first exam was tailored for 2 hours, the second for 4 hours.

F.1. 2012 exam (Wuhan)

Question on Chapter 1: Compute the exponential of $\begin{pmatrix} a & -b \\ b & a \end{pmatrix}$, where $a, b \in \mathbf{C}$.

Question on Chapter 2: Find the unique power series such that $f(z) = (1 - 2z)f(2z)$ and $f(0) = 1$, and give its radius of convergence.

Question on Chapter 3: Prove that $\dfrac{1}{e^z - 1}$ is meromorphic on $\mathbf{C}$ but that it is not the derivative of a meromorphic function.

Question on Chapter 4: Let $A \in \mathrm{Mat}_n(\mathbf{C})$. Show that $\exp(A) = I_n$ if, and only if, A is diagonalizable and $\mathrm{Sp}(A) \subset 2\mathrm{i}\pi\mathbf{Z}$. (The proof that A is diagonalizable is not very easy.)

Question on Chapter 5: Solve $z^2 f'' + z f' + f = 0$ on $\mathbf{C} \setminus \mathbf{R}_-$ and find its monodromy.

Question on Chapter 6: Prove rigorously that $z^\alpha . z^\beta = z^{\alpha+\beta}$.

Question on Chapter 7: What becomes of the equation $z f'' + f' = 0$ at infinity?

Question on Chapter 9: Solve the equation $(1 - z)\delta^2 f - \delta f - z f = 0$ by the method of Fuchs-Frobenius and compute its monodromy. (Just give the precise recursive formula for the Birkhoff matrix and a few terms.)

Question on Chapter 13: Give a necessary and sufficient condition for the differential Galois group of equation $\delta^2 f + p\delta f + qf = 0$, where $p, q \in \mathbf{C}$, to contain unipotent matrices other than the identity matrix.

Question on Chapter 14: Show that the Galois group of a regular singular system is trivial if, and only if, it admits a uniform fundamental matricial solution. Is the same condition valid for an irregular system?

F.2. 2013 exam (Toulouse)

Exercise F.1. Consider the sets

$$\Omega := \{z \in \mathbf{C} \mid 0 \leq \Im z < 2\pi\} = \{x + iy \mid x, y \in \mathbf{R}, \ 0 \leq y < 2\pi\}$$

and, for every integer $n \geq 1$,

$$\Omega_n := \{M \in \mathrm{Mat}_n(\mathbf{C}) \mid \mathrm{Sp}\ M \subset \Omega\}.$$

(Recall that $\mathrm{Sp}\ M$ denotes the set of eigenvalues of M.)

1) Let $A \in \mathrm{Mat}_n(C)$ with additive Dunford decomposition $A = A_s + A_n$ and $B \in \mathrm{GL}_n(\mathbf{C})$ with multiplicative Dunford decomposition $B = B_s B_u$. Prove rigorously the logical equivalence:

$$(\exp(A) = B) \Longleftrightarrow (\exp(A_s) = B_s \text{ and } \exp(A_n) = B_u).$$

2) Prove rigorously that the exponential map from $\mathrm{Mat}_n(\mathbf{C})$ to $\mathrm{GL}_n(\mathbf{C})$ induces a bijection from Ω_n onto $\mathrm{GL}_n(\mathbf{C})$.

3) For every $t \in \mathbf{R}$, write $L(t) \in \Omega_n$ as the unique preimage of the matrix $\begin{pmatrix} \cos t & -\sin t \\ \sin t & \cos t \end{pmatrix}$ under the above bijection (question 2). Compute $L(t)$. What do you notice?

Exercise F.2. Let $a = \sum a_n z^n \in \mathbf{C}(\{z\})$ and let the differential equation

(F.1) $$f' = af$$

be given.

1) Find a nontrivial solution of (F.1) in the form $f = z^\alpha g$, where g is uniform in a neighborhood of 0 in $\mathbf{C}^*$.

2) Show that, if $\alpha \notin \mathbf{Q}$, then f is transcendental over $\mathbf{C}(\{z\})$. (Hint: assuming a nontrivial relation $\sum \lambda_k f^k = 0$, use the monodromy action to obtain a strictly shorter relation of the same kind.)

Exercise F.3. Let $p, q \in \mathbf{C}$ and $\delta := z \dfrac{d}{dz}$ (Euler differential operator). Let the differential equation

(F.2) $$\delta^2 f + p\delta f + qf = 0$$

be given.

1) Write this equation in terms of $f' := df/dz$ and $f'' := d^2 f/dz^2$. Explicitly describe the singularities in $\mathbf{C}$ and their nature (regular or irregular singular points).

2) We now study equation (F.2) near 0. Write the naturally associated differential system in the form:

(F.3) $$\delta X = AX \ , \ A \in \mathrm{Mat}_2(\mathbf{C}).$$

3) Check that A is semi-simple if, and only if, $p^2 - 4q \neq 0$ and give its additive Dunford decomposition $A = A_s + A_n$ in all cases.

4) In this question, we assume that $p^2 - 4q \neq 0$ and we write λ, μ as the eigenvalues of A.
   (i) Compute the fundamental matricial solution $\mathcal{X}$ of system (F.3) such that $\mathcal{X}(1) = I_2$ (identity matrix).
   (ii) Deduce a simple basis of solutions of equation (F.2) in the neighborhood of 1.
   (iii) Compute the monodromy of $\mathcal{X}$ along the fundamental loop.
   (iv) Describe the matricial Galois group of (F.3) with respect to the fundamental matricial solution $\mathcal{X}$. We denote it $G \subset \mathrm{GL}_2(\mathbf{C})$.
   (v) Give a necessary and sufficient condition involving p and q for the inclusion $G \subset \mathrm{SL}_2(\mathbf{C})$ (subgroup of $\mathrm{GL}_2(\mathbf{C})$ made up of matrices with determinant 1).
   (vi) Give a necessary and sufficient condition involving p and q for G to be a finite group.

5) Redo question 4 under the assumption $p^2 - 4q = 0$.

6) We set $g(z) := f(1/z)$. Explicitly describe the linear differential equation of order 2 satisfied by g and check that 0 is a regular singular point of that equation.[1]

---

[1] One then says that equation (F.2) admits a regular singular point at infinity.

Exercise F.4.   1) Let V be a complex vector space of finite dimension n, and let G be a subgroup of $\mathrm{GL}(V)$. Choosing a basis $\mathcal{B}$ of V, we obtain an isomorphism u from $\mathrm{GL}(V)$ to $\mathrm{GL}_n(\mathbf{C})$. Show that the condition "$u(G)$ is an algebraic subgroup of $\mathrm{GL}_n(\mathbf{C})$" is independent of the choice of the particular basis $\mathcal{B}$ of V. We then say that G is an algebraic subgroup of $\mathrm{GL}(V)$.

2) Now let $\phi \in \mathrm{GL}(V)$. Prove the existence of a Jordan decomposition, $\phi = \phi_s \phi_u = \phi_u \phi_s$, such that, writing M as the matrix of ϕ in the (arbitrary) basis $\mathcal{B}$, the matrix of ϕ_s, resp. ϕ_u, in the same basis is M_s, resp. M_u. Deduce from this fact that, for any algebraic subgroup G of $\mathrm{GL}(V)$, if $\phi \in G$, then $\phi_s, \phi_u \in G$.

3) Let $x \in V$. Show that $\{\phi \in \mathrm{GL}(V) \mid \phi(x) = x\}$ is an algebraic subgroup of $\mathrm{GL}(V)$. Deduce from this fact that, if $\phi(x) = x$, then $\phi_s(x) = x$ and $\phi_u(x) = x$.

Exercise F.5.   1) Let $\delta := z\dfrac{d}{dz}$ (Euler differential operator). For every $P \in \mathbf{C}[T]$, compute $P(\delta)(z^n)$.

2) Let $P, Q \in \mathbf{C}[T]$; we shall denote $P = p_0 X^d + \cdots + p_d$ and $Q = q_0 X^e + \cdots + q_e$, with $p_0, q_0 \neq 0$ (so that $\deg P = d$ and $\deg Q = e$). We assume that Q has no root in $-\mathbf{N}$. Let $(a_n)_{n \geq 0}$ be the complex sequence defined by $a_0 := 1$ and the recursive relation: $\forall n \geq 0,\ a_{n+1} := \dfrac{P(n)}{Q(n)} a_n$. Compute the radius of convergence of $F(z) := \sum_{n \geq 0} a_n z^n$.

3) Set $R(X) := Q(X - 1)$ and assume that $R(0) = Q(-1) \neq 0$. Show that $F(z)$ is a solution of the differential equation

(F.4) $(R(\delta) - zP(\delta))F = 0.$

4) Discuss the nature of point 0 for the differential equation (F.4): singular or ordinary? regular or irregular? (You may concentrate on the generic cases and dismiss particular cases.)

5) For $x \in \mathbf{C}$, write $(x)_0 := 1$ and, for n a nonzero integer, $(x)_n := x(x+1)\cdots(x+n-1)$ ("Pochhammer symbols"). Thus, $(1)_n = n!$. Apply the above to the sequence with general term $a_n := \dfrac{(\alpha)_n (\beta)_n}{(\gamma)_n (1)_n}$.

More precisely:
(i) Determine the condition of existence of $F(z) := \sum_{n\geq 0} a_n z^n$ and its radius of convergence.
(ii) Determine the differential equation satisfied by F in the form:

(F.5)
$$\delta^2 f + p(z)\delta f + q(z)f = 0.$$

(iii) Assuming that $\gamma \notin \mathbf{Z}$, determine a solution of (F.5) in the form $G = z^{1-\gamma}H$, where H is a power series with constant term $H(0) = 1$. (Hint: substituting $f = z^{1-\gamma}h$ in (F.5), we are led to a differential equation of the same form but for other values α', β', γ' of the parameters.)
(iv) Show that (F, G) is a basis of solutions of (F.5) and compute the monodromy with respect to that basis along the fundamental loop.

6) Write equation (F.5) in terms of $f' := df/dz$ and $f'' := d^2f/dz^2$. Precisely describe the singularities in $\mathbf{C}$ and their nature (regular or irregular singular points).

Exercise F.6. For the following questions about the algebraic groups $\mathbf{C}$ and $\mathbf{C}^*$, complete proofs are expected. You are asked to determine:

(i) all rational morphisms from $\mathbf{C}$ to $\mathbf{C}$.
(ii) all rational morphisms from $\mathbf{C}$ to $\mathbf{C}^*$.
(iii) all rational morphisms from $\mathbf{C}^*$ to $\mathbf{C}$.
(iv) all rational morphisms from $\mathbf{C}^*$ to $\mathbf{C}^*$.

F.3. Some more revision problems

Some of the following exercises are related to those in the previous section.

Exercise F.7. 1) If $M \in \mathrm{GL}_n(\mathbf{C})$ and $N \in \mathrm{GL}_p(\mathbf{C})$, define $\Phi_{M,N}$ to be the endomorphism $C \mapsto NCM^{-1}$ of $\mathrm{Mat}_{p,n}(\mathbf{C})$.
(i) Show that, if M and N are diagonal, $\Phi_{M,N}$ is semi-simple.
(ii) Show that, if M and N are upper triangular and unipotent, $\Phi_{M,N}$ is unipotent.
(iii) Show that, if $M' \sim M$ and $N' \sim N$ (conjugacy), then $\Phi_{M',N'} \sim \Phi_{M,N}$.
(iv) Show that $\Phi_{M_1,N_1} \circ \Phi_{M_2,N_2} = \Phi_{M_1 M_2, N_1 N_2}$.

(v) Deduce the Jordan decomposition $\Phi_{M,N} = \Phi_{M_s,N_s}\Phi_{M_u,N_u}$ from the above.

2) Using the above questions, show that, if $CM = NC$, then $CM_s = N_sC$ and $CM_u = N_uC$.

3) Prove that, if $\gamma \in \operatorname{Hom}_{gr}(\mathbf{C}^*, \mathbf{C}^*)$ and if $\lambda \in \mathbf{C}$, the above equalities can be generalized as $C\gamma(M_s) = \gamma(N_s)C$ and $CM_u^\lambda = N_u^\lambda C$. Is it legitimate to replace, in the first equality, the group morphism γ by an arbitrary mapping?

Exercise F.8.   1) Let $\alpha \in \mathbf{C}^*$ and let $\Gamma := \langle \alpha \rangle \subset \mathbf{C}^*$ be the generated subgroup.
   (i) Describe the possible structures of the finitely generated abelian group Γ.
   (ii) Describe the possible structures of the algebraic group $X(\Gamma) := \operatorname{Hom}_{gr}(\Gamma, \mathbf{C}^*)$. Call G that algebraic group.
   (iii) Describe the possible structures of the group

$$G^\vee := \operatorname{Hom}_{gralg}(G, \mathbf{C}^*).$$

   (iv) More generally, describe all rational representations of G.

2) Redo all these questions with $\alpha, \beta \in \mathbf{C}^*$ and $\Gamma := \langle \alpha, \beta \rangle \subset \mathbf{C}^*$.

3) Redo all these questions with $\alpha_1, \ldots, \alpha_n \in \mathbf{C}^*$ and $\Gamma := \langle \alpha_1, \ldots, \alpha_n \rangle \subset \mathbf{C}^*$.

Exercise F.9.   1) A subgroup G of $\operatorname{GL}_n(\mathbf{C})$ is called *reducible* if there exists a nontrivial proper subspace of $\mathbf{C}^n$ stable under the natural action of G. Show that G is reducible if, and only if, its Zariski closure $\overline{G}$ is reducible.

2) A representation $\rho : G \to \operatorname{GL}_n(\mathbf{C})$ is called *reducible* if $\rho(G) \subset \operatorname{GL}_n(\mathbf{C})$ is reducible. Show that, if ρ is reducible, there exists a representation $\rho' : G \to \operatorname{GL}_{n'}(\mathbf{C})$ with $0 < n' < n$ and a morphism from ρ' to ρ such that the underlying linear map $\mathbf{C}^{n'} \to \mathbf{C}^n$ is injective.

3) Let $X' = AX$ be a fuchsian differential system over $\mathbf{C}(\{z\})$. Assume its matricial monodromy representation to be reducible. Use the functorial Riemann-Hilbert correspondence to deduce a property of A. Tackle explicitly the case $n = 2$.

Bibliography

[Ahl78]   Lars V. Ahlfors, *Complex analysis: An introduction to the theory of analytic functions of one complex variable*, 3rd ed., International Series in Pure and Applied Mathematics, McGraw-Hill Book Co., New York, 1978. MR510197

[And12]   Yves André, *Idées galoisiennes* (French), Histoire de mathématiques, Ed. Éc. Polytech., Palaiseau, 2012, pp. 1–16. MR2905515

[AB94]   D. V. Anosov and A. A. Bolibruch, *The Riemann-Hilbert problem*, Aspects of Mathematics, E22, Friedr. Vieweg & Sohn, Braunschweig, 1994. MR1276272

[BR78]   Garrett Birkhoff and Gian-Carlo Rota, *Ordinary differential equations*, 3rd ed., John Wiley & Sons, New York-Chichester-Brisbane, 1978. MR507190

[Bir13]   George D. Birkhoff, *The generalized riemann problem for linear differential equations and the allied problems for linear difference and q-difference equations*, Proc. Amer. Acad. **49** (1913), 521–568.

[BorA91]   Armand Borel, *Linear algebraic groups*, 2nd ed., Graduate Texts in Mathematics, vol. 126, Springer-Verlag, New York, 1991. MR1102012

[BorE28]   Emile Borel, *Leçons sur les séries divergentes*, 2nd éd. revue et entièrement remaniée avec le concours de G. Bouligand, Collection de monographies sur la théorie des fonctions, Gauthier-Villars, Paris, 1928, 260 pp.

[Bou07]   Nicolas Bourbaki, *Éléments de mathématique. Théories spectrales. Chapitres 1 et 2* (French), reprint of the 1967 original ed., Springer, Berlin, 2007.

[CR09]   Jose Cano and Jean-Pierre Ramis, *Théorie de Galois différentielle, multisommabilité et phénomènes de Stokes*, unpublished, see `http://www.math.univ-toulouse.fr/~ramis/Cano-Ramis-Galois`, 2009.

[Car63]   Henri Cartan, *Elementary theory of analytic functions of one or several complex variables*, Éditions Scientifiques Hermann, Paris; Addison-Wesley Publishing Co., Inc., Reading, Mass.-Palo Alto, Calif.-London, 1963. MR0154968

[Car97]   Henri Cartan, *Cours de calcul différentiel* (French), nouveau tirage ed., Hermann, Paris, 1997.

[Che99]   Claude Chevalley, *Theory of Lie groups. I*, Fifteenth printing; Princeton Landmarks in Mathematics, Princeton Mathematical Series, vol. 8, Princeton University Press, Princeton, NJ, 1999. MR1736269

[CL55]   Earl A. Coddington and Norman Levinson, *Theory of ordinary differential equations*, McGraw-Hill Book Company, Inc., New York-Toronto-London, 1955. MR0069338

[CH11]   Teresa Crespo and Zbigniew Hajto, *Algebraic groups and differential Galois theory*, Graduate Studies in Mathematics, vol. 122, American Mathematical Society, Providence, RI, 2011. MR2778109

[Del70]   Pierre Deligne, *Équations différentielles à points singuliers réguliers* (French), Lecture Notes in Mathematics, vol. 163, Springer-Verlag, Berlin-New York, 1970. MR0417174

[Del90]   Pierre Deligne, *Catégories tannakiennes* (French), The Grothendieck Festschrift, vol. II, Progr. Math., vol. 87, Birkhäuser Boston, Boston, MA, 1990, pp. 111–195. MR1106898

[DM82]   Pierre Deligne and J. S. Milne, *Tannakian categories*, Hodge Cycles, Motives, and Shimura Varieties, Lecture Notes in Mathematics, vol. 900, Springer-Verlag, Berlin-New York, 1982, pp. 101-228. MR0654325

[DD05]   Régine Douady and Adrien Douady, *Algèbre et théories galoisiennes* (French), 2ème éd., revue et augmentée ed., Cassini, Paris, 2005.

[For81]   Otto Forster, *Lectures on Riemann surfaces*, Translated from the German by Bruce Gilligan, Graduate Texts in Mathematics, vol. 81, Springer-Verlag, New York-Berlin, 1981. MR648106

[Ful95]   William Fulton, *Algebraic topology: A first course*, Graduate Texts in Mathematics, vol. 153, Springer-Verlag, New York, 1995. MR1343250

[God73]   Roger Godement, *Topologie algébrique et théorie des faisceaux* (French), Troisième édition revue et corrigée; Publications de l'Institut de Mathématique de l'Université de Strasbourg, XIII; Actualités Scientifiques et Industrielles, No. 1252, Hermann, Paris, 1973. MR0345092

[GR09]   Robert C. Gunning and Hugo Rossi, *Analytic functions of several complex variables*, reprint of the 1965 original, AMS Chelsea Publishing, Providence, RI, 2009. MR2568219

[Har91]   G. H. Hardy, *Divergent series*, 2nd (textually unaltered) ed., Chelsea, New York, 1991.

[Hil97]   Einar Hille, *Ordinary differential equations in the complex domain*, reprint of the 1976 original, Dover Publications, Inc., Mineola, NY, 1997. MR1452105

[IY08]   Yulij Ilyashenko and Sergei Yakovenko, *Lectures on analytic differential equations*, Graduate Studies in Mathematics, vol. 86, American Mathematical Society, Providence, RI, 2008. MR2363178

[Inc44]   E. L. Ince, *Ordinary differential equations*, Dover Publications, New York, 1944. MR0010757

[IKSY91] Katsunori Iwasaki, Hironobu Kimura, Shun Shimomura, and Masaaki Yoshida, *From Gauss to Painlevé: A modern theory of special functions*, Aspects of Mathematics, E16, Friedr. Vieweg & Sohn, Braunschweig, 1991. MR1118604

[Kol99]   Ellis Kolchin, *Selected works of Ellis Kolchin with commentary*, Commentaries by Armand Borel, Michael F. Singer, Bruno Poizat, Alexandru Buium and Phyllis J. Cassidy; Edited and with a preface by Hyman Bass, Buium and Cassidy, American Mathematical Society, Providence, RI, 1999. MR1677530

[Lan02]   Serge Lang, *Algebra*, 3rd ed., Graduate Texts in Mathematics, vol. 211, Springer-Verlag, New York, 2002. MR1878556

[Mac98]   Saunders Mac Lane, *Categories for the working mathematician*, 2nd ed., Graduate Texts in Mathematics, vol. 5, Springer-Verlag, New York, 1998. MR1712872

[Mal02]   B. Malgrange, *On nonlinear differential Galois theory*, Chinese Ann. Math. Ser. B **23** (2002), no. 2, 219–226, DOI 10.1142/S0252959902000213. MR1924138

[MS16]   C. Mitschi and D. Sauzin, *Divergent series, summability and resurgence I: Monodromy and resurgence*, Lecture Notes in Mathematics, vol. 2153, Springer-Verlag, Berlin, 2016.

[Mor13]   Juan J. Morales Ruiz, *Differential Galois theory and non-integrability of Hamiltonian systems*, Modern Birkhäuser Classics, Birkhäuser/Springer, Basel, 1999. [2013] reprint of the 1999 edition [MR1713573]. MR3155796 [van98]   Marius van der Put, *Recent work on differential Galois theory*, Astérisque **252** (1998), Exp. No. 849, 5, 341–367. Séminaire Bourbaki. vol. 1997/98. MR1685624

[vS03]   Marius van der Put and Michael F. Singer, *Galois theory of linear differential equations*, Grundlehren der Mathematischen Wissenschaften [Fundamental Principles of Mathematical Sciences], vol. 328, Springer-Verlag, Berlin, 2003. MR1960772

[Ram93]   Jean-Pierre Ramis, *Séries divergentes et théories asymptotiques* (French), Panoramas et Syntheses, suppl., Bull. Soc. Math. France **121** (1993). MR1272100

[RS15]   Jean-Pierre Ramis and Jacques Sauloy, *The q-analogue of the wild fundamental group and the inverse problem of the Galois theory of q-difference equations* (English, with English and French summaries), Ann. Sci. Éc. Norm. Supér. (4) **48** (2015), no. 1, 171–226. MR3335841

[Rud87]   Walter Rudin, *Real and complex analysis*, 3rd ed., McGraw-Hill Book Co., New York, 1987. MR924157

[Rud91]   Walter Rudin, *Functional analysis*, 2nd ed., International Series in Pure and Applied Mathematics, McGraw-Hill, Inc., New York, 1991. MR1157815

[Sab93]   Claude Sabbah, *Introduction to algebraic theory of linear systems of differential equations*, Éléments de la théorie des systèmes différentiels. $\mathcal{D}$-modules cohérents et holonomes (Nice, 1990), Travaux en Cours, vol. 45, Hermann, Paris, 1993, pp. 1–80. MR1603680

[Sab02]   Claude Sabbah, *Déformations isomonodromiques et variétés de Frobenius* (French, with French summary), Savoirs Actuels (Les Ulis). [Current Scholarship (Les Ulis)], EDP Sciences, Les Ulis; CNRS Éditions, Paris, 2002. MR1933784

[Sau03]   Jacques Sauloy, *Galois theory of Fuchsian q-difference equations* (English, with English and French summaries), Ann. Sci. École Norm. Sup. (4) **36** (2003), no. 6, 925–968 (2004), DOI 10.1016/j.ansens.2002.10.001. MR2032530

[Sau16]   Jacques Sauloy, *Analytic study of q-difference equations*, Galois theories of linear difference equations: An introduction, Mathematical Surveys and Monographs, vol. 211, American Mathematical Society, Providence, RI, 2016, pp. 103–171.

[Ser78]   Jean-Pierre Serre, *A course in arithmetic*, Translated from the French, 2nd corr. print, Graduate Texts in Mathematics, vol. 7, Springer-Verlag, New York-Heidelberg, 1978. MR0344216

[Sha77]   I. R. Shafarevich, *Basic algebraic geometry*, Springer Study Edition, Translated from the Russian by K. A. Hirsch; Revised printing of Grundlehren der mathematischen Wissenschaften, vol. 213, 1974, Springer-Verlag, Berlin-New York, 1977. MR0447223

[Sin90a]   Michael F. Singer, *Formal solutions of differential equations*, J. Symbolic Comput. **10** (1990), no. 1, 59–94, DOI 10.1016/S0747-7171(08)80037-5. MR1081261

[Sin90b]   Michael F. Singer, *An outline of differential Galois theory*, Computer algebra and differential equations, Comput. Math. Appl., Academic Press, London, 1990, pp. 3–57. MR1038057

[Sin09]   Michael F. Singer, *Introduction to the Galois theory of linear differential equations*, Algebraic theory of differential equations, London Math. Soc. Lecture Note Ser., vol. 357, Cambridge Univ. Press, Cambridge, 2009, pp. 1–82. MR2484905

[Spi65]   Michael Spivak, *Calculus on manifolds. A modern approach to classical theorems of advanced calculus*, W. A. Benjamin, Inc., New York-Amsterdam, 1965. MR0209411

[Ume08]   Hiroshi Umemura, *Sur l'équivalence des théories de Galois différentielles générales* (French, with English and French summaries), C. R. Math. Acad. Sci. Paris **346** (2008), no. 21-22, 1155–1158, DOI 10.1016/j.crma.2008.09.025. MR2464256

[Var96]   V. S. Varadarajan, *Linear meromorphic differential equations: A modern point of view*, Bull. Amer. Math. Soc. (N.S.) **33** (1996), no. 1, 1–42, DOI 10.1090/S0273-0979-96-00624-6. MR1339809

[WW27]   E. T. Whittaker and G. N. Watson, *A course of modern analysis*, Cambridge, 1927.

[Yos87]   Masaaki Yoshida, *Fuchsian differential equations*: *With special emphasis on the Gauss-Schwarz theory*, Aspects of Mathematics, E11, Friedr. Vieweg & Sohn, Braunschweig, 1987. MR986252

Index

Selected Published Titles in This Series

For a complete list of titles in this series, visit the
AMS Bookstore at **www.ams.org/bookstore/gsmseries/**.